Basic Flight Physiology

Basic Flight Physiology

Second Edition

Richard O. Reinhart, M.D.

Boston, Massachusetts   Burr Ridge, Illinois
Dubuque, Iowa   Madison, Wisconsin   New York, New York
San Francisco, California   St. Louis, Missouri

McGraw-Hill

A Division of The ***McGraw-Hill*** *Companies*

Published by The McGraw-Hill Companies, Inc.

12 13 14 15 16 17 18 19 BKM BKM 0 9 8 7 6 5 4

Library of Congress Cataloging-in-Publication Data
Reinhart, Richard O.
Basic flight physiology / by Richard O. Reinhart. — 2nd ed.
p. cm.
Includes bibliographical references and index.
ISBN 0-07-052223-5
1. Aviation physiology. 2. Aviation medicine. I. Title.
RC1075.R45 1996
616.9'80213—dc20 95-50308
CIP

McGraw-Hill books are available at special quantity discounts to use as premiums and sales promotions, or for use in corporate training programs. For more information, please write to the Director of Special Sales, McGraw-Hill, 11 West 19th Street, New York, NY 10011. Or contact your local bookstore.

Acquisitions editor: Shelley IC. Chevalier
Editorial team: Robert E. Ostrander, Executive editor
Norval G. Kennedy, Book editor
Jodi L. Tyler, Indexer
Production team: Katherine G. Brown, Director
Lisa M. Mellott, Coding
Ollie Harmon, Coding
Janice Ridenour, Computer artist
Rose McFarland, Desktop operator
Nancy K. Mickley, Proofreading
Design team: Jaclyn J. Boone, Designer
Katherine Lukaszewicz, Associate designer

TB1
0522235

To my sons Mark and Curt,
who have shared our love of flight

Contents

Foreword *xi*

Foreword from first edition *xii*

Introduction *xiii*

1 Human factors defined *1*

Cause and effect 3
Definitions of human factors and flight physiology 4
The "SHEL" model 5
Incapacitation/impairment 6
Flight surgeons and aviation medical examiners (AMEs) 8

2 Basic human anatomy *10*

The nervous system 11
The musculoskeletal system 14
The gastrointestinal (GI) system 15
The metabolic system 17
The circulatory system 19
Respiration 24

3 The atmosphere *29*

Composition of the atmosphere 30
Physical characteristics of the atmosphere 32
Gas laws 34
Divisions of the atmosphere 37
Physiological divisions of the atmosphere 40
General effects on the human body 41

4 Situation awareness *42*

Situation awareness defined 43
Components of situation awareness 44
Causes and clues of lost situation awareness 45
Prevention of lost situation awareness 45

5 *Altitude physiology* 47

The physiology of oxygen in the body 47
Review of respiration physiology 48
Carbon monoxide and ozone 63
Decompression of cabin altitude 64
Trapped gases 65
Evolved gas disorders 71

6 *Hearing and vibration* 80

Anatomy and physiology of the ear 81
Definition of sound 83
Definition of noise 85
Perception of sound 87
Measurement of hearing 88
Noise and hearing loss 88
Other types of hearing loss 91
Hearing conservation 92
Vibration 93

7 *Vision* 95

The light spectrum 96
Anatomy of the eye 97
Optics and physics of light 99
Correction of acuity 101
Night vision 103
Factors affecting visual acuity 108
Visual scanning 111
Visual illusions and misperceptions 111
Black holes and whiteouts 113
Sunglasses 113
Ultraviolet radiation 115

8 *Orientation* 116

Definition of disorientation 117
Types of disorientation 119
Visual illusions 136
Tolerance to disorientation and illusions 140
Flicker vertigo 142
Motion sickness 143

9 *Self-imposed medical stress* 144

Self diagnosis 145
Over-the-counter medications 148
Caffeine 155
Miscellaneous abuses 158
Physical condition 160
Alcohol and drugs 161

10 Environmental stresses 175

Basic physics of heat 176
The body's temperature control system 178
Coping with heat extremes 181
Coping with cold extremes 185
Dehydration 188
Radiation 189
Cabin air quality 191
Toxic chemicals and fumes 192

11 Sleep, jet lag, and fatigue 194

Sleep 195
Circadian rhythms 204
Fatigue 208

12 Acceleration 216

Types of acceleration 216
Common situations in civilian flying 217
Symptoms of G forces 218
Tolerance to G forces 221
Negative G forces 221

13 Crew resource management (CRM) 222

Definitions in CRM 223
Skills learned in CRM 224
Training in CRM 226
Deterrents to good CRM 227
Stress management 229

14 Human factors of automation 231

Cockpit automation defined 232
Information processing 233
Physiological impairments 234
Quantifying impairment 236

15 Inflight medical emergencies 239

First aid 240
Onboard medical kits 242
Transporting ill passengers 243
Basic survival techniques 243
Reminder 245

16 Health maintenance program 246

Nutrition and diet 247
Hypoglycemia 251
Exercise 252

The medical examination 252
FAA and company flight physicals 262

17 *Medical standards, regulations, and certification* 264

Definitions 265
Typical certification scenario 270
Preferred certification scenario 274
Denial 274
Recertification and special issuance 275

Recommended reading 277

Index 279

About the author 287

Foreword

IN MY MANY YEARS OF ASSOCIATION WITH AVIATION, at NASA, the NTSB, and now Delta Air Lines, my primary concern and interest has been with the human factors of flight. From this broad subject was born CRM, which has evolved into an essential part of a company's culture and its training programs.

The physiology of flight has always been a key component of aviation human factors, but has often taken a back seat to CRM. The physiological status of the human element remains crucial in maintaining safe performance by all aircrew members. One cannot discuss safety without including basic physiological issues such as fatigue, jet lag, hypoxia, self-medication, disorientation, and other related topics.

Until recently, however, the role these factors played in accident causation was not given adequate recognition. During my tenure at NTSB, we were able to identify several transportation accidents caused by fatigue and other physiological factors. Furthermore, many incidents involving one or more of these physiological factors are being reported to the NASA/FAA ASRS. There is a growing understanding of and respect for all facets of human factors, physiological and psychological.

I can't state strongly enough the importance of being familiar with the physiology of flight. As with CRM, you must recognize its importance from the moment you begin flight training and maintain that awareness throughout your career. The safety of flight is influenced by many factors and as our aircraft and their systems become more complex, it is imperative that respect for your fitness to fly remains second nature to you. This should be as much a part of your ritual of flying as putting on your seat belt.

John K. Lauber, Ph.D.
Vice President—Corporate Safety & Compliance
Delta Air Lines

Foreword from first edition

MILITARY AVIATION HAS ALWAYS RECOGNIZED the importance of flight physiology in its training, both initial and recurrent. Periodic trips to the altitude chamber are required for all aircrew to remain on flying status. In my experience with the many aspects of aviation education, flight physiology has always been a topic of great interest to pilots and instructors alike.

Civilian aviation has been slow to incorporate adequate flight physiology into their training, even at the student level. It has been the collegiate community that has taken it upon themselves to provide courses and material to educate future commercial pilots in all factors of this important topic. Some schools require the course, some offer it as an elective, and too many do not offer it at all. Part of the reason for this inconsistency is the lack of educational resources (i.e., textbooks) around which to build an effective course. This text fulfills that need.

I feel strongly that every student must be familiar with the physiology of flight. And this is not limited to the college level. Flight training schools and academies should provide their students the insight to respect the many facets of the physiological deterrents to flight safety. Furthermore, there needs to be a periodic reinforcement of these issues.

This textbook is not meant to make flight surgeons or even flight safety officers of the student. Its objective is to acquaint the pilot with the varied elements of the physiology of flight in ever-changing conditions. This text also becomes a resource of timeless information and needs to be a part of the pilot's personal library. As new-generation aircraft enter this field, the generic topics covered in this text continue to be important.

I encourage the student to take these subjects seriously. In addition to finding them interesting, these topics and issues play an important role in safe flight. I have been in aviation for many years and have seen just about every facet of flying there is, both military and civilian. Aviation is still an exciting and rewarding profession (or hobby), and I remain thoroughly interested in promoting safe flight.

I wish the new aviator success and a fulfilling time in flight.

Kenneth L. Tallman, Lt. Gen. USAF (Ret.)
Former Superintendent, USAF Academy
Past President, Embry-Riddle Aeronautical University

Introduction

TAKE TWO PILOTS, one trained in the military, the other trained as a civilian, and ask them both about hypoxia, trapped gases, decompression sickness, and other flight physiology terms. Chances are they will answer differently. The military pilot will have a quick and concise explanation of each term, whether on active duty or retired. After all, it was a part of his or her initial training, and he or she took altitude-chamber rides every two to three years and underwent periodic safety briefings that frequently covered these topics. The military has been doing this for years.

And the civilian pilot? Often there will be some hesitation in fully explaining each topic. Many pilots are familiar with the topics but have little respect for their significance in flight. Although human factors has become a justifiably important issue upon which to train, flight physiology is often minimized, even though it is a part of human factors. Indeed, more and more human factors insights are being required in training in commercial aviation either as independent topics or as part of CRM (cockpit/crew resource management). Yet the military feels it is as crucial for its pilots to know about flight physiology as it is for their flight-medicine doctors (flight surgeons). Civilian pilots should have no less training or awareness.

Besides complacency, the lack of adequate training and educational resources have been part of the reason all pilots and especially certified flight instructors are not acquainted with, let alone knowledgeable about, flight physiology, hence this textbook. This text is tailored for the beginning pilot, although many civilian pilots are still "beginners" when it comes to flight-physiology education. The pilot readers will not only become knowledgeable about the physiology of flight but will also be able to explain why certain things happen to the mind and body in flight, often only in flight.

When learning to fly, simply knowing how to move the stick and rudder does not make a pilot safe or proficient. The more the pilot recognizes why the stick and rudder work, along with airfoils, engines, weather, hydraulics, and the like, the better and safer that pilot will be. The same holds true with understanding how the body and mind *can* be affected in flight as well as *why* they are affected in flight.

The chapters are organized to lead the reader from the basics of physiology and physics of flight into the rigors of flight that require a healthy, competent, proficient, and "airworthy" pilot. There is obviously much crossover between the topics. Furthermore, it is common for many physiological situations to be occurring at the same time, which is not a very comforting thought if the pilot isn't even aware of what's happening.

The text is not meant to be a comprehensive explanation of all topics. This is not a medical text. After all, medical doctors require many years of education and training just to learn basic medicine. Furthermore, there are many opinions on the same subject. The opinions stated in this text are those shared by many, but there will always be exceptions or disagreement. When there is disagreement or what is stated doesn't ring true with what you feel is correct, then look for other resources to confirm your interpretation. This is medicine; it's not an exact science, it's the "practice of medicine."

The objective for the reader is to get a clear understanding and overview of the many and varied physiological situations that can interfere with safe flight. Bookshelves are full of supplemental resources to further elaborate on any given topic.

Take hypoxia; the basics of atmospheric physics coupled with the essentials of the body's circulation system will give the pilot enough information to be respectful and knowledgeable about hypoxia. Further information on the atmosphere to better quantify how hypoxia can severely affect the body is easily found in weather books. The same is true about the circulatory system. This opens the door for expansion on any topic covered in this text; additional references are listed in the recommended reading appendix.

The military has long been the expert in flight physiology and in providing effective training aids. This text is not meant to "reinvent the wheel" and therefore uses some material already presented in military publications. Civilian pilots are constructed the same as military pilots, so the information is somewhat generic. This text presents the material in a logical order for the classroom and concentrates on the situations that arise in civilian operations.

The reader will also find various "reviews" of flight physiology in many books on human factors, basic flying "how-to," and aviation medicine. Many good articles also appear in trade magazines; however, none seem to take these subjects in depth and relate them to flight for the beginning pilot. Beyond this basic overview of flight physiology, there will be a continuing need to access resources as every pilot matures in flying. This text is also meant to be a resource in a pilot's personal flying library for future situations dealing with human factors and flight physiology. Time constraints in commercial training do not always adequately reinforce the importance of these essential issues. It's often up to the pilot to search out the information that will prevent impairment and reduce the risk of a human factors-related incident or accident.

Flight will be more enjoyable and the pilot will be more confident when he or she knows more about how the body and mind function in flight and the many ways that flying skills and safety can be compromised through unintentional, but controllable, abuses of health and fitness to fly.

Fly Fit, Fly Safe!

1
Human factors defined

Public concern over human error in aviation has been increasing steadily. As a result, FAA and NASA, working in conjunction with the aviation community, have decided to increase their efforts to address human factors. The National Plan For Aviation Human Factors is the first step toward a major program augmentation and is part of the DOT National Transportation Policy. Its purpose is fourfold: (1) To identify the technical efforts necessary to address the most operationally significant human issues in aviation and acquire the necessary resources to respond to these issues; (2) To allocate resources efficiently by coordinating research programs at various government laboratories; (3) To communicate research needs to academic and industrial "centers of excellence"; and (4) To promote the means by which human factors knowledge is transferred to government and industry.

***Federal Register*/Vol. 55/No. 228**

ALL PILOTS WANT TO FLY. A simple fundamental but also a significant statement. Flying means just that: "driving" a flying machine around and through the skies and being in control. Novels have been written solely about the romance and excitement and joy of flight. What pilot hasn't read and even placed on his or her family room wall a copy of "High Flight": "Oh, I have slipped the surly bonds of earth, and danced the skies on laughter, silver wings. . . ."?

Pilots in the movies are fearless, overcoming great odds in the dangerous skies, flying planes that are falling apart as they near their important destination. Pilots are meant to fly. They rise above the inadequacies of their minds and bodies and their vulnerabilities to the physiological and psychological stresses of flying. The macho pilot can drink up a storm, stagger out of the bar, and fly a perfect trip the next day. He or she passes off the effects of fatigue, dehydration, noise, hypoglycemia, and the myriad of other factors associated with flight.

But what about hypoxia, disorientation, jet lag, and vision problems? Some pilots don't even know what these terms mean, many don't care, and most have little access to resources providing them this insight. This situation doesn't mean these issues are less important than flight proficiency or learning how to fly an ILS. They might not be as motivating as sharpening your flying skills, but they all will have a direct and indirect effect on those skills, and the pilot might not even be aware that his or her skills are compromised. You might be the best pilot on the field, but not if you allow yourself to be subjected to the physiological effects that surround all pilots. They became the subtle and unrecognized impairments to your performance that needs your recognition and intervention.

Flight physiology and human factors both have an impact on flight. More than 70 percent of aviation accidents and incidents are in some way related to human factors. No one would deny that he or she is less productive when tired. But does the pilot know what caused the tiredness, how to cope with it, and when to say "I'm not safe."?

That's the purpose of this textbook. To help the pilot understand the other part of flight, the human part, and to recognize when the body and/or mind is not in tune with the aircraft. This insight must be part of a pilot's training, right along with being familiar with emergency procedures and understanding approaches.

We all would like to expect the ideal conditions of good weather, working radios and flight management systems, no delays, functioning instruments, and turning propellers; however, few flights are like that. Add to this the fact that our mind and body might not fully cooperate during problem flights when all our physical and mental skills are needed, and we become a setup for an incident or accident. Flying an approach to minimums with a faltering engine and an intermittent guide slope receiver is stressful enough. Being fatigued, disoriented, and suffering with an ear block makes this approach downright dangerous.

Therefore, these topics are an integral part of any pilot's training. Furthermore, they must be refreshed in your mind periodically. The military has accepted this need and still requires recurrent training in physiology and human factors, despite the moaning of some who say they don't need it. But all will admit that without insight into flight physiology, they are less than safe as well as less proficient.

The FAA is also recognizing that many civilian-trained pilots have a lack of understanding of basic physiology. Regulations are now requiring some knowledge of the physiology of flight, especially at altitude. Part 61, the pilot certification regulations, list hypoxia, decompression, trapped gases, and other physiological topics as a part of training for pilots flying aircraft certified to fly above 25,000 feet MSL. Part 121 and Part 135 of the Federal Aviation Regulations will expect companies to provide training in these topics as well as on fatigue, judgment and communication skills, and other "human-related" issues. Much of past Part 135 operations will be under Part 121 standards where human factors and *cockpit/crew resource management* (CRM) are required. Furthermore, companies who hire pilots will want their new hires to be familiar with human factors and CRM.

Therefore, the beginning pilot needs to become familiar with flight physiology at the same time that proficiency skills are learned and practiced. Reinforcement will come from recurrent training, especially from the company that hires the pilot. Maintaining a high level of awareness of the physiology of flight will become a high priority.

CAUSE AND EFFECT

We often speak of accidents and incidents, and they have their own definitions, requirements for reporting, and impact on the pilot's certification. But despite their differences, they are closely related. No accident occurs without a series of incidents happening beforehand. It's a chain of events that will eventually end with an accident unless someone breaks that chain. It can be said that an accident is one incident too many. The dilemma is that few incidents are reported (although the anonymous Aviation Safety Reporting System (ASRS) administered by NASA is an effective way of accessing these occurrences) and the significance of these incidents is often minimized. Only when a reportable accident occurs does the realization of the role of incidents become apparent. Yet how often do we equate a minor incident to how close we are to an accident?

Another way of describing this relationship is to say that a contributing factor to an accident is probably associated with an incident of varying significance that could have been averted and the accident prevented.

Consider the pilot who is anxious to get home, is tired, had a bit too much to drink the night before (but is legal), has a cold, and is filing a flight plan for a 2-hour trip at FL 230. Is she going to be able to deal safely with changing weather conditions, instrument-approach delays, and a yellow warning light on the panel suggesting the oil pressure is not stable? Is she going to be able to break a chain of events that she might not even be aware of because of her condition? We would all like to think that we can rise above these challenges. But would you ride along as this pilot's passenger?

Another relationship that needs to be understood is the terminology used when describing what caused an accident (again, accidents alone generate an accident report by the NTSB and FAA) and the contributing factors. Statistics will often include cause in their reports but rarely contributing factors. Here's an analogy to keep in mind. Cancer does not kill people; it's not a causative factor in death. *Cancer*, by definition, is an abnormal cell that multiplies rapidly and spreads to and invades other essential organs. One or more of those organs are then unable to function properly, leading to heart failure and death. The cause of death will probably be reported as being cancer, even though it technically was a *contributing* factor—and we place great emphasis on cancer as a cause.

But in aviation, only the cause is usually reported as the pilot being unable to fly the plane safely, for example, "controlled flight into terrain." But rarely are contributing factors discussed when placing blame, even though they are mentioned in accident reports. Unlike in medicine, contributing factors (the incidents in a chain of events) are minimized, and we tend to focus on who was in error, not why that pilot erred during the preceding events. Always look for the contributing factors because you can control them. The final event and the accident cannot be reversed.

DEFINITIONS OF HUMAN FACTORS AND FLIGHT PHYSIOLOGY

Human factors has a variety of definitions, depending on whom you ask. For example, an engineer will consider the positioning of controls and instruments within the cockpit a human factor—the man-machine interface. The psychologist will define it as how the pilot deals with stress, how he or she communicates, and how he or she manages available resources (CRM).

A pilot might say that it means the ability to perform under extreme conditions, how he judges what action to take when things turn sour. A physiologist or physician will consider the effects of fatigue, hypoglycemia, illness, noise, and other "medical" issues. The result is the same: how to get the human pilot from point A to point B in a flying machine without a human factor interfering.

Flight physiology is how the body and mind work in the flying environment. It includes such topics as understanding how our organs function, what keeps them from functioning, and what the pilot can do to protect these functions before and during flight. Hypoxia, dehydration, fatigue, vibration, visual illusions, noise, disorientation, jet lag, self medication, alcohol, smoking, and many more topics are included in the list. Furthermore, how to define health and then how to protect health becomes an important goal.

Flight physiology is not dull. It is not just "nice to know." It *is* essential to safe flight.

However, we tend to take these issues for granted because our bodies and minds are usually very tolerant and forgiving of the abuses to which we subject ourselves. "Human factors happen to the other pilot," not me. Denial of the importance of human factors compared to flying proficiency begins to take over our reasoning. The near-miss or dumb incident usually gets our attention and brings us back to the reality of the significance of human factors. Now, these are the pilots who don't have to be told to learn because they know its importance, at least until they fall back into old habits. We experience this all the time when driving; the near-collision with another car gets us back to driving safely again.

Flight physiology, therefore, is an integral part of human factors and safe flight. We expect our aircraft to be airworthy. We look to our mechanics to keep the aircraft airworthy. Before we fly, we preflight the machine to ensure that it is airworthy. Our expectations are high for that aircraft. We should have the same expectations for the pilots and the other flight crew. They must also be airworthy—medically airworthy. This text will concentrate on keeping the pilot medically airworthy while discussing human factors in general. Obviously, the pilot's airworthiness is directly related to whether or not there are any human factors in flight. Flight physiology is the most important part of human factors. It's the human element of human factors and safe flight, and it has a direct effect on performance.

Throughout this text, the term "medical airworthiness" will be used. As stated above, this will be the single designation to summarize the physiological and psychological state of the pilot. It will also relate to any illness and medical or physical impairment that might be present. It will be used in the same context as with expressing the flying status of an aircraft; is the pilot "medically airworthy" before being subjected to the stresses inherent in flight?

Safety is the prime goal of all flight for all pilots. The best trained pilot, however, isn't safe if he or she isn't medically airworthy. Experience will *not* overcome any impairment. Knowing flight physiology, being aware of its effects on performance, and maintaining a high index of suspicion when performance becomes substandard will continue to make anyone a better and safer pilot. Without this knowledge, the pilot is practicing "crisis management," assuming that a problem can be dealt with if it arises but indifferent until then.

THE "SHEL" MODEL

The International Civil Aviation Organization (ICAO) has used the concept of the SHEL model (developed by Edwards and Hawkins) to better define the role of human factors in aviation and how they interrelate. It is helpful in recognizing the importance of every factor when looking at the big picture of flight (Fig. 1-1).

The term comes from the basic components:

- Software (procedures, documentation, symbology, etc.)
- Hardware (machines and equipment)
- Environment (internal and external)
- Liveware (human element)

Liveware becomes a component as well as the central figure upon which each component will have an effect; thus, we can talk about the "human-machine" interaction (pilot moves a control), for example, while keeping in mind that there are other interactions (turbulence caused by weather). It is this model that helps to visualize how and why CRM works and that CRM is not limited solely to "human-human" interactions (see chapter 13).

The model is also important in explaining why the physiology of flight is important since the human element is obviously crucial and central and every aspect of physiology will affect every other interaction as defined in the model. In other words, if the pilot isn't medically airworthy, the rest of the system model will be affected. If the pilot is impaired, his or her medical airworthiness will affect all the interfaces. Human factors is how these interfaces and interactions ultimately affect performance; human factors is a dynamic process. There are other models for explaining these interactions, such as "Reason's Model." It is not within the scope of this text to explain Reason's Model, but it is worth mentioning and can be researched independently.

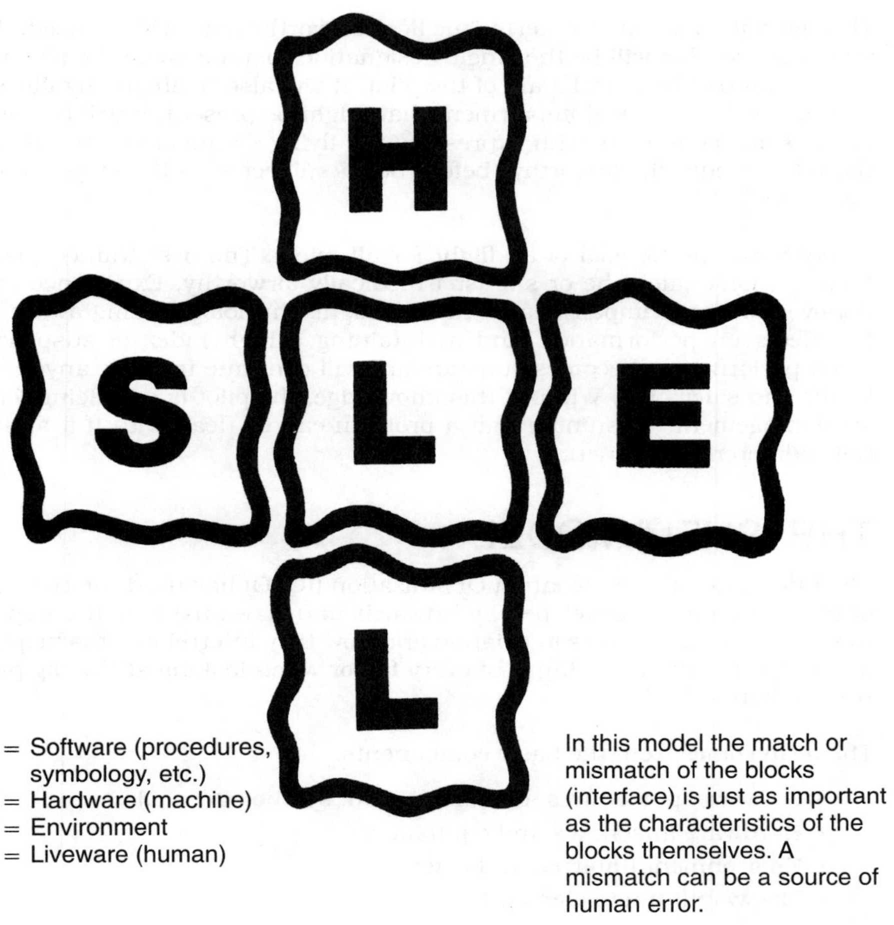
H
S
L
E
L
S = Software (procedures, symbology, etc.)
H = Hardware (machine)
E = Environment
L = Liveware (human)
In this model the match or mismatch of the blocks (interface) is just as important as the characteristics of the blocks themselves. A mismatch can be a source of human error.

1-1 The SHEL model examines the interrelationship of human factors and the aviation environment.

INCAPACITATION/IMPAIRMENT

The ultimate objective of flight physiology awareness and comprehension is the prevention of incapacitation or impairment, whether physical or mental. *Incapacitation* is defined as being incapable of performing expected normal activity. Mental incapacitation is the mind's inability to use proper judgment, reasoning, and decision making. Beyond that, mental incapacitation turns into "neurological" incapacitation whereby the signals from the brain fail to use the sensory information and data from the eyes, ears, touch, smell, and the like.

Physical incapacitation refers to the body's inability to function in an expected way. The end result in any of these incapacitating or impairing situations is an unsafe and poorly performing pilot. A variety of classes of incapacitation all lead to the same unacceptable result, an unsafe pilot.

Sudden

This is obvious. The pilot collapses or slumps over the yoke. He or she might cry out in pain or have a seizure. The solo pilot is committed to whatever the plane now chooses to do. With a crew, the others can take immediate action to *keep the plane flying*, which is the first rule of safe flight in an emergency. Because the situation is so obvious, there is no doubt that something has to be done and quickly. Get the troubled pilot away from the controls and take over the flying, then declare an emergency, ask for help from onboard or from the ground, and land.

Subtle

This is less obvious. In fact, it can go unnoticed by the pilot or her crewmates, which makes this a potentially dangerous situation. It's like driving your car and not knowing your brakes are out until you need them. This pilot might be dazed, semiconscious, or unable to move (as with a stroke). The pilot might even know she is incapacitated but is unable to convey the problem to anyone, even to ATC.

This is such a major concern that some flight training ground schools, especially CRM courses, actually simulate an incapacitated pilot in a simulator setting to determine how the rest of the crew recognizes the presence of the problem and determines what action they take. One of the solutions is a continuous observation of each crewmember's actions to ensure that each one is doing or saying something all the time. It's a confirmation that all crewmembers are in the loop of responsibility. Any deviation from expected actions needs to be challenged with the suspicion of being incapacitated or impaired.

Total and partial

This is self explanatory. *Total* means the pilot is totally out of it, up to and including dead. There is no doubt that a problem exists that needs immediate attention. Partial is more troublesome. Like subtle, *partial* means that the pilot might not be aware of a deficiency or relates the reason he isn't performing up to expected standards to being fatigued, sick with the flu, or something else. In any case, the pilot might not accept the realization that he is partially incapacitated and pride forbids admitting this to himself, his crewmates, or ATC.

Distraction

Somewhere amidst these terms is a place for the healthy pilot to be "kind of incapacitated" as a result of being distracted. This distraction might be worrying about a problem at home, concentrating on a yellow warning light, watching another plane, listening to a conversation, and many other scenarios that virtually take the pilot out of the flying loop. Some call this a lack of *situation awareness* (or *situational awareness*; both are accepted terms, as explained in chapter 4), not knowing or recognizing what's going on around himself or herself.

In a way, the pilot is incapacitated because he or she is not doing the expected job. This became such a problem on air carriers (distraction caused by unnecessary cockpit conversation) that the sterile-cockpit rule came into effect: no unnecessary talking during crucial phases of flight.

This took care of the dangerous distractions problem, but interestingly, another "distracting incapacitation" is taking place. Pilots in glass cockpits with cathode ray tubes (CRTs), liquid-crystal displays (LCDs), and numerous black-box computers are "playing" with these fascinating devices beyond what is necessary for flight. Their heads are down instead of keeping in touch with the outside. Once again, by being distracted, they are out of the performance loop by focusing on things not directly related to their duties.

Recognized and unrecognized

Again, by definition this is fairly obvious. But this is where the problem becomes insidious, especially in a crew situation. Does the troubled pilot know she is having problems and is she dealing with them? If not, what should the other pilots do? Some forms of incapacitation are clearly recognized and one can then take action, either for himself or herself or for the other person. Unrecognized becomes more of a challenge.

Ego and pride also get in the way. Few pilots are willing to admit that they are having a problem if they think that they can deal with it themselves, which is another breakdown in the use of cockpit and crew resources. Another unsafe situation is where the pilot is unwilling to challenge another crewmate when substandard performance is noticed as possibly being a result of some physiological problem. Assertiveness and self-confidence become important personality traits in good pilots and are key topics taught in CRM courses.

The reason for identifying these variations on the same theme is to advise you that the chances of any or all of these conditions existing at the same time in some form is real and it happens. And now what do you do? You can certainly stay on top of unexpected scenarios that could lead to disaster, if you are aware that this can and does happen. Then you define what to look for and recognize specific actions to take to determine the extent of incapacitation. Then and only then can appropriate action take place to intervene and keep the plane flying.

FLIGHT SURGEONS AND AVIATION MEDICAL EXAMINERS (AMEs)

Flight surgeons and aviation medical examiners (AMEs) are medical professionals, often physicians, who either specialize or concentrate part or all of their medical practice on aviation medicine and flight physiology. An understandable function related to these doctors is that they do flight physicals and can keep pilots flying or ground them. This is a big part of their responsibility no matter where their practice is located.

But there is some difference between the military and civilian aeromedical doctors. This will be discussed later, but for now keep in mind that the military flight surgeon is often considered a resource for human factors education and consul. FAA's AMEs could be this resource in civilian aviation, but they rarely get involved in education of human factors; they are satisfied to just do physicals. Most civilian pilots rely on the FAA, company, or organizational sources for any information on flight physiology, if it is even available or readily accessible.

The civilian pilot is virtually on his own to gather insight on human factors and, unfortunately, often puts these topics on the bottom of his educational priority list unless it's required by the FAA, company, or school. As time passes and the realization that the aviation community and the public they serve cannot depend on individuals to educate themselves, more and more training will be required, as it should be. Current and future requirements confirm this trend.

Why? As stated before, the pilot should expect to be as airworthy as his or her aircraft. In the military, the flight surgeon is responsible for providing this information at safety briefings and in unit publications. Because this is not usually the case in the civilian aviation world, it is up to you, the pilot, to seek out aeromedical information on human factors. AMEs and flight surgeons can be valuable resources and should be used.

Some companies have a safety department. The military has a flight-safety officer (FSO) for its flying units. They both provide the same insight to their pilots and aircrew—how to keep flying safely and productively. These resources, however, are uncommon in the rest of the aviation community.

As you review the information in this text, the intent is not to make you flight surgeons or AMEs: FSOs maybe. But certainly each pilot should be aware of what's going on and what can happen in flight. Recognizing before a flight what physiological events can affect your performance will result in a safer and more enjoyable trip, whether for fun or for pay.

2
Basic human anatomy

Our body is a well-set clock, which keeps good time, but if it be too much or indiscreetly tampered with, the alarm runs out before the hour.

Joseph Hall, English bishop (1574–1656)

It is shameful for a man to rest in ignorance of the structure of his own body, especially when the knowledge of it mainly conduces to his welfare, and directs his application of his own power.

Philip Melancthon, German reformer (1497–1560)

IN ORDER TO DESCRIBE the basics of how and why the body works, this discussion begins with the premise that the main part of the body is the brain and that all other parts and physiology support this one organ. The brain, therefore, is the ultimate organ, wherein control of all body functions occurs to satisfy the physiological and psychological needs of the brain. For the sake of this text, we will concentrate on keeping the body healthy so as to "service" the brain such that the brain can perform as a master computer in control of the body parts and system organs to allow us to fly in an unhealthy and often hazardous environment.

We will discuss each major anatomical system: brain, musculoskeletal, gastrointestinal, metabolic, and circulatory. During the discussion of each, the healthy body will be described. Next will be a brief overview of the "abnormal body," including discussions on infection, injury, tumor, inflammation, and the like. This will provide the foundation for the subsequent chapters where we will subject this body to the additional stresses of flight. Such knowledge will provide the pilot some insight as to how the body should work under ideal and controllable situations, and raise the level of awareness of what can, and often does, happen in less then ideal conditions. Then and only then can the pilot take action to either avoid unsafe situations or be prepared to cope with the physiological and psychological challenges of flight as well as be better prepared medically. We often have little control over many of these flight stressors, and we can't make them go away; however, with insights about flight physiology, the responsible and safe pilot can find ways to cope.

When considering the abnormal body, we usually think of *pathology*, which is the medical term referring to anything that is not normal. One frequent and common form of pathology of any part of the body is an inflammation.

This can result either from an infection, irritation, or injury. Any part of the body can be potentially inflamed. If a body part ends in *-itis*, then we are talking about the pathology of inflammation of that part: tendinitis, tonsillitis, appendicitis, arthritis (arthro=joint), meningitis (menin=the covering of the brain), carditis (cardi=heart), and so on. The intensity of the inflammation and the degree of involvement of body parts determines how impaired the pilot will be.

Although inflammation is the most common general pathology, there are others, of course, and they will be defined with the specific body system.

THE NERVOUS SYSTEM

The brain

The brain or central nervous system (CNS) is the central part of our total nervous (or neurological, or nerve network) system. It controls all functions—physiological, mechanical, and mental—by sending electrical signals to the various parts of the body, much like a phone system or computer. Instead of wires, nerves are used through which a signal is transmitted electrically through the nerve via a series of nerve cells and biochemically between nerve cells. A physical row of these cells is what we call a nerve; the nerve varies from a single cell in diameter to a bundle of nerves the diameter of a pencil.

The brain is divided into three main parts, the largest being the *forebrain*, or *cerebrum*. This contains much of our "gray matter," or thinking cells. The forebrain is further divided into two hemispheres, right and left. The surface of these hemispheres is called the *cerebral cortex*. Each hemisphere is further divided into four lobes, each supplying its own functions.

The *frontal lobes* are responsible for complex thoughts, decisions, and judgments. From these lobes, signals or "messages" are transmitted via nerves to muscles, telling them what to do. Next are the *parietal lobes* where some of the senses send information for processing. The *temporal lobes* are where the speech center is located and where the brain computes information and data and assists in written and spoken communications. The *occipital lobes* are where information from the eyes is processed.

The second part of the brain is the *midbrain*, which contains the hypothalamus that produces hormones that affect temperature, growth, and other physiological activities. The *hindbrain*, which is the third part of the brain, is the center of regulation of many of the body's basic functions, including breathing, blood pressure, heart rate, and many others.

Various parts of the brain can become infected or inflamed by afflictions such as meningitis and encephalitis, which can result in permanent damage to brain tissue. The same is true of injury. Because the brain is in a closed container (the skull), any impact to the skull can cause torn blood vessels, bruised brain cells, or swelling of the brain itself. To explain how the brain is damaged when the head is hit, some have used the descriptive analogy of what happens to a fresh tomato that just fits inside a jar, and then someone drops the jar. The tomato "bounces" back and forth (a rebound effect) within the jar with predictable results. A human would have multiple areas of injury to all the brain; therefore, any closed-head injury is considered serious.

A *concussion*, by definition, is any period of unconsciousness caused by a blow to the head. If one experiences any kind of loss of consciousness (LOC) for any reason, there is usually short- or long-term brain damage or some other underlying problem. This brain damage is a potential risk for problems, especially in flight. Such problems range from a simple headache to convulsions. Hypoxia is known to trigger serious symptoms in someone with a history of any brain pathology, whether from injury, surgery, burst blood vessel (stroke), or infection; therefore, any brain pathology is a potential risk to flying until it can be medically proven that there is little chance of a problem developing in flight.

The spinal cord

Signals from the various parts of the brain are transmitted to the rest of the body by a cable of nerves called the *spinal cord*, located in the bony protection of the spine. This is the link between the central nervous system (the brain) and the peripheral nervous system, becoming the direct linkage to muscles, internal organs, and our sensory organs. All nerves to everything in the body come from this cord; therefore, we experience pain in the foot as it is transmitted to the brain from nerves that leave the spinal cord at the end of the spine, which is the lower back.

In addition, the spinal cord brings back signals from "end organs" (organs that process and execute the body's functional activities) as a form of feedback, which the brain processes with other data needed for overall function. For example, the brain will tell the pilot to turn the ignition on with his fingers, and when the sound of the engine starting reaches the ear, that signal is transmitted back to the brain to tell the pilot to release the ignition switch, all via the spinal cord.

Any pathology in the spinal cord will interfere with these critical transmissions. Injury to the cord results in total loss of neurological control to any part of the body that is connected to nerves distant from the brain and closest to the injury. The same is true if there is a tumor pressing on the cord or an infection or inflammation affecting even a small portion of the cord. It's similar to cutting a telephone or computer cable, except the "human cable" usually can't be replaced and probably can't be repaired, nor will it likely grow back; however, current surgical and medical technologies are accomplishing remarkable results of nerve regeneration and repair.

The peripheral nervous system

The signals to and from the brain are transmitted to the various organs and parts of the body via the series of smaller single nerves in the peripheral nervous system. Each muscle and organ has one or more nerves controlling its functions. Comparing again to a telephone cable, just as every house has its own telephone line, every part of the body, including selected cells, has its own nerve.

Control of body functions is more complex than just a single nerve telling an organ or body part what to do when the person or the brain decides what action it wants to take. Breathing, digestion, heart rate, blood pressure, internal temperature control, and many more functions are automatic and do not require conscious effort to initiate action. The feedback system from various organs, like an air conditioning thermostat, tells the brain what it needs and then the brain responds without our awareness. This process is often called the *autonomic* or *sympathetic nervous system.*

This is also true in the "fight-or-flight" situation. When one is suddenly alarmed, scared, facing a dangerous activity, or in a situation requiring quick response, the brain senses the urgency and, in addition to sending electrical signals, also releases chemicals and hormones (commonly adrenaline), all of which increase the body's metabolism to be able to respond to the emergency. Many times this is instantaneous. Everyone has experienced the rush of adrenaline with the associated rapid heart rate, tingling sensation, and other symptoms in a potentially dangerous situation, like a near-collision in your car. The body instinctively responds without your intervention.

Control of our senses is also automatic. We see, hear, taste, smell, and feel without having to consciously ask the brain to begin. The switch is always on and we are continuously receiving cues from our sensory organs. What we do once we begin processing these signals now becomes a conscious activity. Pain is a good example. We can reach for a hot pan on the stove and pick it up through conscious commands. If it's too hot, we automatically and instinctively let go very quickly and pull our hand back, probably also dropping the pan. This happens so suddenly that we are usually unable to intercede and hang onto the pan before dropping it. What we do next to recover from that reflex requires a conscious effort.

Training could, in effect, allow us to overcome the reflex so as to hold on to the pan even though it is hot. The same holds true in flying. Many actions and functions of flight are contrary to what the body thinks it ought to do without our input. The brain tries to protect us from this perceived threat. This is why it is so important to be familiar with those flight conditions not usually tolerated by the body and to train and practice until the body computer, or "thermostat," can be reset to accommodate the flight environment. Lack of currency in flying skills in instrument conditions is a good example of slipping back into unsafe old habits and conditioned responses.

The peripheral nervous system must also be intact to transmit the various signals to and from the brain. A nerve can be infected and inflamed like any other body part. Add *itis* to *neuro* (nerve) and you have *neuritis*. The symptoms of neuritis depend on what body part or organ the nerve is serving. If it works on the skin, there is a tingling or painful symptom. If it goes to the muscle, there could be twitching. Injury to the nerve, like to the spinal cord, cuts off any input to or output from the organ, making that organ useless. Cut the nerve to a finger, and the finger goes limp. Peripheral nerves do have the property of growing back over a long period of time (months) if they are injured and then heal properly. This is another area where surgery, specifically microsurgery, is accomplishing incredible results with sewing nerves back together.

THE MUSCULOSKELETAL SYSTEM

In addition to supplying the needs of the body, the musculoskeletal system allows us to participate in life's activities. It's what allows us to fly airplanes. Most parts of this system are used in some way during flight, but they are not all essential. Given the appropriate tools and controls, and the proof of performance as a result of training and flight testing, many people can fly without a perfect body.

The body framework is made of bones (skeleton), tendons, and muscles. The covering, or skin, is actually a body organ, the largest in the body. The skull houses the brain and most sensory organs, the chest contains the lungs and heart, and the pelvis protects part of the abdomen and supports the body by being attached to the legs. The rest of the bones are used to manipulate the body and perform a variety of functions (walking, sitting, picking up things, flying airplanes, etc.).

The marrow in the center of bones produces the blood cells that carry oxygen, carbon dioxide, and nutrients.

Mobility of all the skeleton requires joints to function. Most are like simple hinges; others, like the hip and shoulder "ball-and-socket" joints, have a wide range of motion. All joints are held together with *ligaments*, which are tough connective tissues that don't stretch. In addition, there is a capsule surrounding each joint, within which there is a thick viscous fluid (*synovial*), which lubricates the joint just like any mechanical joint.

Tendons are the linkage between a bone and muscle. One end of the muscle is attached directly to the bone, usually away from a joint. The other end of the muscle, using the tendon, is attached to a variety of locations on the bone, often closer to the joint. Contraction of the muscle bends the joint.

Bones and joints do little by themselves except provide support and protection. Muscles make the skeleton functional. Muscles are controlled automatically as well as consciously, allowing for a wide range of motion and activity. Muscles can do only one thing, contract. Contraction is an active event; relaxing is a passive event. Yet even with this simple property, the body and its skeleton can stand, run, jump, make delicate adjustments to controls, pick up a pin, and assume a variety of positions few machines can duplicate. This all comes from a combination of perfectly structured bones that are connected to each other by unique joints that are controlled by a variety of small or large muscles that are controlled by the brain.

We are born with the same amount of muscle tissue that we have now. We do not add muscles when we exercise; we increase the size of each muscle cell. The efficiency of a muscle group is dependent on its physical condition and whether or not it gets adequate oxygen and nutrients.

Injuries are the most common pathology to the musculoskeletal system. The body is remarkably capable of rebuilding these tissues if healing is allowed to happen. We often tend to get anxious and put an injured part back to work only to be reinjured after premature use. Inflammation is another very common problem. Tendinitis, arthritis (artho=joint), and myositis (myo=muscle) are the most common. Tendons and ligaments are particularly slow to heal because of their low blood supply. Muscles, on the other hand, heal fairly quickly because they have a large blood supply needed to provide sustenance for their functions. Muscles, therefore, are easily bruised (a *bruise* is bleeding into the muscle tissues under the skin).

As one ages, the musculoskeletal system stiffens and becomes more brittle, and the muscles become weaker. Inflammation, especially in the joints, becomes more frequent. It also takes longer for the body to heal and it is less tolerant of the abuses to which we subject it. Some people find this difficult to accept. Staying in shape with a good exercise program minimizes these aging effects, but they still occur.

THE GASTROINTESTINAL (GI) SYSTEM

The purpose of the gastrointestinal (GI) system is to digest and provide nutrients and fluids for metabolism into the tissue cells. Foods and fluids that we ingest are transformed into energy and used for tissue rebuilding and maintenance of body functions.

The *GI system*, or *tract*, is essentially one long tube beginning at the mouth and ending at the anus. Within this tube, fluids are quickly absorbed and transferred to the bloodstream. Food is first physically broken up, then converted into basic components (mostly glucose and amino acids) through digestion. All waste products (undigested components) are passed out of the body.

When food enters the mouth, it is chewed, and saliva is made to moisten the food. Saliva also begins the digestive process with an initial breakdown of simpler foods. This partially processed food is then swallowed via the esophagus and into the stomach.

The stomach is the first major location for the breakdown of food. Within the stomach are *hydrochloric acid* and *pepsin* chemicals that begin to transform all food (protein, fat, and carbohydrate) into usable products. Food might remain in the stomach for as long as 4 hours, allowing the material to be digested into a semiliquid pulp. This entire mass is churned around inside the stomach by the process of *peristalsis*, which occurs when muscles within the stomach wall and the rest of the GI tract contract in a coordinated manner so as to mix the contents, much like squeezing a toothpaste tube to work out the toothpaste. The partially digested material is now squeezed into the first part of the small intestine.

The *duodenum* further digests material from the stomach by using four different chemicals: bile, trypsin, lipase, and amylase. This process results in products (simple sugars and amino acids) that can now be absorbed by the rest of the small intestine (the *jejunum* and *ileum*) and passed into the bloodstream. Material that is not digested, such as fiber, continues to be passed down the GI tract through peristalsis.

Throughout this whole process thus far, fluids are being absorbed and secreted. The absorbed fluids replace those lost through metabolism, sweating, and urination. Some are returned to the GI tract to assist in digestion.

The large intestine (colon and cecum) is where undigested material is collected and compacted and further broken down with bacteria. As this material, now called *fecal matter*, continues down the tract, fluids are once again exchanged to maintain a soft consistency. Without this process, constipation (or hard fecal matter with infrequent bowel movements) results. Fluids and fibers are important to maintain a consistent soft fecal matter that passes easily.

A bowel movement allows this material to pass from the large intestine, through the rectum, and out the anus. Peristalsis does most of the work, especially if the stool (fecal material) is soft enough and stimulates the action of the large bowel. Sometimes active straining is necessary.

This whole process of digestion within the GI system is another part of our diurnal and circadian rhythms. In other words, our internal rhythm tells the body when to expect food, which triggers hunger and stimulates the digestive juices. Peristalsis also begins, even though there might be no food yet (the "growling" stomach). This activity extends to the large bowel, which assumes there is food coming and it needs to rid itself of accumulated fecal material. If this whole process does not take place at the usual time, the system is fooled, like any disruption of a circadian body rhythm. Hunger patterns and signals become minimized and constipation is common.

Other pathology includes common inflammation: esophagitis, gastritis (gastro=stomach), ileitis, colonitis, and proctitis (procto=rectal). Depending on the location, this change can cause a variety of symptoms, ranging from nausea and vomiting to diarrhea. Gastritis can evolve into an ulcer of the lining of the stomach and duodenum. Further ulceration into the wall of these sections results in bleeding. This bleeding can be very dramatic with blood being vomited, often without pain.

More often there is pain in the upper abdomen with black tarry looking stools ("digested" blood mixed in with fecal matter). The concern with any of these problems is the unpredictability of recurrence. The GI tract is a relatively common place for tumors and polyps, which are sometimes malignant. Significant tumors can cause obstruction of the GI tract with few symptoms leading up to the recognition of even a partial obstruction.

THE METABOLIC SYSTEM

The metabolic system makes everything work. The nervous system supplies the information and direction, the musculoskeletal system is the container and does the mechanical work, the GI system processes the fuel and building material, and the metabolic system converts the resources into substances, chemicals, energy, and the like, which support the activities of the body.

There is a close relationship between all functions of metabolism that require continuous communication between the user (the end organs) and the decision maker (the brain) to determine on what functions the metabolic system must concentrate. The entire process is extremely complex and interrelated. Thick textbooks are written on portions of metabolic processes. Experts agree that there is still much to be learned about how our bodies function.

The organ carrying most of the workload is the liver, located in the upper righthand corner of the abdomen, partially under the rib cage. This organ (sometimes called a chemical processing plant) is an essential part of metabolism, changing the simple chemicals from the GI tract and blood stream into usable products such as vitamins, enzymes, bile, and cholesterol. In addition, it detoxifies drugs, medications, and alcohol. It makes glycogen (the body's storage form of glucose for future use), processes fats and proteins, removes waste products, and plays a role in the generation of heat for the body. The liver will regenerate itself if part of it is healthy. An inflamed liver is *hepatitis*, and this can be a serious illness, sometimes resulting in recurrence and converting liver tissue into fibrous nonfunctioning tissue (cirrhosis).

The gallbladder, which stores bile made by the liver, can form stones that block bile entering the GI tract. This backup can interfere with hepatic function. Malignant tumor cells from other locations and organs often migrate to the liver, taking over healthy liver cells and shutting down its activities. Since alcohol is a toxin to the liver, there is often an increased production of certain enzymes that can be measured in blood tests, indicating the liver is being affected.

The kidneys (located on both sides of the spine just below the back of the rib cage) act like a filter for the bloodstream and take out impurities and some waste products. The kidneys also regulate the volume of blood through conservation or excretion of water and adjust the composition and density to ensure the necessary balance of essential products. An important function is to maintain a fluid balance in the body. If there is not enough fluid (water) within the blood, the kidney will concentrate the urine it makes for removing unneeded body chemicals.

Dehydration is noted by a darker yellow color in the urine. If more water is ingested than is needed, the kidneys will allow more to pass out of the body in the form of urine. Now the urine is very pale. Because the kidneys are capable of this increased activity, one can rarely drink too much water, but there can be too little. Urine produced by the kidneys passes through tubes (ureters) and into the bladder where it is stored until there is an urge to urinate. The bladder can hold a lot of urine, which is why you can sleep through the night while the kidneys continue to produce urine.

Another problem relatively common in aviation is kidney stones. Here, concentrated urine from periodic dehydration over many months tends to form small sandlike "stones." If small enough, they pass without any symptoms; however, if a stone gets stuck in the ureter or at the entrance to the bladder, severe pain can suddenly result. Some people are more prone to making stones, and once one stone is made, that person is at higher risk of making others. Maintaining a high level of hydration, especially on long trips, is essential in preventing stone formation.

Situated to the left of the liver, the pancreas is an important organ. Its primary function is to produce insulin and some other enzymes and hormones. All body tissues require insulin for metabolism of glucose. The amount of insulin made (not stored) by the pancreas is dependent on the feedback system described earlier. There is also a circadian component in that at certain times of the day, the pancreas is "programmed" to begin producing insulin to metabolize expected food being transformed into glucose.

Diabetes is a disease where insulin is not produced enough or at all. Insulin-dependent diabetics need to inject insulin into their bodies to maintain a balanced use of glucose. Some diabetics have the capability of making some insulin but cannot without the help of an oral medication. This is not oral insulin but a chemical that stimulates the pancreas to make and excrete insulin, and even more important, helps the body cells utilize insulin better. Because the pancreas is stimulated by the presence or absence of the right kind of food, diet becomes a special problem. Exercise also plays an important role in preventing a diabetic situation. One of the most devastating cancers occurs in the pancreas. The problem is that this is a fast growing malignancy, rarely with symptoms until it is too late.

The thyroid is a gland located at the middle front of the lower neck. It makes two different hormones that play a significant role in *metabolic rate*, or the speed with which your body functions and responds to activities. Too high a rate makes you hyper, resulting in weight loss, nervousness, and a sense of feeling too warm. Another hormone assists in the metabolism of calcium and bones. The thyroid hormone is essential, but it can be replaced with oral medication; therefore, pathology that interferes with thyroid function often is treated by stopping the thyroid's function and replacing the hormone orally.

Females have unique physiological events that can affect safe flight. While most females have little trouble with their menses (or periods), some do experience both psychological and physiological problems, usually minor;

however, these periodic variations must be considered in the same category as other possible causes of subtle impairments.

Pregnancy has also been an issue for some individuals and some companies. In most cases, allowing a pregnant pilot to fly is left up to her doctor; there are no regulations guiding either party. Considerations must include radiation, "morning sickness," complications, and comfort as the pregnancy progresses, to name a few. Be sure the doctor is aware of the conditions of flight when requesting when it is safe to fly.

THE CIRCULATORY SYSTEM

Although the metabolic system is crucial to make everything work, the circulatory system carries the blood, which in turn transports oxygen, carbon dioxide, nutrients, and waste products. This system is comparable to any closed hydraulic system in an airplane, which includes fluid (blood), a pump (heart), tubing (blood vessels), and the part requiring support of the system (an organ). This system is of most interest to pilots because any change in oxygen levels to the cells immediately changes the performance of many organs, especially the brain; therefore, anatomy will first be described and then the physiology will be explained in more detail.

The heart

The heart is the pump in this human hydraulic system, which is a closed system with blood flowing from the heart, into arteries, then through capillaries that are spread around the tissues and individual cells (Fig. 2-1). From there, the blood continues into the venous system, through the lungs, and back to the heart.

This pump is divided into four chambers that essentially take blood from a major vein and pump it into arteries. The two smaller chambers on top (the *atria*) take in blood from the veins and "squeeze" it into the larger *ventricles* below. The ventricles then pump blood to the lungs and the rest of the body. The heart is a muscle and responds just like any other muscle in the body. Every muscle fiber or cell must contract at the same time, which squeezes the blood through the one-way heart valves. The blood moves forward under pressure but not backward, so there is a moment immediately after contraction when blood pressure goes down, but not to zero.

Flowing blood is not under the same pressure throughout the system, which is one of the differences compared to an aircraft hydraulic system. This "sinus wave" pressure is why we can feel a pulse. When the heart contracts, the pressure goes up, and we feel that rise. When the heart muscle relaxes, getting ready for the next contraction, there is a decrease in pressure. The heart contracts again before the pressure gets too low.

Like any other muscle, the heart needs a blood supply, and this is supplied via the coronary arteries. These arteries become blocked in heart disease, leading to heart attacks or poor blood perfusion (*ischemia*) of the

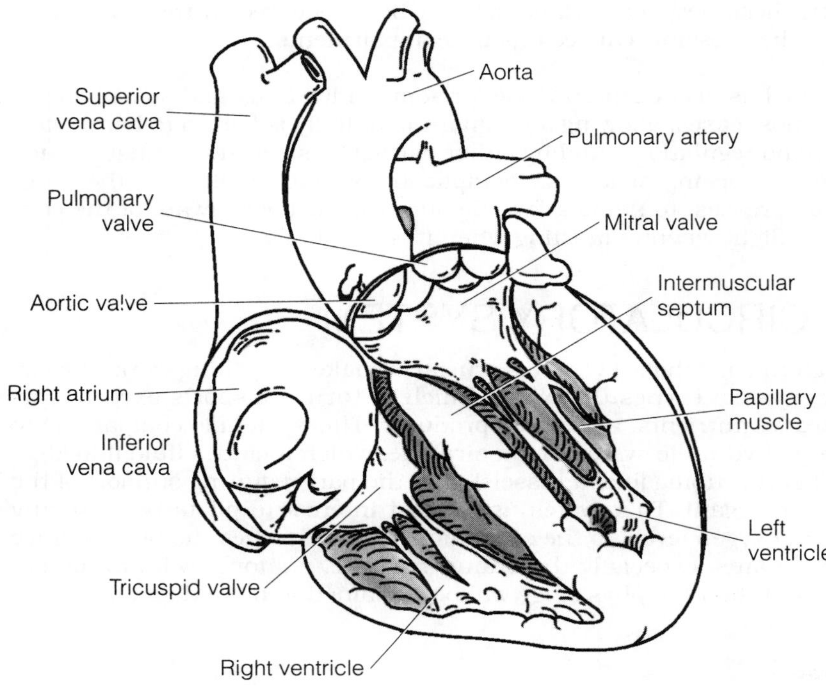

2-1 The heart is the pump of the circulatory system.

muscle cells. The heart muscle requires oxygen for energy; the heart can fail or become less effective if oxygen does not reach the heart muscle. Total blockage of blood to the heart muscle leads to death of some of the tissues devoid of blood. Dead heart tissue is called a *myocardial infarct* (myo=muscle, cardia=heart, infarct=dead), or *MI*. This myocardial infarction, therefore, is another name for *heart attack*.

If the heart-muscle fibers do not contract simultaneously, there are changes in the rhythm and contractility. *Fibrillation* occurs when the muscle fibers are contracting at different times and not in synchrony, looking like a can of live worms. Rhythm problems also can develop that compromise the efficiency of the heart's ability to move blood. Any of these situations leads to heart failure.

The amount of blood that flows out of the heart per minute is dependent on the heart rate (pulse), the volume of blood ejected from the heart, and the force of the contraction. Much of this is controlled by the autonomic nervous system that senses the need for more blood to a specific area of the body. An increase in carbon dioxide and other metabolites also makes the heart go faster to compensate for metabolic needs.

The flow of blood will become somewhat turbulent if it is interfered with by either a constriction of a blood vessel or a heart valve that is too small or leaks. This turbulence can be heard with a stethoscope and is called a *murmur.* A murmur is not pathology; it's an audible sign of potential pathology such as a calcified heart valve or an artery that is narrowed from arteriosclerosis. That cause can impair the performance of blood flow, weaken the heart, and decrease the perfusion of tissues in the organs. Like any hydraulic system, the end result is no better than the weakest part of that system.

The lungs

The lung organs exchange oxygen and carbon dioxide from the ambient air drawn into the air sacs (*alveoli*) with the same gases in the blood (Fig. 2-2). Air enters the lungs through the mouth or nose, into the pharynx, trachea, and then to the main bronchus. This then splits into two bronchi, which further split into thousands of smaller bronchioles and end up in tiny alveoli (tiny rubberlike balloons).

Capillaries surround these alveoli, and because both walls are so thin, gases can pass back and forth (diffuse), out of and into the air or into the bloodstream. This diffusion and gas exchange at the cell level will be further explained in the section on physiology of respiration.

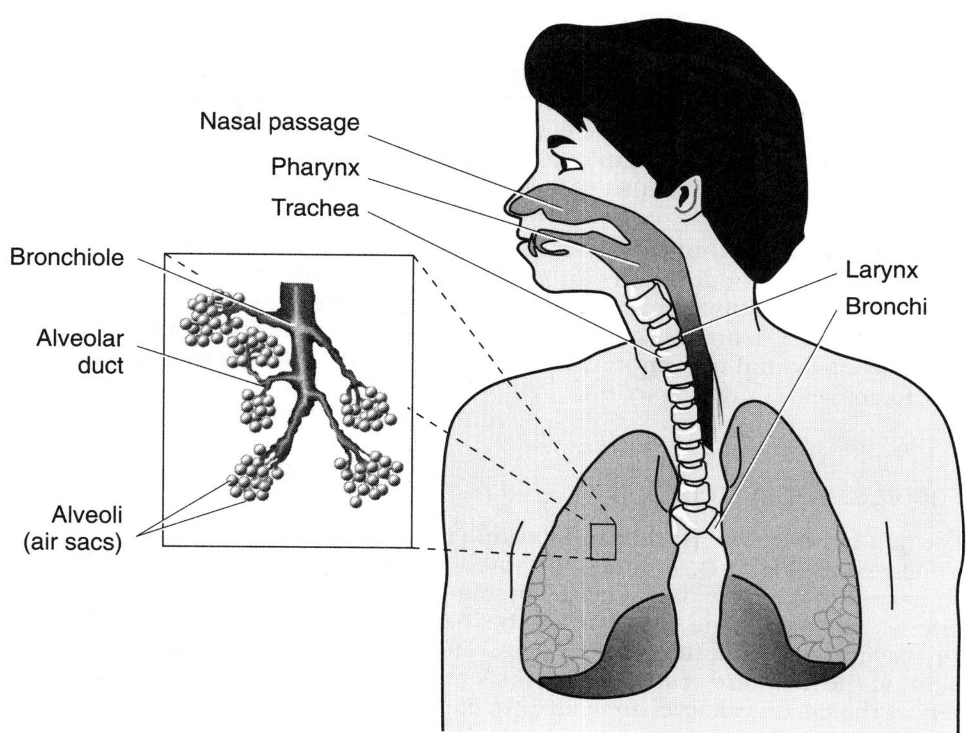

2-2 The lung resembles an upside-down tree.

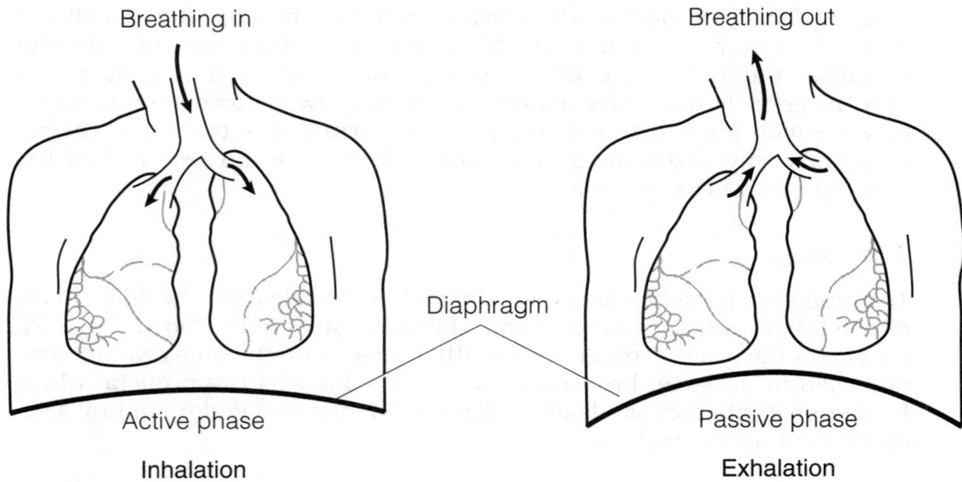

2-3 When the diaphragm descends and the chest cavity expands, air inflates the lungs (left). Air is forced out of the lungs when the diaphragm and lungs relax (right).

Air is pulled into the lungs by decreasing the pressure within the chest cavity (Fig. 2-3). This is accomplished by the diaphragm, a muscular, dome-shaped covering of the lower part of the rib cage, forming a closed container with one opening. When this muscle contracts, it flattens out, creating lower pressure and bringing in air to fill the alveoli. This inhalation is an active event (a muscle contracting), whereas exhaling is passive when the muscular diaphragm relaxes and returns to its dome shape. The muscles between the ribs also supplement this activity.

This process of breathing is also an autonomic function, controlled by the CNS and feedback data from various organs telling the breathing rate to change. With exercise, both breathing and heart rate are selectively increased to meet demands without the person having to think about it. We can also easily and actively control our breathing rate, whereas it is difficult to actively control heart rate.

The vascular system

The link between the heart, lungs, brain, and other parts of the body is the blood vessels (Fig. 2-4), and they maintain an uninterrupted blood supply to all tissues of the body. It is a closed system of arteries and veins. Basically, *arteries* distribute the blood from the heart to various organs, getting smaller (*arterioles*) as they branch out. The collecting vessels that return blood to the heart are *veins*; smaller veins are called *venules*. Very small vessels at the tissue-cell level are called *capillaries*.

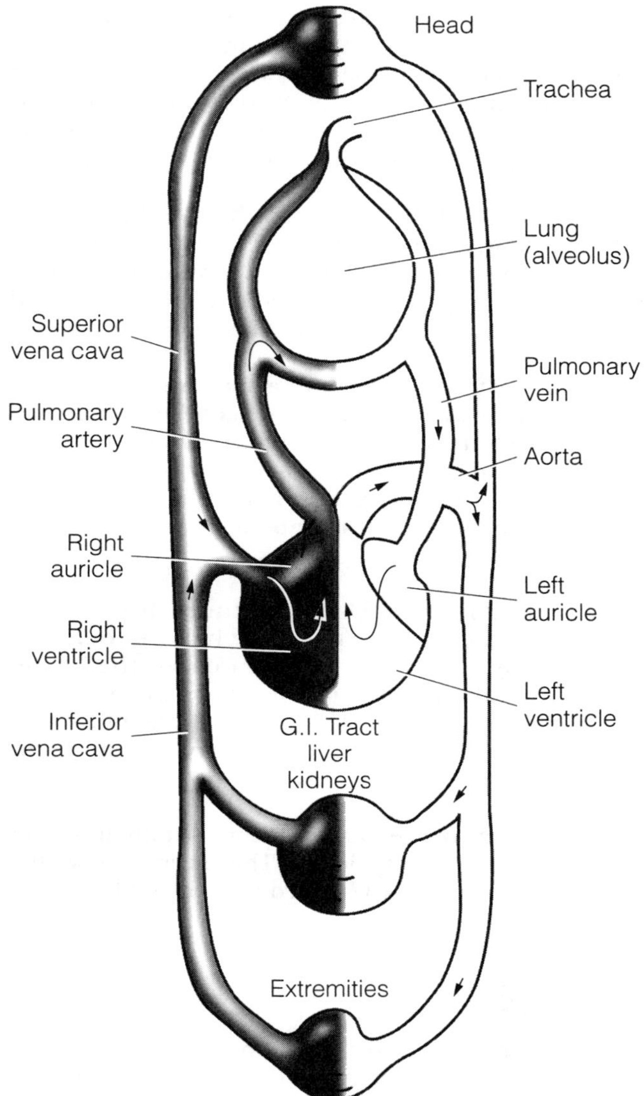

2-4 Essential oxygen and nutrients are transported throughout the body by the circulatory system.

Pulmonary arteries carry oxygen-poor blood from the heart to the lungs. Pulmonary veins carry oxygen-rich (reoxygenated) blood from the lungs to the heart. Oxygen-rich blood is then pumped through the arteries of the circulatory system to the tissues and individual cells. Arteries are unique from other blood vessels in that they have muscle and elastic cells within their walls, allowing them to dilate or constrict by muscle contraction or the inherent elasticity.

This is an effective way of increasing or decreasing (shunting) blood flow to various parts of the body. For example, during exercise, the leg muscles need more blood, so arteries to these muscles automatically dilate to their full size. Arteries to other parts, such as the digestive tract, are constricted by the wall muscles, shunting blood away from that area and into the legs. The same process happens in control of body temperatures.

The elasticity of the artery wall helps keep the blood pressure more constant during the period when the heart muscle relaxes between beats (contractions). Like blowing up a balloon, the pressure within the artery resulting from the elasticity enhances and prolongs the blood pressure that is generated by the heart during the relaxed phase.

Veins are simple tubes without muscles or elastic tissues. Some veins in the arms and legs have one-way valves that prevent blood from flowing backward during the pause or relaxed phase after it has been pumped forward during a heart contraction. This is necessary because blood pressure in the veins is very low, less than 10 millimeters of mercury (mm Hg) (as compared to pressure within the major arteries of about 120 mm Hg at the height of a heart contraction).

Contracting arm and leg muscles that surround veins compress the vessel's wall and squeeze the blood back to the heart in only one direction because of the one-way valves. This is often called the "muscle pump." Such action does not take place when you have been sitting for a long period of time and blood has pooled in the extremities (*stagnant hypoxia*). Veins carry blood that is deficient in oxygen and high in carbon dioxide.

Tissues and cells get their oxygen and nutrient supply from blood flowing next to the cells in the smaller capillaries. These very thin-walled tubes allow diffusion of gases through the walls from the blood to the cell and back again.

RESPIRATION

Respiration is the exchange of gases between the body and its tissues and the outside ambient air. The main objective of respiration is to add oxygen and remove carbon dioxide. Part of this activity requires transferring the gases from another gas mixture (the atmosphere) to a liquid (blood). It is more than just dissolving a gas in a liquid.

Blood is essentially cells and a liquid called *serum.* About 55 percent of blood is serum and about 90 percent of serum is water. One of the blood cells is a red blood cell, which physically carries oxygen molecules attached to a chemical on the cell called *hemoglobin.* The comparison of partial pressures and the percentage of hemoglobin saturation is called the *dissociation curve* (Fig. 2-5). Inadequate oxygen to the cells (hypoxia) will occur when the saturation goes below 72 percent, where the curve drops off quickly with decreasing oxygen partial pressure.

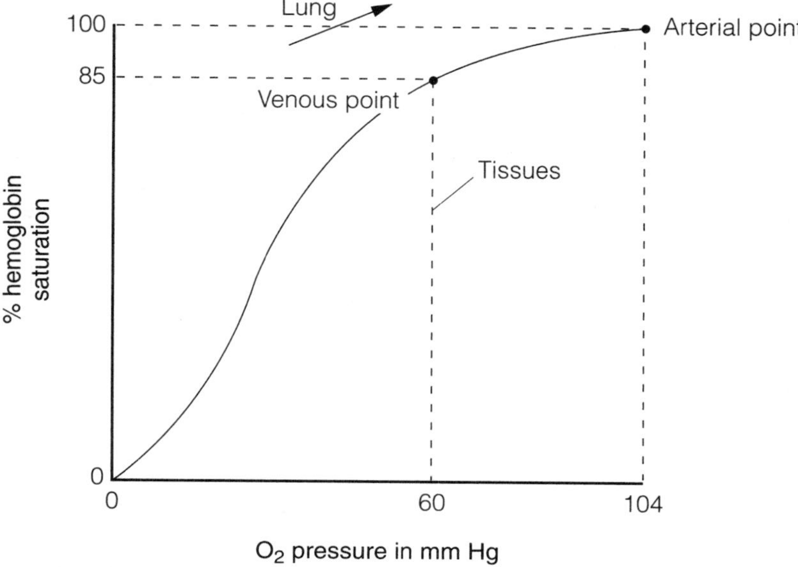

2-5 Arterial and venous saturation points for the normal sea-level conditions are shown.

The red blood cells transport oxygen through the use of the hemoglobin. Serum can carry only a small amount of oxygen in solution. The difference between the body's needs and what the blood can carry is made up by the hemoglobin's ability to carry oxygen. Any change in the hemoglobin, the number of red blood cells, or the diffusion of gases will cause some form of hypoxia (see chapter 5); therefore, respiration involves taking in air, transferring it to blood, and then transporting it via arteries to the cells. The whole process is reversed for removing waste gases via the veins.

Red blood cells are unique in how they carry oxygen. The bright-red color of arterial blood results from the combination of oxygen with hemoglobin. The darker color of venous blood is an indication of hemoglobin that has minimal oxygen. The amount of oxygen carried (or saturated) by the red blood cells is normally (and needs to be) about 95–98 percent.

This relationship between oxygen available and its partial pressure (see chapter 5) is seen in the *oxygen dissociation curve* and indicates that oxygen saturation of the blood falls rapidly when the partial pressure of oxygen goes below 60 mm Hg. The curve describes how venous blood carries less oxygen than arterial blood. The lower availability of oxygen is when hypoxia develops.

Physiology of respiration

Respiration is the process of exchanging gases in an environment with the gases within the tissues and cells of a living organism (Fig. 2-6). This process can be divided into two activities: external respiration and internal

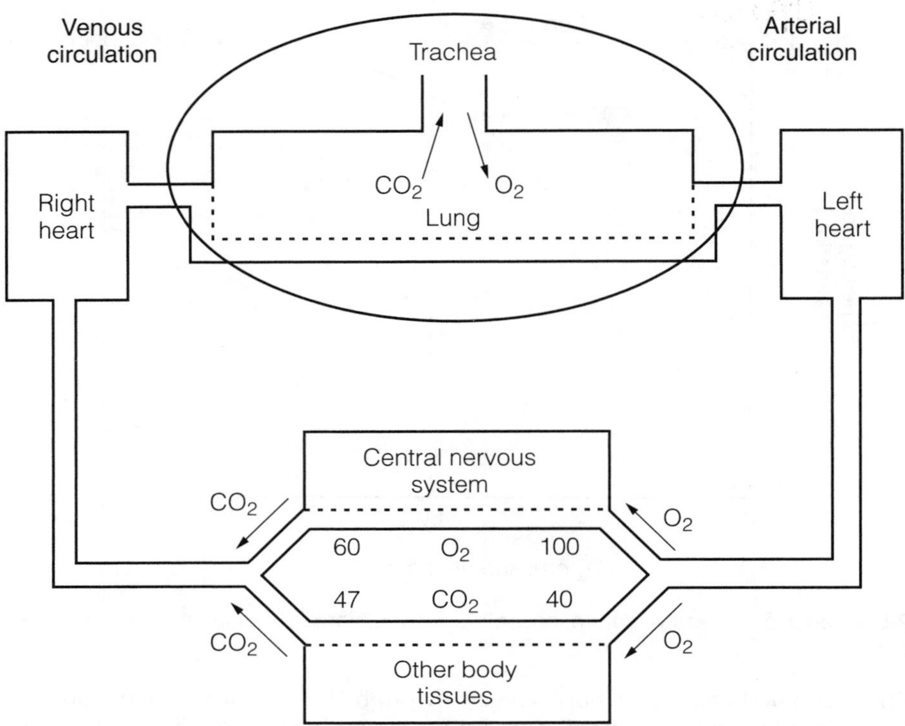

The problem area in hypoxic hypoxia is circled.

2-6 Gas exchange within the body.

respiration. External respiration occurs in the lungs, where air is inhaled and exhaled and gases are transferred (diffused) through the lungs and into the bloodstream. The internal process refers to the transport of gases to and from body cells and tissues by the blood and red blood cells.

The function of respiration is to get oxygen into the body and to the cells and to take carbon dioxide from the cells and remove it from the body. Body temperature is also controlled somewhat by the transfer of heat that takes place during this same circulatory process (chapter 10).

By reviewing the actions of the lungs, chest wall, and diaphragm, one can see how external respiration takes place. It becomes more complex when considering how the gases get into the bloodstream. The alveoli within the lungs are very thin walled (one cell thick) and are surrounded by the small capillaries, also with walls one cell thick. These capillaries are so small that only one red blood cell can pass through, one behind the other; therefore, the gas molecules and the blood cell are separated by only two permeable walls each one cell thick.

Gases move across these membranes by the process of *diffusion*, which lets a gas move from an area of high pressure to an area of lower pressure (Fig. 2-7). The oxygen, when it reaches the alveoli, has a partial pressure of about 95–100 mm Hg (Fig. 2-8). Within venous blood, the partial pressure is 40 mm Hg. Diffusion takes place as long as there is a differential. Carbon dioxide diffuses in the same manner. The partial pressure of carbon dioxide in venous blood is 47 mm Hg; in the arteries, it's 41 mm Hg; and in the alveoli, it's 40 mm Hg. The same process of diffusion occurs at the tissue and cell level within the body.

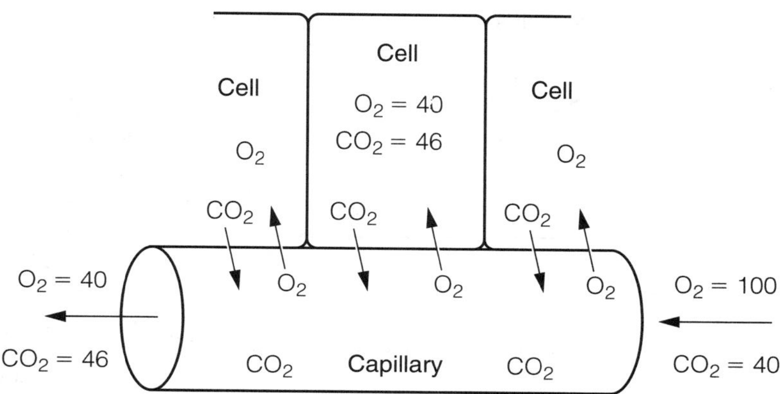

2-7 Tissue-capillary diffusion of oxygen and carbon dioxide (in mm Hg.)

The body has a variety of receptors and feedback mechanisms that can detect when oxygen supply and carbon dioxide levels are incompatible with metabolism. This control and regulation determines how quickly and deeply we breathe, how hard the heart pumps, where blood is shunted, and how metabolic processes are controlled. A description of this process is beyond the scope of this text, but it should be recognized that this regulatory system determines our level of impairment in flight.

Obviously, the body is far more complex than what is described here; however, in order to understand how the physiology of flight can affect performance, you must have a basic understanding of how the body works. Then some of what you hear and read about in terms of impairment and associated medical standards will make sense; the regulations and company insurance policies will also make better sense. Misinformation abounds regarding why these standards, policies, and regulations exist, and only people who understand and respect flight physiology can judge their significance in safe flying. Furthermore, anything that interferes with a normally functioning body—your health—can also determine if you can meet FAA medical certification standards.

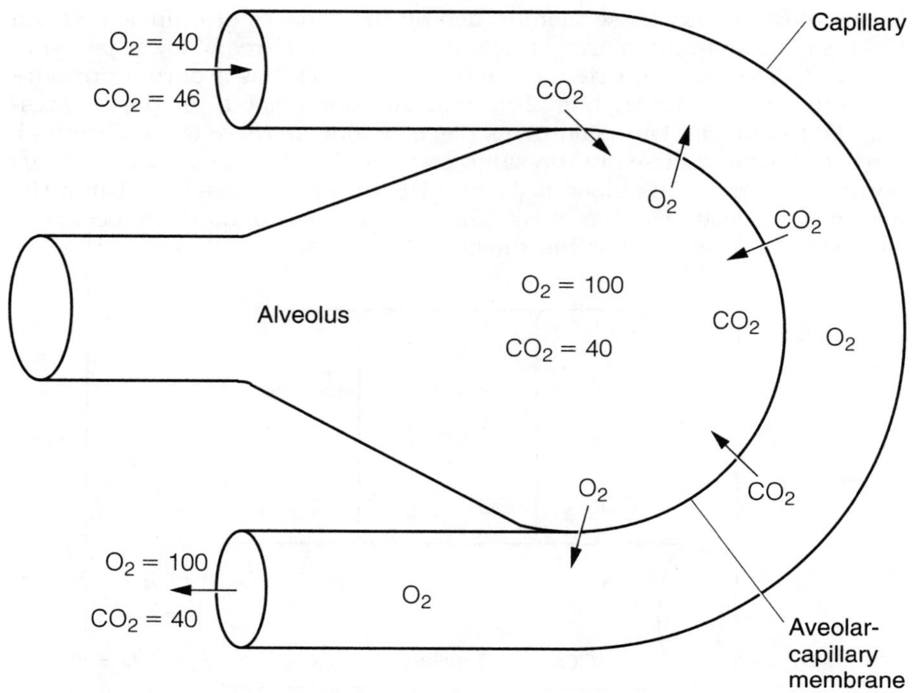

2-8 Alveolar-capillary diffusion of oxygen and carbon dioxide.

3
The atmosphere

The air, like the sea, can be lethal to the human animal stripped of his artificial protections. Adapted to breathing air at sea level density and pressure, he loses consciousness and dies at great altitudes where the pressure of the atmosphere is too low to force life-giving oxygen through his lungs and into his blood stream. In the realm of the 'special senses,' man is found deficient. God may not have built man to fly; yet today, borne aloft by the power of thousands of horses, he soars through the heavens as one returned to an ancient domain, the land of fantasy now made real.

Douglas H. Robinson, M.D. "The Dangerous Sky," 1973

THERE ARE THREE DISTINCT PARTS TO FLIGHT: the plane (or "aerospace vehicle"), the pilot, and the environment in which the pilot lives and flies (the atmosphere). All three are interrelated, but the atmosphere affects both the plane and the pilot. The airplane might be right at home in thin air; the pilot is not. An aircraft engine might be more efficient at altitude; the human body is not.

Therefore, it is imperative that all aircrew have a complete understanding of the makeup of the atmosphere from a physiological perspective. Knowledge of the pilot's working environment will help her recognize the potential dangers and subtle changes in herself and her crewmates and maintain a high index of suspicion and potential problems in her performance. Only through awareness of the atmosphere's impact on a pilot will she have a true respect for her vulnerability. The safe pilot knows where she is working.

This chapter will explain the atmosphere. This insight will be a base of needed information for later subjects that impact most physiological events. Plan on using this chapter as a frequent reference. Subjects covered will include the composition of the atmosphere with elements that have a direct effect on our health and performance, as we all know but often minimize. This composition is affected by pressure and temperature, and one needs to know when and where.

Additionally, a variety of independent layers are within the atmosphere (troposphere, stratosphere, etc.), much like layers of water and oil in a standing bottle. The composition of air within these atmospheric layers is different. These layers also can be divided physiologically to identify where the dangers to humans lie. All of these factors are directly related to known gas laws, especially how our body is affected.

COMPOSITION OF THE ATMOSPHERE

The atmosphere (atmos = vapor) is the gaseous envelope of air that surrounds the earth and extends to about 25,000 miles (40,000 km). It rotates with the Earth and is continuously changing with temperature and pressure. There will be constant parameters and many variables, often occurring in unpredictable situations. From a biological and physiological point of view, however, the composition of the atmosphere is constant.

Because of this overall dynamic state, the composition of the atmosphere is expressed in percentages and not absolute amounts. In other words, the percentage of oxygen is about 21 percent within the entire sphere of gases covering the Earth, but the number of oxygen molecules, the amount of oxygen available to the human body, decreases with altitude and is affected by temperature, the presence of water vapor, and other variables. Furthermore, these percentages are for dry air. The presence of water vapor is also described, but it is not considered part of the gaseous composition of the atmosphere. Rather, its presence varies and decreases the percentages of the other gases proportionately and to a minor degree (less than 5 percent).

Of primary importance to the pilot are oxygen and nitrogen and to a lesser degree carbon dioxide, ozone, and water vapor. Too much or too little of any of these gases could compromise health and safe performance. The body, especially the brain, is not very tolerant of wide variations of gas availability. Such extremes will be covered more extensively in chapter 5 on altitude physiology. Suffice to say, understanding the role of the atmosphere's composition is as important as the physics of why an airfoil flies. You can fly without knowing what is happening, but you will not be a competent, knowledgeable, or safe pilot when something unexpected happens to disrupt the flight.

Oxygen

A colorless, odorless, tasteless elemental gas, oxygen (O_2) is the most abundant of all elements on Earth and occurs as a free element in the atmosphere. Oxygen comprises 20.95 percent, or roughly one fifth, of the total gases in the atmosphere. The source of oxygen in our atmosphere is generally from the process of plant photosynthesis where it is generated from carbon dioxide and water.

Of all the gases, oxygen is the most crucial in maintaining life. As explained in chapter 5 on altitude physiology, oxygen is essential for life as we know it. Any level of oxygen deprivation is incompatible with safe performance, even if only a subtle unrecognized minor impairment.

Nitrogen

Another colorless, odorless, and tasteless elemental gas that is also free in the atmosphere is nitrogen (N_2). It is the most plentiful gas in the atmosphere, comprising 78.08 percent of the total gases. Consequently, nitrogen is responsible for the major portion of the total atmospheric pressure, or weight, upon the Earth's surface and the human body.

This gas is an essential building block of all living things on Earth, yet it is not necessary to be used by man in the metabolic process; however, nitrogen saturates the cells and tissues of the body and, like other gases, too much in the tissues means trouble (i.e., the bends). This also will be explained in chapter 5.

Carbon dioxide

Although only 0.03 percent of the atmosphere, carbon dioxide (CO_2), also colorless, odorless, and tasteless, creates some of the most disabling physiological problems, especially in flight. A by-product of respiration, carbon dioxide is somewhat heavier than air. As a result of fossil-fuel combustion, carbon dioxide is the only gaseous constituent of the atmosphere that has increased significantly in amount during the past century.

Inert gases

Comprising less than 1 percent of the atmosphere, inert gases include argon (0.93 percent), neon (0.0018 percent), helium (0.0005 percent), krypton (0.0001 percent), hydrogen (0.00005 percent), and methane (0.0002 percent). These atmospherically rare gases have little impact on the body's physiology.

Water vapor

As previously stated, water vapor is not considered one of the components of the atmosphere. When we talk about humidity and dew point, we are talking about the presence of water vapor in the ambient (surrounding) air within which we live and breathe. The content of water in the air depends on how much water remains as a gas as the temperature changes. Even at 100 percent humidity at 100°F, there is less than 5 percent water vapor. If all the water vapor in the atmosphere were to be removed at once, it would represent about 1 inch of rainfall all over the Earth's surface. The percentage of water vapor decreases dramatically as we ascend. Compared to the troposphere, the stratosphere is virtually dry. Water vapor has a variety of well-known sources, usually evaporation from large bodies of water. The impact of high and low humidity on health and performance will be described in chapter 10 on environmental stress.

Ozone

Ozone (O_3) is an unstable pale-blue gas with a characteristic pungent odor often noticed during thunderstorms and around electrical equipment that produces sparks. It is formed by photochemical action on atmospheric oxygen and is found in high concentrations at atmospheric levels in excess of 5 miles. Lesser concentrations are found at lower levels and have in recent years become involved in the physiology of flight. (See chapter 5 on altitude physiology.)

The ambient ozone concentration is dependent on the season, latitude, and altitude as well as changing weather systems. Although present mainly in the stratosphere, ozone is not considered a component in the composition of the atmosphere. In the 1980s, this ozone layer became a controversial issue relative to the layer's shielding the surface of the Earth from radiation. Radiation is further discussed in chapter 10. Ozone is also recognized for causing discomfort and possible impairment primarily of the pulmonary system, especially for cabin crews and in high concentrations when exposed during extended high-altitude flight.

PHYSICAL CHARACTERISTICS OF THE ATMOSPHERE

As with any gas, especially a mixture of gases, physical properties are universally applied (Fig. 3-1). The atmosphere is basically one large volume of gases surrounding the Earth, and it has the same characteristics as a volume of gas within a closed container in a physics lab. Each of these gases has a direct impact on how the body functions in flight.

Altitude		**Pressure**		**Temperature**	
(Feet)	**(Inches Hg)**	**(mm/Hg)**	**(PSI)**	**(°C)**	**(°F)**
0	29.92	760.0	14.69	15.0	59.0
10,000	20.58	522.6	10.11	–4.8	23.3
18,000	14.95	379.4	7.34	–20.7	–5.3[1]
20,000	13.76	349.1	6.75	–24.6	–12.3
25,000	10.51	281.8	5.45	–34.5	–30.1
30,000	8.90	225.6	4.36	–44.4	–48.0
34,000	7.40	187.4	3.62	–52.4	–62.3[2]
35,332	6.80	175.9	3.41	–55.0	–67.0
40,000	5.56	140.7	2.72	–55.0	–67.0
43,000	4.43	119.0	2.30	–55.0	–67.0
48,000	3.79	96.0	1.86	–55.0	–67.0[3]
50,000	3.44	87.3	1.69	–55.0	–67.0
63,000	1.86	47.0	0.980	–55.0	–67.0[4]
80,000	0.882	20.8	—	–55.0	–67.0
100,000	0.326	8.0	—	–55.0	–67.0

1—½ atmosphere
2—¼ atmosphere
3—⅛ atmosphere
4—1/16 atmosphere

3-1 Standard pressure/temperature values at latitude 40°, temperature 36°F (2°C).

In addition, because of the instability and dynamic properties of the atmosphere, it is difficult to predict at any given time what the atmosphere will be like as we fly through it. The only way to be prepared is to have a familiarity with these gases, how they relate to each other, and how they affect the physiology of the body in flight.

Temperature

Solar radiation heats the Earth's surface, which in turn reradiates the heat to the atmospheric air. Radiation causes only minimal heating to the atmosphere as it passes through. As we ascend from the surface, the temperature falls steadily and predictably with altitude throughout the troposphere. This *lapse rate* is 3.56°F (1.98°C) per 1,000 feet (305 m). This temperature lapse rate levels off at the tropopause, around 35,000 feet. The tropopause is higher over the equator and this results in lower temperatures at higher altitudes (up to 80,000 feet). The temperatures also will vary according to the time of the year because the tropopause is higher in the summer than in the winter.

Pressure

Barometric (atmospheric) pressure is the weight per unit area of all the molecules of the gases (the air) above the point at which the measurement is made; therefore, this weight, or atmospheric pressure, decreases with altitude. The pressure drop, however, is not linear because air is compressible relative to the weight above. Density now plays a role. The air near the surface of the Earth is more dense, having more molecules of gases within a given volume; therefore, a greater pressure change, or pressure differential, occurs as one nears the Earth's surface.

A variety of physical properties of the atmosphere will change this pressure, in addition to its weight. This results from temperature changes secondary to the seasons, weather systems, and location on the globe (the latitude and longitude of a particular point). To establish a base pressure from which to work, a standard atmosphere is defined. The most commonly used base is the *international standard atmosphere* (ISA), which establishes a mean atmospheric pressure of 29.92 inches of mercury (in. Hg), or 760 millimeters of mercury (mm Hg) at a temperature of 15°C (59°F) in dry air at sea level.

This pressure is also stated as 14.7 pounds per square inch (PSI) or 1,013.2 millibars (mb) at the same temperature. Key landmarks, stated in atmospheres are: Sea Level/760 mm Hg = 1 atmosphere, FL 180 (18,000 feet) = ½ atmosphere, FL 430 = ¼ atmosphere. Compared to scuba diving, 33 feet of sea water is twice sea level pressure, or 2 atmospheres.

Ninety-nine percent of the Earth's atmosphere, in terms of the quantity of gases, exists below 20 statute miles (32 km). The greatest pressure differential encountered during an ascent is from sea level to about 5,000 feet (1,500 m); therefore, even in a pressurized aircraft, the problems associated with changes of pressure must still be considered because pressurized cabins often get up to 6,000 feet (1,800 m) or more.

From a physiological point of view, most pressures are expressed as standard atmospheric with the base at sea level; however, for the pilot at varying altitudes, it is important to recognize that pressure is not constant and is a variant of many physical properties, as defined in traditional gas laws. This will be especially true as we discuss the physiology of oxygen/carbon dioxide exchange in the body, pressure breathing at altitude, decompression sickness, and ear blocks. Thus, human performance is measured by the pressure altitude, which is shown by the actual reading on the cabin altimeter, not corrected for temperature. This is different for the airplane, which is dependent on the variations of the atmosphere, especially density altitude, which is related to temperature and humidity.

GAS LAWS

The composition of the atmosphere is a mixture of gases and is subject to the same laws of physics as you would see in a physics lab. If you are to understand how and why these gases affect the body, you must also have a basic understanding of the classic gas laws.

Boyle's Law

This law states that the volume of a gas is inversely proportional to the pressure (temperature remaining constant). This applies to all gases. $V_1/V_2 = P_2/P_1$ (V_1 is the initial volume of the gas, V_2 is the final volume, P_1 is the initial pressure on the gas volume, and P_2 is the final pressure). In other words, if the pressure of the gas decreases with the temperature unchanging, then its volume increases and vice versa (Fig. 3-2).

In dealing with gas expansion in the body, a correction must be made for the ever-present water vapor; therefore, the formula now becomes: $V_1/V_2 = (P_2\text{-}47 \text{ mm Hg})/(P_1\text{-}46 \text{ mm Hg})$ (Fig. 3-3). Water-vapor pressure at body temperature is 47 mm Hg. Such characteristics applied to the body ex-

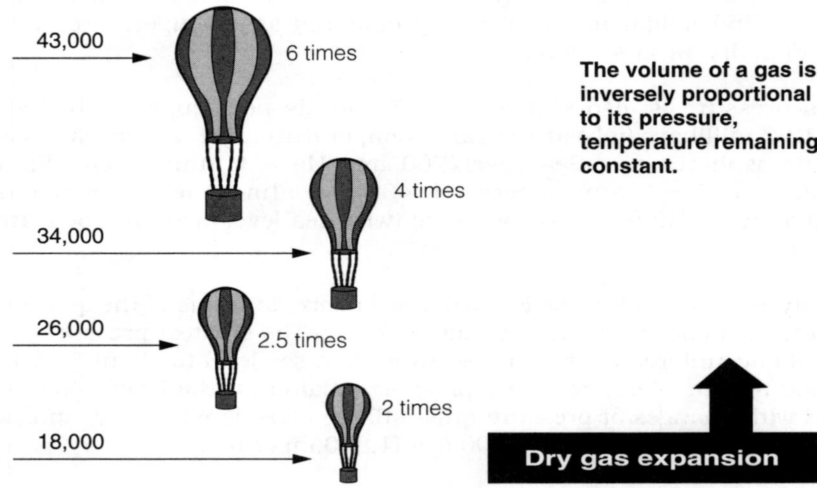

3-2 Comparative volumes of dry gases at increasing altitudes and decreasing pressure.

plain the expansion of gases trapped within such moist areas as the middle ears, sinuses, stomach, and intestines. These are all actual or potential cavities within which moist air is present and can become trapped and expand like any other gas; hence, the physiological topic of "trapped gases," which will be discussed in chapter 5 regarding altitude physiology.

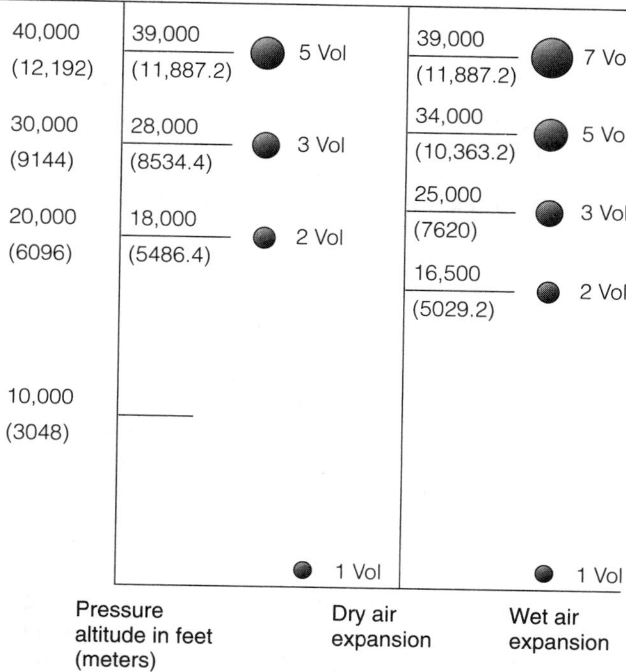

3-3 The effect of water on gas expansion.

Charles' Law

Charles' Law states that the volume of gas is directly proportional to the temperature (pressure remaining constant). This applies to all gases. This law has no direct physiological significance because body temperature remains fairly constant. It does, however, explain the fact that pressure within supplemental oxygen containers will decrease if the ambient temperature surrounding the storage container decreases, even when no oxygen has been used, such as at altitude.

Dalton's Law

Since the atmosphere is a mixture of gases, and each gas has its own pressure at any given temperature within a given volume, it is important to also be familiar with the physics of the combined pressures. Dalton's Law states that the total pressure of a gas mixture is the sum of the individual pressure (also called partial pressure) that each gas would exert if it alone occupied the whole volume.

Or expressed mathematically: $PT = P_1 + P_2 + Ps + Pn$; where PT is the total pressure of the mixture of gases and the P value is the partial pressure of each gas, which is determined by multiplying the percentage of the individual gas times the total pressure. Each kind of gas exerts its own pressure, depending on the percentage of that gas in the mixture; thus, even though the percentage of oxygen in the atmosphere is constant (about 21 percent), its partial pressure will decrease proportionately as atmospheric pressure (expressed as barometric pressure) decreases.

The body is affected by the pressure of gases available. In other words, as we ascend, the percentage of each gas in the atmosphere remains the same, but there are fewer molecules at less pressure that the body can use (or get rid of). This decrease of available molecules of oxygen at a pressure required to pass to a blood cell, for example, is what leads to hypoxia in the body.

Henry's Law

Henry's Law states that the amount (or weight absorbed) of gas in solution, not chemically combined, varies directly with the partial pressure of that gas over the solution. $P_1A_2 = P_2A_1$; where the A value is the amount of gas in solution initially. In other words, when the pressure of a gas over a certain liquid decreases, the amount of that gas dissolved in the liquid will also decrease (and vice versa); therefore, when equilibrium is attained, the dissolved gas tension will equal the partial pressure of the gas in the atmosphere to which a solution is exposed. As the pressure "falls" (during ascent), the amount of gas that can be held in solution is reduced.

This is demonstrated every time you open a carbonated beverage. Once the seal is broken and the gas that has been under pressure escapes, gas that has been dissolved within the beverage begins to escape by forming bubbles on the side and rising to the surface (Fig. 3-4).

In the human body, however, additional factors influence and modify the process of gas uptake and elimination, such as the varying types of fluids in the body, the circulation rate and volume of the blood, and the amount of hemoglobin (where the gas is chemically attached), among others. These will all be discussed in subsequent chapters.

Graham's Law

Graham's Law of gaseous diffusion states that a gas of high pressure exerts a force toward a region of lower pressure, and that if an existing membrane separating these regions of unequal pressure is permeable or semipermeable, the gas of higher pressure will pass (or diffuse) through the membrane into the region of lower pressure. This process, which occurs in milliseconds, continues until the unequal regions are nearly equal in pressure.

For example, this law explains the transfer of oxygen from one part of the body to another as it passes within blood vessels and then through the ves-

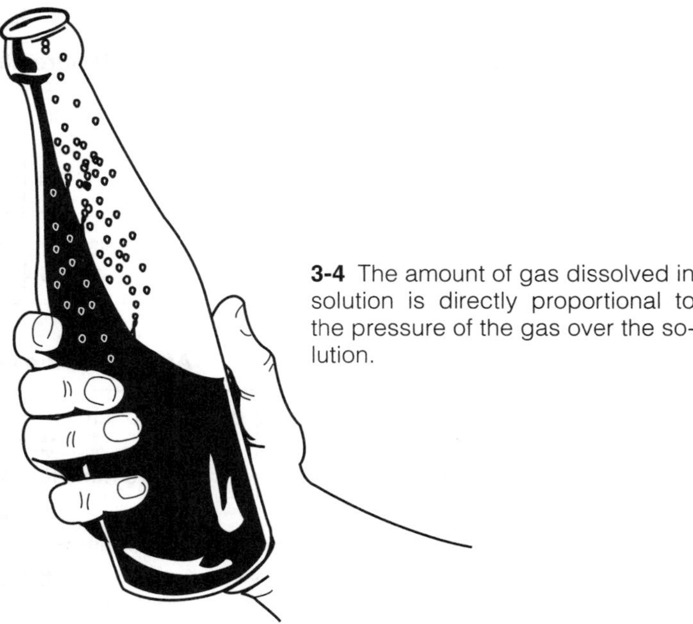

3-4 The amount of gas dissolved in solution is directly proportional to the pressure of the gas over the solution.

sel wall to the adjacent cell. Each gas behaves independently of the other gases in a solution and might move in opposite directions to another gas of differing partial pressures (Fig. 3-5). Figure 3-6 summarizes the gas laws.

DIVISIONS OF THE ATMOSPHERE

The atmosphere is divided into layers around the Earth, similar to the layers of an onion (Fig. 3-7). These envelopes of air are composed of several gases described earlier. *Sphere* signifies the actual layer of air, whereas *pause* is the outer boundary of that layer, separating the next layer. The density of the atmosphere decreases with height because it is a compressible gas, and the molecules are pressed closer together near the Earth's surface from the sheer weight of these gases. Gas is more dense, therefore, at ground level.

The troposphere

The troposphere is the layer that lies in direct contact with the surface of the Earth and is characterized by the presence of:

- Water vapor (humidity).
- A constant decrease in temperature with increasing altitude.
- Large-scale vertical currents that keep the gaseous component remarkably constant through continuous mixing. These vertical currents also lead to turbulence.

Most weather systems occur in this layer.

A

Before diffusion

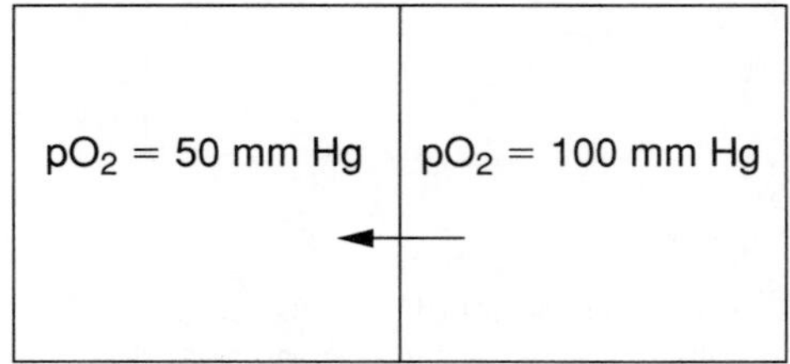

After diffusion (at equilibrium)

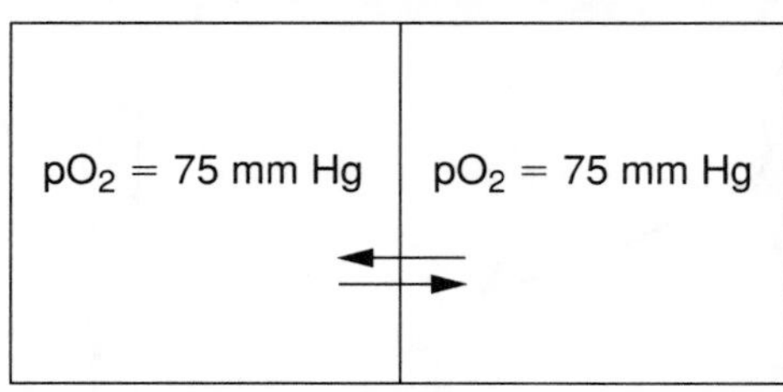

B

Before diffusion

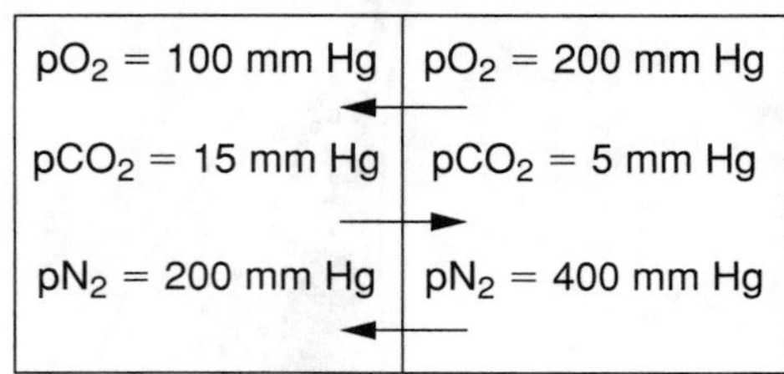

After diffusion (at equilibrium)

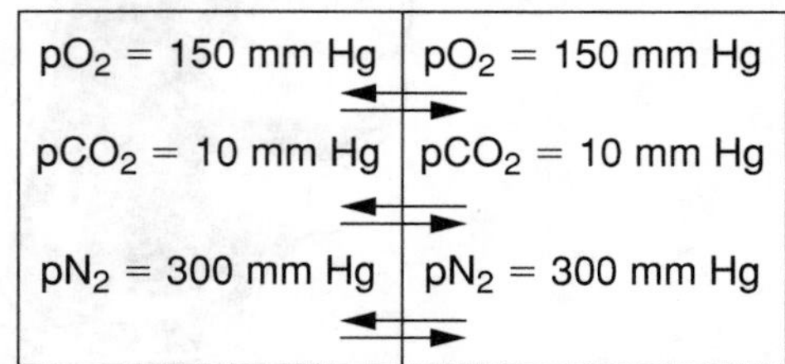

3-5 Gases diffuse across membranes from a higher partial pressure to lower and vice versa.

Gas law	Formula	Statement of law	Physiological significance
Boyle's	$P_1V_1 = P_2V_2$	Volume of gas is inversely proportional to its pressure (temperature constant).	Trapped gas in the body
Dalton's	$P_T = P_1 + P_2 ... P_n$	Total pressure of a mixture of gases equals the sum of the partial pressure of each gas in the mixture.	Hypoxia
Henry's	$P_1A_2 = P_2A_1$	Amount of gas in solution is directly proportional to the partial pressure of that gas over the solution.	Evolved gas in the body
Charles'	$P_1T_2 = P_2T_1$	Pressure of a gas is directly proportional to its temperature (volume remaining constant)	Storage of oxygen in containers
Gaseous diffusion		A gas will diffuse from an area of high concentration (or pressure) to an area of low concentration.	Transfer of gases in body (O_2 and CO_2)
	P = Pressure V = Volume A = Amount T = Absolute temperature		

3-6 Review of gas laws as used in flight physiology.

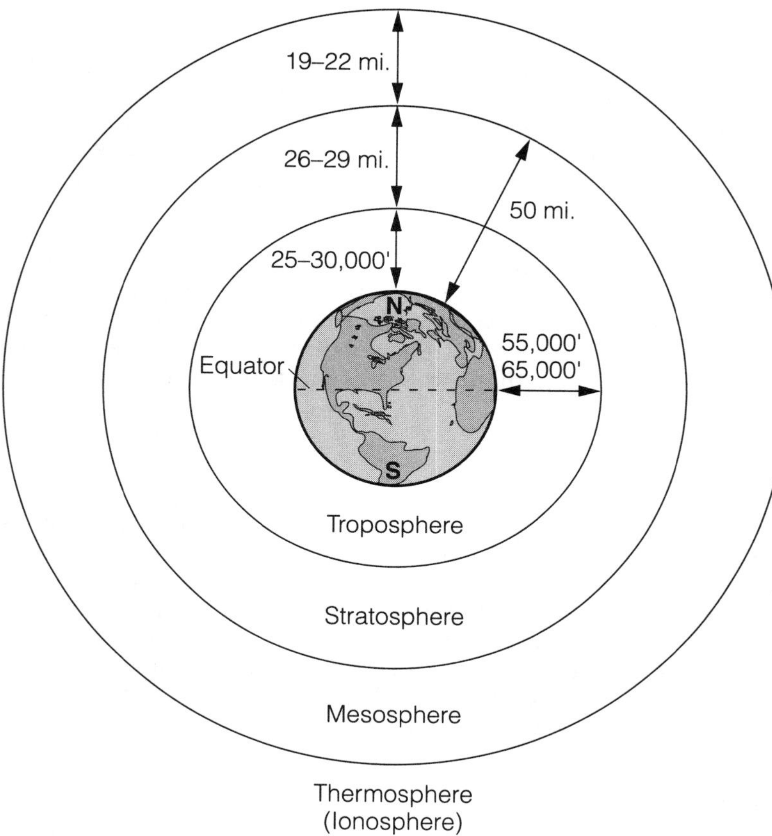

3-7 The atmosphere is divided into layers base on temperature.

Recall the lapse rate previously mentioned in this chapter; the rate at which temperature decreases with altitude averages 3.6°F (2°C) per 1,000 feet in still air. When the tropopause (the outer boundary) is reached, the temperature remains constant. This layer reaches an average of 30,000 feet (9,000 m) at the poles to about 60,000 feet (18,000 m) above the equator (as a result of the rising warm, heated air at the equator). Almost all commercial flights take place within the troposphere.

The tropopause

The tropopause is affected by the heat within the troposphere and the seasons as well as its position relative to the surface. For example, more solar energy is received by the Earth in the region of the equator than at the poles, and this directly determines the height of the tropopause. This level between the troposphere and stratosphere is a region of temperature stability.

The stratosphere

The stratosphere is known for an almost complete absence of water vapor and a fairly uniform temperature (–69.7°F, –56.5°C) that changes little at different altitudes. It reaches an additional 46,000 feet (14,000 m) above the tropopause. In some classifications of the stratosphere, it is extended to a height of 50 miles (80 km), and this additional area is sometimes called the *mesosphere*.

This higher layer has a temperature that first rises and then falls to a minimum temperature of about –162°F (–107°C). The stratopause is the upper boundary of this layer. Contrails (water-vapor trails) of high-flying aircraft indicate the plane is below the tropopause. If it were in the stratosphere where there is no water, there would be no trails, and only the aircraft could be seen.

The ionosphere

The gases present in the ionosphere, the layer above the stratosphere, are ionized as a result of photochemical and photoelectric reactions involving ultraviolet radiation from the Sun; thus, this is where ozone is most prevalent and in sufficient amounts to shield the Earth from ultraviolet radiation. This layer, with its ozone, acts as a reflector for other long-wavelength electromagnetic radiation and is further affected by solar disturbances. This will be more fully described in chapter 10. The Aurora Borealis and Aurora Australis are found in this layer.

The exosphere

Above the ionosphere is deep space, or the exosphere, the outer boundary of the atmosphere of the Earth.

PHYSIOLOGICAL DIVISIONS OF THE ATMOSPHERE

In addition to its pressure and temperature divisions, the atmosphere is also differentiated into zones that can affect us medically and physiologically. Knowing that you are flying in any of these areas should raise the level of suspicion of any performance that is unexpected or not considered acceptable in yourself and others. The reason might be physiological changes in the varying atmosphere.

Physiological-efficient zone

The physiological-efficient zone extends from sea level to approximately 10,000–12,000 feet and represents where the human body is adapted or can adapt. The barometric pressure drops from 760 mm Hg at sea level to 483 mm Hg at 12,000 feet. The oxygen levels are usually sufficient to maintain an otherwise healthy pilot without supplemental oxygen. Decompression sickness is also rare in this zone; however, rapid changes after a prolonged time at altitude can be incapacitating, such as ear blocks.

Physiological-deficient zone

In the physiological-deficient zone, there is increased risk of problems, especially hypoxia, trapped-gas, and evolved-gas situations. This zone extends from the top of the physiological-efficient zone (12,000 feet) to about 50,000 feet. Above this altitude, a human cannot exist unless he or she is in a pressurized suit or cabin; therefore, within this zone, the body responds as if it were in deep space.

For example, body fluids, such as blood, boil at about 63,000 feet (also called *Armstrong's Line*), where the boiling point of water is 98.6°F. Despite this dangerous environment, high-tech aircraft such as supersonic transports (SSTs) as well as some corporate aircraft and undoubtedly many military "aerospace vehicles" are routinely operating in this zone. Above 50,000 feet is the *space-equivalent division*.

GENERAL EFFECTS ON THE HUMAN BODY

Recognizing and understanding this hostile environment should allow the pilot to respect his working conditions and the associated problems. Rather than just accepting that it is "dangerous up there" because someone said so, aircrew can now reason for themselves what the risks are and anticipate potential problems. Ear blocks and hypoxia are common, although rarely reported into a database. Often the pilot is not aware of the problem simply because as long as he feels okay, he does not suspect anything. This is a poor indicator of unsafe performance for a responsible pilot.

Despite war stories of braving the elements at all altitudes, the body still responds to traditional gas laws and is affected by predictable gas concentrations; therefore, if you are in the atmosphere anywhere above the level at which you live, you are at risk, no matter how you perceive your performance. Simply "being at any altitude"—it doesn't have to be "high altitude"—invites potential and often undetected changes in piloting skills. Flying at any altitude is safe and tolerated by the protected body, but the atmosphere is not forgiving if taken for granted. Subsequent chapters will give you the insight to stay fit and safe.

4

Situation awareness

On June 8, 1992, a Beechcraft C99 was cleared by ATC for the ILS approach to Runway 5 at Anniston, Alabama. The flightcrew turned the airplane toward the north away from the airport in the erroneous belief that the airplane was south of the airport. In actuality, the airplane had intercepted the back course localizer signal for the ILS approach and the flightcrew had commenced the approach at a high airspeed about 2,000 feet above the specified altitude for crossing the final approach fix. The airplane continued a (rapid) controlled descent until it impacted terrain. The captain and two passengers received fatal injuries.

NTSB accident report

THE NTSB DETERMINED THAT THE FLIGHTCREW FALSELY BELIEVED that the flight was receiving radar services from ATC and commenced the approach from an excessive altitude and at a cruise speed without accomplishing the published procedure specified on the approach chart. They did not know their aircraft's position. One of the probable causes, according to the NTSB, was "failure of the aircrew to use approved instrument flight procedures, which resulted in a loss of situational awareness and terrain clearance."

Obviously there were other contributing and causal factors that led to the accident. The key term here is "situational awareness." And this and related terms are showing up more and more in accident reports. For example, a Hibbing, Minnesota, accident in December 1993: The aircraft flew into terrain on an ILS approach. It was probably caused by "loss of altitude (situation) awareness by the flightcrew during an unstabilized approach in night IMC."

But why, you might ask, is situation awareness a part of a flight physiology textbook, especially close to the beginning? Because situation awareness (SA), or situation*al* awareness (both terms are acceptable and used interchangeably), has become a descriptive title for many factors that can lead to an accident or incident. It's in the same category as impairment and incapacitation, defined in an earlier chapter. They all are human factors and can add links to the growing chain of events (or situations) that end up being described in an accident report.

The reason it's brought up along with the physiology of flight is that how we feel affects how we think, mentally process information, and then react. Everyone, several times a day, while at work, driving, puttering around the house, and the like, will "disconnect" from what's happening around them because they are tired, ill, distracted, disoriented, or a multitude of other physiological factors; hence, you will notice SA and the word "situation" mentioned frequently in the upcoming chapters. As you study each of the physiological and psychological stressors in flight, you can define a relationship between the topic and its effect on maintaining SA. How many times have you been lost, communicated on the tower frequency instead of the departure frequency, or made an approach to the wrong runway or airport? Those events don't happen because you are dumb or untrained. You lost SA because you were probably subtly impaired, distracted, or disoriented.

SITUATION AWARENESS DEFINED

Despite everyone's acceptance of the existence of SA, it's often difficult to define it specifically. Much like explaining safety or human factors, it's a broad topic and means different things to different people in different roles in aviation. First off, you will hear both situation and situational. Depending on who you talk to, you will notice both terms being used. They mean the same thing but might not be grammatically correct in the context of being used.

The next issue is that we know what it means, but how do we explain it to someone else? And it's important to recognize this because SA impacts our performance and decision making; therefore, we must educate and train flightcrew in SA. Furthermore, with a clearer, albeit varied, description, there can be several educational and training scenarios within which SA can be discussed, including safety, CRM, human factors, and, of course, flight physiology.

The military has a definition, stating that it is the accurate perception of what is going on with you (and your crew, wingman, lead, student, instructor, etc.), the aircraft, and the surrounding world, both now and in the near future. Someone said that "when perception matches reality, you are situationally aware."

Here's another variation from a CRM expert: "SA is the continuous ability of a crew, acting as a single entity, to accurately perceive the relationship of themselves and their aircraft to all factors that could influence successful completion of their mission and to forecast as well as execute tasks based on that perception."

You can come up with your own definition based on your experience. If you acknowledge the fact that situational awareness exists and its maintenance must be respected, then you are a step closer to a safe flight.

NASA's explanation seems to be more specific and proactive. That is, it includes how to deal with loss of SA as well as define it. NASA's definition includes "the awareness of acknowledging and assessing; and is the basis for choosing courses of action of the situation both now and in the future." The key point is "assessing" and how to assess the situation. It follows and is equally important: How do we train pilots to assess what's happening in an environment filled with many variables and many different situations? Then, how does our fitness to fly impede SA?

Probably the best example of loss of SA is *controlled flight into terrain* (CFIT), commonly found in accident reports and statistics. It's also another example of how easy we can label what caused an accident or, as stressed throughout this text, an incident. But little is stated as to the underlying or contributing causes, which are those early links in the accident chain that are not identified and therefore not broken. This is where hypoxia, fatigue, self medicating, stress, and other human factors must be included when it comes to training pilots in CRM and prevention of pilot error. CFIT prevention, as a very visible topic as well as SA, is usually expected, often required, in a pilot's initial and recurrent training. And it must be introduced at day one of a pilot's career.

COMPONENTS OF SITUATION AWARENESS

Given the various definitions stated above, SA can be broken into several performance components:

- Maintaining awareness of the situations occurring in flight requires a continuous sensitivity to location, communications, surrounding traffic, weather, and crew performance, to name a few.
- Also, each crewmember must be tuned into the ever changing performance of the aircraft and its multiple systems, using a variety of cues, instrumentation, observations, and other sources of data to monitor that performance.
- Furthermore, each pilot must have an awareness of "the big picture," that is, what's going on around him or her inside the aircraft and within the airspace and environment ahead.
- Finally, the crew must be able to assess all theses parameters and have the ability to anticipate decisions rather than waiting for the situation to change and then react in a crisis-management mode.

Note that for each of these components there is some element of flight physiology that affects the results. Virtually all of our senses are tasked: vision, hearing, smell, taste, touch, proprioceptive, and the like. Our mental capacity is challenged with multiple tasks, which means nothing must impair the brain's ability to function as from hypoxia, fatigue, medication, and the like. You might be the best pilot with a keen sense of situational awareness, but if you are tired, in a noisy cockpit, not wearing your glasses, and at a cabin altitude of 7,000 ft., you will be compromised and eventually lose SA. What happens next depends on your assessment of the situation just before an incident evolves into an accident.

In fact, complacency, a trait common with anyone when one feels overconfident, becomes the greatest deterrent to maintaining SA. Assuming you are functioning physiologically because you feel OK leads to thinking that you can handle problems when they arise, rather than planning ahead by anticipating the situations that could develop. Complacency lulls us into thinking that every flight does not require all our skills and experience. Losing SA is our first clue that we are complacent. And as you study further, you will find that complacency becomes a symptom of fatigue and hypoxia, which becomes a vicious circle.

Reading transcripts of NTSB accident reports becomes a new and revealing educational exercise when you factor in the crew's complacency about their awareness of their deteriorating situation that they are in, often compromised by their physiological status. More and more accident and incident reports (and submissions to the Aviation Safety Reporting System) are noting these contributing factors. Although the probable cause might be CFIT due to loss of SA, there has to be an explanation as to why situational awareness was lost.

CAUSES AND CLUES OF LOST SITUATION AWARENESS

Here is a partial list of causes for losing situational awareness; there are many more, depending on "the situation." Note that they are all "CRM related" and usually fall into that educational category; however, keep these in mind as you continue your studies and, more importantly, as your flights become more complex and multitasked. Then carefully consider how your physiological status could make situations worse.

- Failure to stay ahead of the aircraft (not anticipating activities)
- Increased number of repeated ATC communications
- Ambiguity (disagreement of information from different sources)
- Not meeting expected checkpoint times or other expected events
- Fixation and channelized thinking (symptoms also of fatigue and hypoxia)
- Confusion or uncertainty about clearances, performance parameters, and the like
- Violation of minimums (even the mere *consideration* of taking a chance)
- Taking shortcuts from standard operating procedures (The leading cause of accidents and incidents is not following procedures.)
- Operating aircraft outside its published limitations
- No one looking out the windows
- Pilot unsure of condition of aircraft

PREVENTION OF LOST SITUATION AWARENESS

The most obvious controllable preventive measure that all pilots can take is increased awareness of SA: maintaining a higher index of "suspicion," especially when less than fit to fly. Noticing two or more clues that SA is "drifting away" should get your attention; although circumstances might justify missing one clue. Communication is probably the leading resource to help you avoid getting into trouble. This is where CRM comes in: how to manage the resources you have available to prevent loss of SA or get out of a jam when you recognize that you're "behind the power curve."

Maintaining situational awareness is crucial in the ongoing decision-making process. When workload increases, this process can become impaired. Losing SA, even minimally, adds to the chances of making the wrong decision. All this is assuming that the pilot is free from physiological impairments. Under the usual conditions of flight, no pilot is free of these impairments, but they are often minimized or go unrecognized. Raising the awareness of the impact of flight physiology on SA and decision making becomes a primary objective of a flight-physiology course.

A mental checklist that can be very helpful is the DECIDE model. It's been around for years and is still worthwhile to keep handy. It can be your final "checklist" before you begin a flight, to remind you of the process to determine the safe action.

D = Detect the fact that a change has occurred that requires attention.

E = Estimate the significance of the change to the operation.

C = Choose a safe outcome for the operation.

I = Identify plausible actions and their risks to control the change.

D = Do the best option. This letter "D" can also mean "discuss" options with crewmates.

E = Evaluate the effect of the action on the change and on progress of the operation.

Couple this model with an inventory of your fitness and a healthy respect for the effects of the physiology of flight and you can decrease the odds of becoming an accident or incident.

5
Altitude physiology

A 46-year-old airline pilot was test flying a new widebody airplane before acceptance by his company. Part of the evaluation included some engine-out performance that included loss of cabin pressure. Some minor aircraft problems kept the pilot working for several minutes at a cabin altitude of 12,000 feet. After the flight, he mentally reviewed what had happened and recognized he had felt strange. This concerned him enough to seek medical attention. He called his doctor about his feelings of lightheadedness and the headache he developed after the flight along with some vision disturbances. The pilot was concerned he might have had a mild stroke. The doctor diagnosed the symptoms as those of hypoxia, something the pilot never even considered at the time of the flight.

HIGH ALTITUDE IS A MISNOMER commonly used to label those conditions that affect the body as a result of altitude above where one's body usually lives. The perception is that only at high (whatever high means) altitudes should one be concerned about hypoxia, ear blocks, and the like. In actuality, anytime the human body remains at any level above where that body is used to being, there is a risk of problems; therefore, this chapter is not limited to airline pilots or anyone flying above 18,000 feet. The contents to follow affect all pilots, at all levels, and include recreational flying, helicopter activities, and low-level commercial flight.

The most common problems directly associated with the physical properties of the different layers of the atmosphere include hypoxia, trapped-gas situations, and decompression sickness often called the "bends." In addition, scuba diving has become very popular for all ages and is a major factor in safe flying. Being able to identify the symptoms of any of these conditions in yourself and in your crewmate is essential. The safe pilot will respond with immediate and appropriate action to prevent incapacitation and minimize impairment.

THE PHYSIOLOGY OF OXYGEN IN THE BODY

By definition, the lack of adequate oxygen in the body's metabolism is called *hypoxia* (or *anoxia* for the total absence of oxygen). It is the lack of sufficient oxygen to the body's tissues and cells. This lack of oxygen to the cells exists for reasons in addition to the availability or lack of oxygen in the air we breathe. These reasons include the blood's ability to carry oxygen and the interference of drugs, alcohol, carbon monoxide, smoking, and illness, to name a few.

Many incidents and some accidents are "officially" related to the pilot's inability to detect that he or she was hypoxic, even mildly hypoxic, and was therefore unsafe because of compromised skills/judgment, subtle pilot error, and lack of respect or understanding of hypoxia. Hypoxia is often not considered when a pilot is not functioning as expected. To fully understand the challenge in getting oxygen to the cell, we must understand the physiology of oxygen. Then we can understand and respect hypoxia, no matter what the cause.

REVIEW OF RESPIRATION PHYSIOLOGY

Oxygen in the ambient air must reach the cell and does so because of gas laws, the lungs, circulatory system, blood, and hemoglobin. An interruption of that process from a single problem or a combination will result in hypoxia, each situation with a different name. Ambient atmospheric partial pressure of oxygen decreases during ascent, which translates into a lower pressure gradient in the lungs for diffusion to take place across the alveolar membrane (Dalton's and Graham's gas laws). Hemoglobin (Hgb), the primary transporter of oxygen within the blood, does not have access to adequate amounts of oxygen to attach. Or there isn't enough Hgb available. Oxygen molecules are biochemically attached to the hemoglobin molecule, which is affected by the surrounding partial pressures of oxygen (and carbon dioxide) which, in turn, transfers these gases to and from the tissue cells. The oxygen hemoglobin dissociation curve indicates a rapid decrease of saturation—and transfer—when oxygen partial pressure drops below 60 mm Hg. Ideally, arterial pressure should be near 80–90 mm Hg or a saturation of 87–97 percent.

The tissue cell might be unable to take on the oxygen because it is impaired or diseased. Also, the amount of blood flow to the tissue cells is compromised if the circulatory system is compromised.

The entire process is reversed for removing carbon dioxide from the tissue cell and bringing back oxygen-poor hemoglobin. This is critically important since both directions of this transfer process affects the feedback "signals" to the brain, which adjusts the entire system to meet the needs of the body. The biochemical process is quite complex, and further information is available in the resources appendix.

Classification of hypoxia

No matter what the cause of hypoxia, the symptoms and effects on flying skills are basically the same. The following classification is therefore intended only to review these causes so that your degree of suspicion of being hypoxic (or observing hypoxia in crew or passengers) is raised when you are subjected to these varied situations. How you feel when hypoxic and what to do about it will be discussed next (Fig. 5-1).

Phase of respiration	Condition	Specific cause	Type of hypoxia
Ventilation	Reduction in alveolar PO_2	Breathing air at reduced barometric pressure Strangulation/respiratory arrest/laryngospasm Severe asthma Breathholding Hypoventilation Breathing gas mixtures with insufficient PO_2 Malfunctioning oxygen equipment at altitude	Hypoxic hypoxia
	Reduction in gas exchange area	Pneumonia Drowning Atelectasis Emphysema (chronic lung disease) Pneumothorax Pulmonary embolism Congenital heart defects Physiologic shunting	Hypoxic hypoxia
Diffusion	Diffusion barriers	Hyaline membrane disease Pneumonia Drowning	Hypoxic hypoxia
Transportation	Reduction in oxygen-carrying capacity	Anemias Hemorrhage Hemoglobin abnormalities Drugs (sulfanilamides, nitrites) Chemicals (cyanide, carbon monoxide)	Hypemic hypoxia
	Reduction in systemic blood flow	Heart failure Shock Continuous positive pressure breathing Acceleration (G forces) Pulmonary embolism	Stagnant hypoxia
	Reduction in regional or local blood flow	Extremes of environmental temperatures Postural changes (prolonged sitting, bed rest, or weightlessness) Tourniquets (restrictive clothing, straps, and so forth) Hyperventilation Embolism by clots or gas bubbles Cerebral vascular accidents (strokes)	Stagnant hypoxia
Utilization	Metabolic poisoning or dysfunction	Respiratory enzyme poisoning or degradation Carbon monoxide Cyanide Alcohol	Histotoxic hypoxia

5-1 Causes and types of hypoxia. R.L. DeHart. *Fundamentals of Aerospace Medicine.* Philadelphia, Lee & Febiger, 1985. Reproduced with permission.

Hypoxic (altitude) hypoxia

This term is commonly used when talking about hypoxia associated with lack of available oxygen, as experienced when flying at altitude in an unpressurized cabin. It means that there aren't enough oxygen molecules available to breathe with sufficient pressure, as when we ascend. The number of molecules of oxygen decrease, despite the fact that the percentage remains the same. This situation is particularly evident in the physiologically deficient zone of the atmosphere.

Strictly speaking, one is hypoxic even a few hundred feet above the ground. In actuality, the symptoms do not become a significant factor until about 5,000 feet, especially at night. The significance of the symptoms is related to many factors, which will be explained later. Suffice to say, hypoxia must be considered as being present at all flight levels, including high cabin altitudes.

From a gas-law perspective, hypoxic hypoxia exists when the partial pressure of oxygen in the atmosphere or the inhaled ambient air is reduced. This reduced partial pressure is also present in the inspired air as it travels into the bronchial tree and into the lungs. In other words, the partial pressure of oxygen as it is presented to the blood within the lung is too low to effectively carry and transfer enough oxygen to the cells of the tissues.

Hypemic (anemic) hypoxia

This occurs when the blood's ability to carry oxygen molecules is the problem, even though there is adequate oxygen available in the air to breathe and exchange. This happens for a variety of reasons.

Anemia, or a reduced number of healthy, functioning red blood cells for any reason (disease, blood loss, deformed blood cells, etc.), means less capacity for blood to carry oxygen. Hemoglobin (Hgb) physically carries 75 times more oxygen molecules than are dissolved in solution. Anything that would interfere or displace oxygen that is attached to the hemoglobin would also reduce the oxygen available to the cell. This occurs most commonly when carbon monoxide is inhaled unknowingly along with the cabin air. Hemoglobin accepts carbon monoxide 250 times more than oxygen and therefore competes with adequate oxygen transfer. Other chemicals such as sulfa drugs and nitrites (as in food preservatives), have the same combining activity to hemoglobin, thus competing with oxygen.

Stagnant hypoxia

If the blood flow (the circulation of the oxygen carrying hemoglobin) is compromised for any reason, then sufficient oxygen cannot get to the cells and tissues. *Stagnant* implies a diminished flow, not necessarily a complete stoppage. Such decrease in blood flow results from the heart failing to pump effectively, an artery constricting and cutting off or reducing the flow, and venous pooling of blood because of gravity, such as in varicose veins of the legs.

Another cause, unique to flying, is during positive-G maneuvers—pulling Gs and long periods of pressure breathing at extreme cabin altitudes where oxygen masks are required. Another situation is in cold temperatures where blood supply to the extremities is decreased so as to shunt blood away to crucial organs. All these situations can lead to stagnant hypoxia.

Histotoxic hypoxia

Histotoxic means the cell expecting and needing oxygen is abnormal and unable to take up the oxygen that is present. This abnormality has been created as a result of a toxin or toxins present/absorbed by the cell. In other words, the oxygen might be inhaled and reach the tissue or cell in adequate amounts, but the cell is unable to accept and use the available oxygen. This can occur when alcohol is present in the blood or in the cell and prevents the essential use of oxygen by the cell. Alcohol becomes a toxin to the cell. The same is true for some narcotics and certain poisons, such as cyanide.

Stages of hypoxia

No matter what the reason for oxygen not getting to the cell or being used in its metabolism, the lack of oxygen—hypoxia—results in a variety of subtle and not so subtle symptoms: any single or concurrent combination of several situations causing some degree of incapacitation. The danger of hypoxia is that the pilot is probably unsuspecting that she is hypoxic.

The key to flying safe at altitude is to recognize:

- The conditions under which you could be hypoxic.
- The physical and mental symptoms that indicate you are hypoxic.
- When a crewmate is susceptible to hypoxia in those conditions.

Because the nervous system tissues have a heavy requirement for oxygen, especially the brain (and eyes), most hypoxic symptoms are directly or indirectly related to the nervous system. If hypoxia is prolonged, serious problems develop with ultimate death (Fig. 5-2). In extreme cases (prior to death), some brain cells are actually killed, and they cannot be regenerated. It bears repeating. The single most dangerous characteristic of hypoxia is that if the crewmember is hypoxic and engrossed in flight duties, the pilot might not even notice the impairment. Only through education, continued awareness, and actual exposure to hypoxia in controlled conditions (as in an altitude chamber) can the pilot truly respect this insidious hazard.

Because hypoxia can and often does develop gradually, the pilot must recognize its various stages, allowing some degree of anticipation if symptoms in the early stages are identified. The earlier that hypoxia can be recognized, the sooner that corrective action can be taken before the pilot becomes unable to act appropriately. Keep in mind that although altitude is the common denominator in being hypoxic, the health of the pilot can affect his tolerance. Other abuses, such as smoking, alcohol, and stress, can reduce this tolerance in the otherwise healthy pilot; therefore, hypoxia can be very unpredictable. Saying you had no problem at 10,000 feet last week does not mean you will be safe today at the same altitude.

Indifferent stage

One of the earliest symptoms of hypoxia is its effect on the eye. Vision, especially night vision, will deteriorate even at altitudes less than 5,000 feet. And you won't know you are having problems. Other classic symptoms can be present at lower altitudes if the body is less tolerant, as will be explained later. Suffice to say, at any altitude, at night, be aware that your

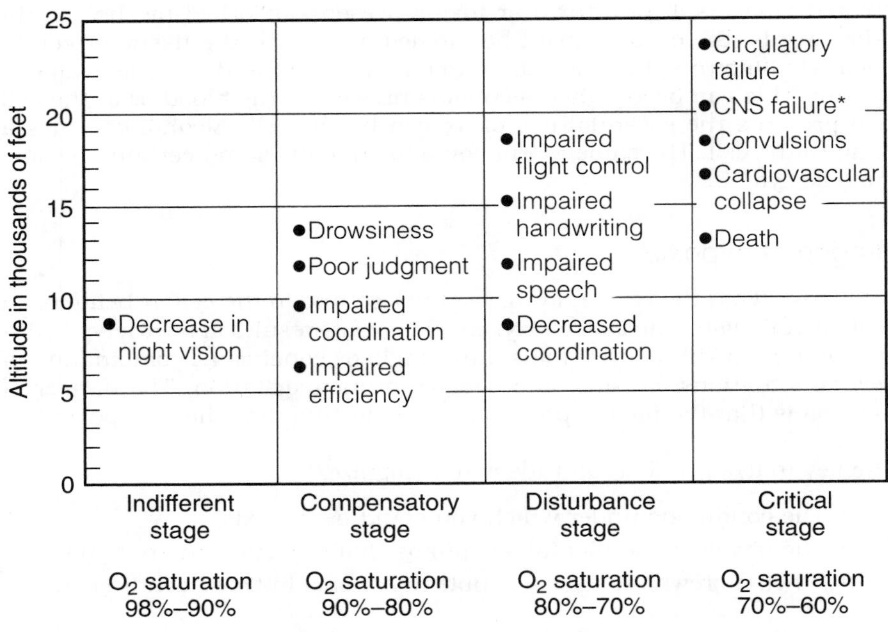

5-2 Stages of hypoxia related to arterial blood oxygen levels.

vision is compromised. For example, night vision is lost by 5–10 percent at 5,000 feet, 15–25 percent at 10,000 feet, and 25–30 percent at 12,000 feet.

Compensatory stage

During this stage, the body and mind can be severely affected and in increasing and subtle ways; however, the circulatory system and, to a lesser degree, the respiratory system can provide some defense against hypoxia while in this stage. This happens as a result of increased heart rate, enhanced circulation, and a more productive cardiac pumping of blood. Respiration (breathing rate and depth) also increases. Although these body responses are automatic, one should not assume recovery without taking immediate conscious corrective action whenever you suspect hypoxia.

At 12,000 to 15,000 feet the effects of hypoxia on the nervous system can become increasingly incapacitating, especially for the unacclimated. As time at this altitude continues (as little as 10–15 minutes), impaired skills are very evident. A variety of symptoms develop (a complete list follows), many being unsafe. Such symptoms as drowsiness, poor judgment, and frequent subtle errors in flying skills become apparent. More dangerous is a feeling of well-being and indifference (euphoria).

Once again, the crucial characteristic about hypoxia, especially in this stage of potential recovery and compensation, is the recognition of an increasing oxygen deficit before you are unaware you are in trouble. The same holds true in your crewmates. Keep checking on each other especially in an environment conducive for hypoxia. Beyond this stage, you will not suspect hypoxia and be so impaired that you won't be able to correct the situation by descending or donning an oxygen mask.

Disturbance stage

This is the stage when chances of recovery are greatly diminished. Symptoms become more severe with headaches, hyperventilation, impaired peripheral vision, marked fatigue, sleepiness, and especially euphoria. By now you might even recognize you are hypoxic, but you don't care and have little incentive nor energy to take corrective action such as getting to a lower altitude or going on oxygen.

Critical stage

This is when you have lost it—you're unconscious. All this can happen within 3–5 minutes after you failed to recognize you were hypoxic during the compensatory and disturbance stages. Some will faint as a result of circulatory failure, or the more serious will occur, failure of the central nervous system, which also results in unconsciousness. Convulsions can also occur before or after unconsciousness.

Symptoms of hypoxia

The following are some of the more common symptoms experienced by pilots with hypoxia (not necessarily in any order of severity):

- Change in peripheral vision, even noting "tunnel vision."
- Visual acuity impairment, images slightly blurred, can't focus.
- Difficulty in visual accommodation, focusing from near to distant and back.
- Weakness in muscles, more difficult to change the airplane seat.
- Feeling very tired and fatigued, sleepy for no reason (not boredom).
- Sense of touch is diminished, the controls feel different.
- Sense of pain is diminished (the aching sprained ankle is better).
- Headache, especially if hypoxic for a long period (2+ hours).
- Lightheaded and mild dizziness, reacting poorly in tight turns.
- Tingling in fingers and toes.
- Muscular coordination decreased, sloppy at controls.
- Stammering, can't get the right words out to ATC.
- Cyanosis, bluish lips and fingernails (notice in your crewmate).
- Impaired judgment, doing dumb things, slow thinking.
- Loss of self-criticism, no longer care if you're a great pilot.
- Overconfidence, "No problem!", taking risks not normally acceptable to that individual.
- Overly aggressive, too much in charge, challenging ATC.
- Depression, small irritants become perceived as major problems.
- Altered respiration, breathing faster, shallower.

- Reaction time decreased, you've lost your touch.
- Greatly reduced color discrimination and night vision (even at 5,000 feet).
- Euphoria, you settle for less, who cares?

Note that many are very subjective; that is, most symptoms are not definitive of how hypoxic you are. For example, a change in heart rate is objective evidence but it can also change with exertion, stress, and other situations. One might not think of being hypoxic. A common symptom is fatigue, which is a very subjective sign. How fatigued must you be for you to consider hypoxia? Are you fatigued from hypoxia or from the length of the flight? And if you are indeed hypoxic, you will be indifferent to any degree of fatigue. Also, these symptoms are not listed in order of importance or ease of recognition. Some pilots experience a specific kind of headache, often lasting for a very long time; others only feel fatigued before they become incapacitated and lose control.

Hypoxia is an easy trap to fall into, and it happens to all pilots, many times, in varying degrees. It is unfortunate, but the fact is that the human body does not have an effective warning system to alert one to the onset of hypoxia. Hypoxia is painless. Each pilot will react differently to the same degree of hypoxia, and he or she, in turn, will experience different symptoms even under similar conditions. These individual symptoms can also change as one gets older. It is for these reasons that the military continues to expose its flight crews to an altitude-chamber ride periodically. The main reason is for the individual pilot to experience what his or her symptoms are for hypoxia before he or she reaches the disturbance stage and then cannot (or will not) recover or take appropriate action. No amount of lecturing or reading will prepare you for how to detect your own level of incapacitating hypoxia; therefore, it is strongly encouraged that at least once in your flying career you take a chamber ride, preferably early in your flying education. This will be discussed later in this chapter.

In summary, the signs and symptoms of hypoxia are many and varied, and they differ from individual to individual and are unpredictable at any given time and altitude. Be aware, be suspicious, and watch your crewmate.

Factors influencing tolerance to hypoxia

Recall that it is impossible to predict exactly when, where, or how hypoxic reactions will occur in an individual. One of the reasons is because individuals vary widely in their susceptibility to oxygen deficiency. This susceptibility is related to many factors now to be defined that are controllable by the pilot, in most cases. Avoidance of these factors becomes part of the pilot's responsibility in being safe. In other words, given a cabin altitude that the pilot will be at, his tolerance to oxygen deficiency will be diminished by any of the following.

Self-imposed factors

Although the cabin might be at less than 10,000 feet (pressure altitude), there is a "physiological altitude," which is an altitude that the body "feels" it is at. The presence of these self-imposed factors effectively raises this physiological altitude. Consequently, the mind and body react accordingly with some incapacitation. So instead of the body responding to a 10,000-foot altitude, it responds as if it were at 13,000 feet.

Alcohol in the body can result in histotoxic hypoxia because the alcohol is a toxin. It has been observed that 1 ounce of alcohol can equate to about an additional 2,000 feet of physiological altitude. It interferes with oxygen uptake and metabolism at the cell level, and, of course, is dependent on the amount of this toxin in the body. Furthermore, the usual depressant effects of alcohol on behavior can cloud the pilot's recognition of his own hypoxia, adding to his decreasing tolerance. It's possible to still have alcohol in your system and be legal but still be impaired.

An individual who is mentally or physically fatigued tolerates hypoxia poorly because that person is already bordering on performance decrement. And since fatigue is a symptom of hypoxia, it becomes difficult to discriminate how hypoxic one is. The pilot will often erroneously reason that his fatigue is not a symptom of hypoxia and not consider any preventative actions.

The carbon monoxide in cigarette smoke is a great threat to the smoking pilot (Fig. 5-3). Carbon monoxide has an affinity for hemoglobin 210–250

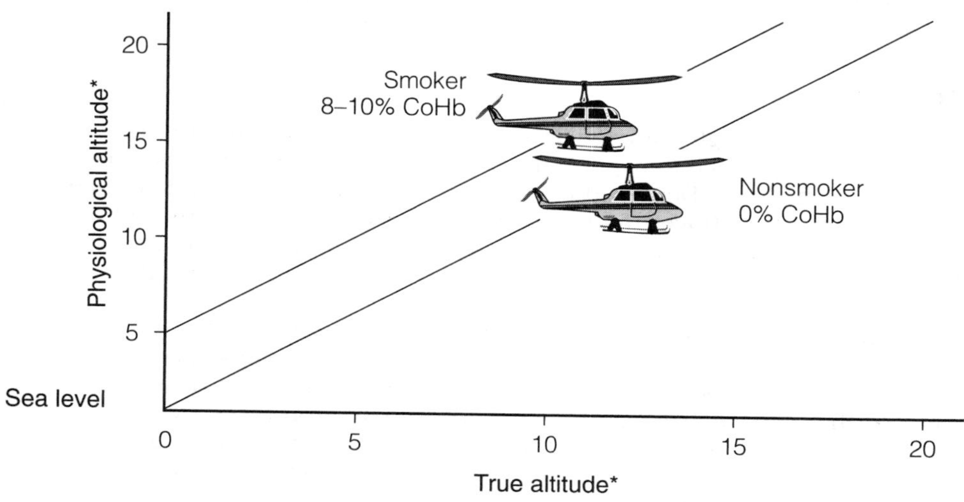

5-3 Smoking increases a pilot's physiological altitude.

times more than oxygen, and this results in hypemic hypoxia. As with alcohol, it has been observed that smoking three cigarettes in rapid succession or smoking 20–30 cigarettes in a 24-hour period prior to flight can saturate from 8–10 percent of the hemoglobin in the body. In addition, approximately 20 percent of a smoker's night vision is lost even at sea level. This can translate into a physiological altitude of an additional 3,000–5,000 feet.

The pilot who is in good shape physically (not overweight, has been exercising, and has a fairly nutritious diet) will be much more tolerant of the effects of low oxygen. The symptoms and potential incapacitation from hypoxia are still present, but the pilot will find them less severe, and more important, she will be able to take appropriate action to recognize and avoid increasing hypoxia before she is truly at risk. Another way of looking at this is that the physiological ceiling is reduced. Also, keep in mind that it might be more than just improved tolerance. The healthy body itself might be more efficient in the use of oxygen and therefore requires less oxygen in its metabolism. In any case, being in good medical condition is a true investment in safe flight. Being in less than optimum shape becomes a self-induced deterrent to tolerance.

Other factors affecting response to hypoxia

The following are conditions that determine the degree of hypoxia one can expect. One has some control over these factors but to a lesser degree since they are a part of our flying environment and working conditions. When tolerance is described, it refers to how susceptible the body is to low levels of oxygen. In other words, as a result of the following, one could become hypoxic at lower altitudes sooner than expected.

Acclimatization

People who live at high altitudes develop an increased tolerance to the conditions that would lead to hypoxia in people living at lower altitudes; therefore, pilots who live in Denver (the "Mile-High City"), are more adapted when flying at, say, 18,000 feet than someone who lives in Los Angeles. By the same token, the LA-based pilot who lands in Denver and has an overnight layover will already be slightly hypoxic when she takes off the next day.

A relatively common occurrence in the winter months is the influx of skiers from the "lowlands" to the higher elevations of the ski slopes. By the time they are at the ski lodge, they probably have ascended at least 5,000 feet and then are carried higher via the ski lift to the top of the slope. Usually the evening is filled with good times, wine, and a big meal. Sleep patterns are often disturbed that night because it is difficult for the hypoxic skiers to fall asleep and stay asleep.

The following morning often begins with one heck of a hangover, not just from the wine, but also due to the combination of being out of shape, too much sudden and extended exertion, and, of course, being hypoxic. Severe cases of hypoxia are called *altitude sickness* and are often medical emergencies. The same physiology is in action when flying to cabin altitudes of more than 5,000 feet. If the pilot is not acclimated to a higher altitude, she is at risk of becoming hypoxic at lower altitudes than the acclimated pilot.

Absolute altitude

This is rather self-explanatory. The higher the altitude of the cabin (whether pressurized or not), the higher the risk of becoming hypoxic. Be advised that an airliner's cabin altitude often reaches 8,000 feet. Since the partial pressure of oxygen is decreased with ascent, the pilot must find ways to get more oxygen to his cells. This is the only way that hypoxia can be prevented.

When the absolute altitude reaches a dangerous level, the pilot must consider either lowering the cabin altitude or donning a space suit. The higher one gets, the less one's inherent tolerance is helpful. You might get by at 15,000 feet, but no way at 20,000 feet, even though you think you can.

Rate of ascent

The quicker the ascent, the less effective your individual tolerance will be. Often, as a result of rapid ascent, the pilot is even less aware of approaching hypoxia. The lack of time to experience changing suspicious symptoms often leads the pilot into a sudden incapacitating series of symptoms.

Duration of exposure

Staying at 8,000 feet for several hours without supplemental oxygen could result in the same symptoms of compromised skills as staying at, say, 16,000 feet for half an hour. Symptoms are time related but very unpredictable, and the higher the altitude, the shorter the exposure time before symptoms of hypoxia occur.

Physical activity

Increased activity obviously requires more oxygen for the body to function. One can expect a higher risk of becoming hypoxic during a flight where, for example, the autopilot isn't working and you are flying by hand in turbulent conditions requiring more activity. This situation, however, is more of a concern to individuals who exert more energy during flight, such as flight attendants and military load masters.

Ambient temperature

Temperature extremes in the cockpit, especially in smaller recreational aircraft and poorly air-conditioned pressurized cabins, will make the body less tolerant to diminished oxygen levels. Shivering when cold and coping with excessive heat and humidity use body energy and are forms of increased activity. The body's circulatory system is working harder to maintain its "core temperature." Concurrently, the metabolic needs of the cells are changing, generally requiring more oxygen. All these situations decrease the pilot's tolerance to hypoxic conditions, and she might become hypoxic at lower altitudes.

Effective performance time (EPT) or time of useful consciousness (TUC)

The effective performance time (EPT) is the period of time that a pilot has from the time oxygen becomes less available until the time when he or she loses the ability to recognize and take action, such as getting on oxygen or descending to a lower altitude. As described earlier, a major symptom of hypoxia is euphoria; therefore, if one goes beyond the EPT, then he or she isn't even aware of the problem and will not take corrective action. The key word is *effective or useful*. He or she might be conscious, but not making expected useful or effective decisions (Fig. 5-4).

Altitude (m)	Altitude (ft)	Effective performance time
5,486	18,000	20 to 30 minutes
6,706	22,000	10 minutes
7,620	25,000	3 to 5 minutes
8,534	28,000	2.5 to 3 minutes
9,144	30,000	1 to 2 minutes
10,668	35,000	0.5 to 1 minute
12,192	40,000	15 to 20 seconds
13,106	43,000	9 to 12 seconds
15,240	50,000	9 to 12 seconds

5-4 Effective performance time decreases as altitude increases.

The EPT decreases with increasing altitude and is dependent on the same tolerance factors defined earlier. The EPT can become very short but only at altitudes not frequently experienced; however, decompression (sudden or slow loss of cabin pressure) still occurs in flight, and the pilot will be flying at altitudes in excess of 20,000 feet without realizing the onset of the shorter EPT. In fact, in a more rapid decompression, the EPT will be shortened by ⅓ to ½.

Of greater concern, and more likely to occur, is subtle decompression, where the pilot is unaware of increasing cabin altitude and a shortening EPT. For example, when flying without supplemental oxygen at 18,000 feet where the EPT is about 25 minutes, the pilot can easily get into trouble, especially if very busy, not keeping track of the time at that altitude, and not anticipating hypoxia (Fig. 5-5).

Prevention of hypoxia

There is nothing remarkable about preventing hypoxia. Avoidance of those conditions that increase the risk of becoming hypoxic is paramount. It's

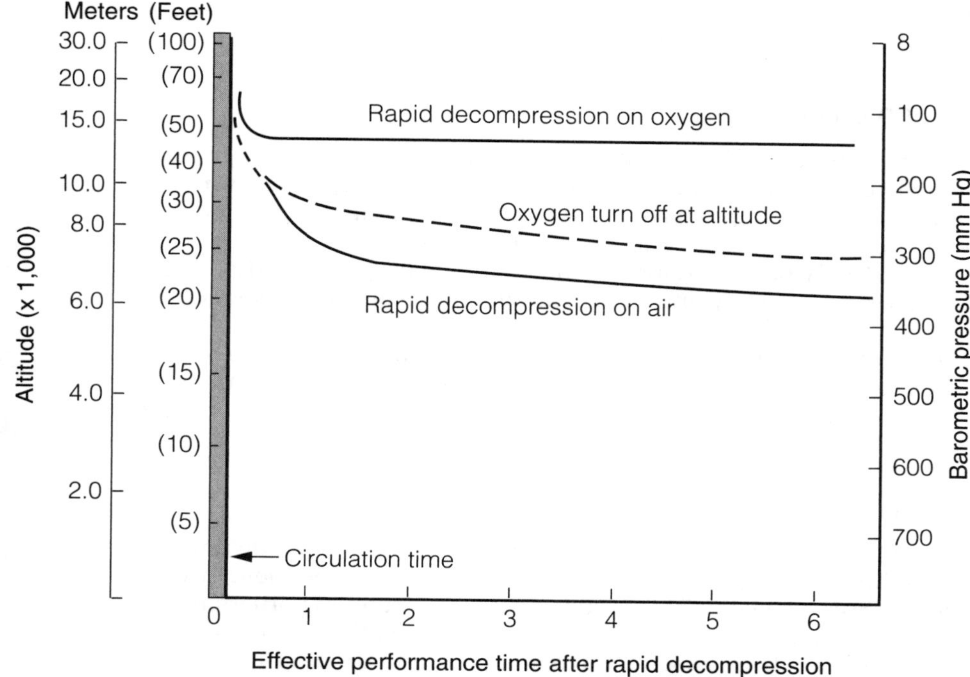

5-5 A pilot flying at an altitude of 22,000 feet has twice as long an effective performance time as a pilot flying just 3,000 feet higher.

also prevented by maintaining a high level of suspicion anytime the cabin altitude is above 5,000 feet, especially at night. Like maintenance on the aircraft, preventive management is more productive than crisis management. (Waiting for symptoms of hypoxia to be observed before taking action is crisis management.) Prevention of hypoxia, therefore, becomes anticipation of being hypoxic where hypoxia can happen. If you are a smoker, moderate drinker, and physically out of shape, you are at greater risk than a physically fit person who does not smoke or drink.

The obvious solution is to be in a pressurized cabin at an altitude of less than 10,000 feet or using supplemental oxygen by mask. Although according to the regulations you are legal without oxygen at 10,000 feet, you can still be subtly incapacitated.

Treatment of hypoxia

If there is any suspicion of hypoxia in yourself or a crewmate, the immediate use of 100-percent oxygen is essential. Aside from the rare occurrence of oxygen toxicity, you cannot hurt anyone by giving them oxygen. If hypoxia is occurring, then there will be almost immediate response. If there isn't this immediate improvement, you must think of other causes for the symptoms. If supplemental oxygen is not available, then getting the airplane or cabin altitude below 10,000 feet is mandatory. Hypoxia is a valid reason to declare an emergency to ATC.

Hyperventilation

Hyperventilation is by definition simply a matter of breathing too rapidly, which results in exhaling carbon dioxide during respiration in disabling amounts. The symptoms are seldom completely incapacitating, but they can produce, at the very least, a potentially serious distraction in often very active flying situations. Then the problem is only further aggravated by increasing anxiety and breathing rate. It becomes a vicious circle.

The body's respiratory activity reacts automatically under conditions of stress and anxiety, such as those associated with flight situations that are deteriorating. If the pilot has a sense of losing control, hyperventilation is a possibility.

Causes

Fear, stress, and anxiety can cause an individual to override normal automatic functions such as breathing rate. As we become more stressed, we tend to breathe faster. This is apparent when watching others in stressful situations. In fact, you are apt to observe this in others more than in yourself. This is another reason for depending on and trusting your crewmates.

In situations of increasing hypoxia, the body is attempting to circulate more oxygen, which leads to increased heart and respiratory rates. This might bring more oxygen to the cells, but hyperventilation now is evident and becomes counterproductive. Additionally, at high altitudes with pressure breathing or even wearing a regular oxygen mask with usual air pressures, one tends to breathe more rapidly. Just thinking about breathing often leads to an increased rate.

Symptoms

Unfortunately, many of the symptoms of hyperventilation are similar to those of hypoxia, fainting, and decompression sickness (Fig. 5-6). They include, but are not limited to, tingling sensations in the fingers, toes, and around the lips; small muscle spasms; hot and cold sensations; visual impairment; dizziness; and, in severe cases, unconsciousness.

The challenge is to make the proper diagnosis based on the symptoms and then take appropriate action. For example, hyperventilation rarely is painful, nor is feeling faint; decompression sickness can be painful. Decompression sickness often is *asymmetrical* (symptoms limited to one arm and/or leg and not the other), whereas hyperventilation is symmetrical, affecting both sides.

If hyperventilation is suspected, trying to get voluntary control of the breathing rate is the most effective method of treatment. Normal breathing is about 12–16 breaths per minute. Recognizing that carbon dioxide is

Signs and symptoms	Hyperventilation	Hypoxia
Onset of symptoms	Gradual	Rapid (altitude-dependent)
Muscle activity	Spasm	Flaccid
Appearance	Pale, clammy	Cyanosis
Tetany	Present	Absent
Breathlessness	X	X
Dizziness	X	X
Dullness and drowsiness	X	X
Euphoria	X	X
Fatigue	X	X
Headache	X	X
Judgment poor	X	X
Lightheadedness	X	X
Memory faulty	X	X
Muscle incoordination	X	X
Numbness	X	X
Performance deterioration	X	X
Respiratory rate increased	X	X
Reaction time delayed	X	X
Tingling	X	X
Unconsciousness	X	X
Vision blurred	X	X

5-6 Comparison of hyperventilation and hypoxia.

blown off by the rapid breathing, using the "paper bag method" will also help. Breathing into the bag gets the person to concentrate on slowing the rate as well as rebreathing carbon dioxide.

Since hypoxia and hyperventilation have similar symptoms and hypoxia can cause one to hyperventilate, the initial treatment is the same: oxygen. A good rule of thumb, therefore, is to go on 100-percent oxygen and get the person working on something or mentally figuring out a problem—get his mind off his anxiety. Even singing or talking has been suggested. Most symptoms should go away.

Final thoughts on hypoxia

For both hypoxia and hyperventilation, it is essential that there be a mutual understanding and respect by all aviators. You should expect your crewmate to watch you just as you are obligated to keep your suspicions up in hypoxia-generating conditions. This is true for all facets of aviation, including helicopter flight where there is a false sense of security as a result of the perception that these pilots aren't going to get hypoxic. Helicopters are flying higher all the time, sometimes in excess of 20,000 feet.

This is a common occurrence in mountainous areas where the pilot can leave sea level and quickly be at an altitude in excess of 15,000 feet. Although flying below 10,000 feet AGL is the norm, remember that the physiological altitude can be much higher, especially if you are at the higher altitude for extended periods; therefore, all aviators must be familiar with hypoxia.

Oxygen support systems

The amount, or percentage, of oxygen required to maintain normal oxygen saturation levels varies with altitude. At sea level, 21-percent ambient-air oxygen concentration is necessary to maintain the oxygen saturation of 96–98 percent. At cabin altitudes of 20,000 feet, however, a 49–percent oxygen concentration is required to maintain the same saturation. The upper limit of continuous-flow oxygen is reached at about 25,000 feet, where the partial pressure of oxygen just equals that within the circulatory system. Above this altitude, positive pressure breathing (pressure higher than ambient pressure) is necessary to maintain adequate oxygen saturation.

Oxygen regulators will usually begin some positive pressure above 25,000 feet to keep oxygen saturation above 90 percent and will noticeably force oxygen into the lungs above 40,000 feet. At 50,000 feet, even with pressure breathing, the oxygen saturation is only 70 percent; therefore, for flights above 50,000 feet cabin altitude, pressure suits are required.

Supplemental oxygen can usually be delivered two ways: gaseous and liquid. Gaseous aviation oxygen (Grade A, Type 1) must be free of water (99.9 percent pure) to prevent freezing at high altitude. Liquid oxygen (Grade B, Type 2) is very efficient, with 1 liter generating about 850 liters of gaseous oxygen. Solid chemical generation of oxygen is becoming more practical, especially in commercial aircraft.

Whatever the source, the oxygen is delivered by a mask. In fact, in nonpressurize cabins, supplemental oxygen is required in the United States (FAA) after 30 minutes of exposure to cabin altitudes of between 12,500 and 14,000 feet, immediately on exposure above 14,000, and encouraged to use above 10,000 feet. This is a very conservative requirement.

Oxygen mask styles range from a nasal cannula or simple mask delivering a continuous flow of 100-percent oxygen to mix with ambient air to pressure-demand masks that are more common in the military.

Pressurized cabins

Pressurization (forcing air into the cabin by means of compressing mechanisms) maintains a cabin/cockpit altitude below 10,000 feet, often in the 6,000–8,000-foot range; however, maintenance of this safe level is dependent on aircraft equipment (its airframe integrity and the pressure differential it can tolerate) and pilot control of cabin pressure. This control is often automatic and is dependent upon settings of rates of descent by the pilot, which will depend on how fast the aircraft will be changing altitudes.

Another factor to be considered is the fact that cabin humidity is extremely low as a result of this pressurization. Aircrew are still hypoxic in pressurized cabins, especially considering the varying physiological altitudes.

Positive-pressure breathing

This use of an oxygen mask that has a high-air pressure is necessary only when the cabin altitude exceeds 34,000 feet. Here, the partial pressure of oxygen within the human respiratory system is too low to be transferred to the cells, even when breathing pure oxygen. The oxygen is literally forced into the lungs by the delivery system and regulator. The altitude-chamber ride is the best place to experience positive-pressure breathing for the sake of familiarization.

CARBON MONOXIDE AND OZONE

Introduction to carbon monoxide

The effects of carbon monoxide (CO) are subtle and deadly. Carbon monoxide poisoning is more common than any other toxic gas. It is colorless, odorless, tasteless, and slightly lighter than air. Like hypoxia, carbon monoxide poisoning is painless and can incapacitate a pilot without his or her knowing the problem existed. Carbon monoxide is present in every smoker's body. It can enter the airplane by a leaky exhaust system when cabin airflow passes over manifolds for heating. Review chapter 2 on basic human anatomy and chapter 10 on toxic chemicals to better understand how carbon monoxide affects the body's delivery of oxygen.

Effects of exposure

Relatively low concentrations of carbon monoxide in the air you breath can, in time, produce very high blood concentrations. It is a cumulative problem. Such accumulations can occur in less than 30 minutes at carbon monoxide levels in the air of only one half of one percent. Remember that carbon monoxide competes with oxygen for attachment to hemoglobin (250 times more). This situation leads to hypemic hypoxia even at lower altitudes.

Symptoms

Like hypoxia, the symptoms of carbon monoxide intoxication are often subjective. They include headache, weakness, nervousness, tremors, and ultimately unconsciousness. A loss of visual acuity also occurs; especially affected are night vision and peripheral vision. More on this is found in chapter 7 on vision.

Treatment

Again as with hypoxia, the treatment is immediate 100-percent oxygen; however, this treatment is an extremely slow process, high carbon monoxide concentrations taking several days to be reduced. Treatment, once carbon monoxide poisoning is diagnosed, is often in a hyperbaric chamber using 100-percent oxygen. Although the symptoms of hypoxia and carbon monoxide poisoning look alike, going to 100-percent oxygen will not immediately relieve carbon monoxide poisoning symptoms like it would with hypoxic hypoxia. This is another clue to determining if a pilot is impaired.

Introduction to ozone

General

Ozone is a gas by-product of the ionization of oxygen within the ionosphere. Problems for aviation are seasonal, with more concerns in late winter or early spring. Areas of increased ozone directly affecting air travel are predictable, which means that high flying aircraft can avoid most ozone problems by watching forecasts for where the ozone concentrations are. Until the last decade, ozone was not considered a factor in medical problems; however, new evidence clearly identifies significant symptoms directly related to increased ozone levels within an aircraft cabin.

Symptoms

Most symptoms are related to respiratory-tract irritation with cough and mild chest pain. Sometimes headaches and fatigue are also reported. These symptoms are aggravated by increased physical activity, which is one reason flight attendants are particularly susceptible. There is no evidence that exposure to ozone interferes with blood oxygenation.

It would appear that there is some adaptation to repeated exposures over a period of several weeks. Aircraft flying in high ozone concentrations now can have a filtration system installed for cabin air cleansing.

DECOMPRESSION OF CABIN ALTITUDE

Loss of cabin pressure for any reason, suddenly or slowly, is called *decompression*. The cabin altitude decompresses to the ambient (surrounding) pressure. In other words, if the cabin is pressurized to 6,000 feet and you are flying at 20,000 feet and a leak develops, your cabin can suddenly be at 20,000 feet.

Some of the characteristics of a rapid decompression include a loud popping sound. Concurrently there is usually flying debris, paper and dust mostly, with some items being sucked out of the aircraft as the pressurized air rushes out the hole into lower pressure. Sudden fogging is very common because more water can remain in gaseous form at higher pressure (and temperatures) but condenses out of the air as particles of water when the pressure suddenly decreases. The temperature also suddenly drops to equalize the outside temperature, often a significant change because ambient temperature at 35,000 feet is about −67°F.

Explosive decompression

An explosive decompression occurs so quickly that the lungs decompress and can be damaged. It usually takes about two-tenths of a second (0.2) for the lungs to release their air. A decompression that takes less than a half second (0.5) is considered explosive. Usually occurring in smaller high flying aircraft, this is not a common event, and there is little you can do about it except immediately get on oxygen.

Rapid decompression

A rapid decompression occurs more slowly than an explosive. Rapid decompressions are more common and do not usually result in lung damage. Such an event is easy to recognize because there is usually a "bang" and sudden fogging of the air. Because this is often confused with smoke, fire is first suspected before a decompression. This form of decompression is usually experienced in larger aircraft.

Subtle decompression

Subtle decompressions could be considered the most potentially dangerous in the sense that you might not be aware that your cabin altitude is going up. Not all pressurized airplanes give any warning about cabin-ascent rate. In fact, subtle decompression has become such a potential factor that the flight profile in the altitude chamber has been revised to take into account that this kind of situation needs to be recognized during the chamber ride. The exercise is to take off your mask while the cabin altitude slowly goes up. The pilot in the chamber must recognize when to suspect hypoxia and put on the mask before reaching the euphoric stage or exceeding the EPT.

Acute effects of decompression on the body

The most noticeable effect of all but the subtle decompressions is the sudden rush of air from the lungs. There might be trapped gas problems, but they probably are not significant. Having a cold or flu will increase the risk of problems. The primary problem is the sudden exposure to extreme hypoxic conditions, especially at high altitudes. The treatment remains the same: get on oxygen, get to a lower altitude, and, of course, fly the airplane. This is a flight emergency.

The ultimate insight can be gained from an altitude-chamber ride. Here you can experience the actual event and realize that decompression is not that threatening and that the pilot does have time to take action, providing he or she recognizes what is happening. If you know what to expect, identifying the real cause of the events as being a decompression and not smoke, for example, you can deal with the problem. Get on oxygen immediately, declare an emergency, and get to a lower altitude.

TRAPPED GASES

Mary has a checkride tomorrow and hadn't been able to schedule an airplane until today. But she also had a cold that had hung on for several days. She thought she would be okay because she was feeling better; however, about 45 minutes into her practice, she began to feel a headache over her eyes. It got worse every time she made her approach. Finally, on the ground, the headache got progressively worse until it was bad enough for her to see her doctor. She had a sinus block severe enough that she couldn't fly for another week, even using decongestants. Her doctor shared her frustration because the cold had cleared but there was still congestion in her sinuses, enough to cause the block on the descents.

The body can withstand tremendous changes in pressure as long as air pressure in the body cavities is able to equalize with the ambient air pressure. In other words, the pressure inside any part or organ of the body where gas is present does not generate any problems unless that pressure is not the same as the pressure surrounding that cavity. Problems occur when the expanding (or contracting) gas within the body cannot escape to allow ambient and body pressures to equalize. When gases within body fluids, such as blood, escape from the fluid and enter the body as gases, this also causes problems, often called *the bends*. The bends will be discussed later in this chapter under the section on decompression sickness.

Review of gas laws

Boyle's Law needs to be reviewed because it is the basis for virtually all of trapped-gas problems. This law states that the volume of a gas is inversely proportional to the pressure exerted upon it; therefore, as we ascend and the pressure outside the aircraft decreases, the air expands. If that air were in a balloon, the balloon would get larger. When considering expanding gases in the body, these gases also have water vapor included, which expands more than dry air.

Dysbarism is a synonym often used for the bends and sometimes associated with trapped gas. This is no longer an accepted term since it is nonspecific. It is better to use the terms *trapped gas* and *ear* or *sinus blocks*.

As a result of a greater differential in pressure per thousand feet at lower altitudes, the risk of developing any kind of block is actually higher at the lower altitudes (below 15,000 feet). In other words, a rapid descent from 30,000 feet to 20,000 feet will often cause little or no discomfort where a similar rate of descent from 15,000 feet to 5,000 feet will cause great distress. The greatest pressure change (differential) is from sea level to 5,000 feet. This effect is also noticed in scuba diving where the most noticeable pressure changes occur in the first 15 to 30 feet.

Review of pertinent anatomy

Those air-filled cavities in the human body most susceptible to trapped-gas problems are the sinuses, middle ear, and occasionally the gastrointestinal system. Rarely, dental cavities will be involved. The sinuses are cavities in the bony skull with usually one narrow opening into the nasal passages. The sinuses, including the openings, are lined with moist mucous membranes. The middle ear is another air-filled cavity with only one opening, the eustachian tube, with a lining of mucous membranes. The gastrointestinal system is not a common problem location, but because the bowels are potential air-filled cavities, there can be a bloated feeling with passage of air at both ends.

Middle ear block (barotitis)

During ascent, as the ambient air pressure is reduced, the expanding air within the middle ear is intermittently released through the eustachian tube (Fig. 5-7). During descent the same process should take place, but in the opposite direction. This equalization of pressure on either side of the eardrum is essential for proper hearing because the flexibility of the eardrum must not be compromised. As the pressure increases in the mid-

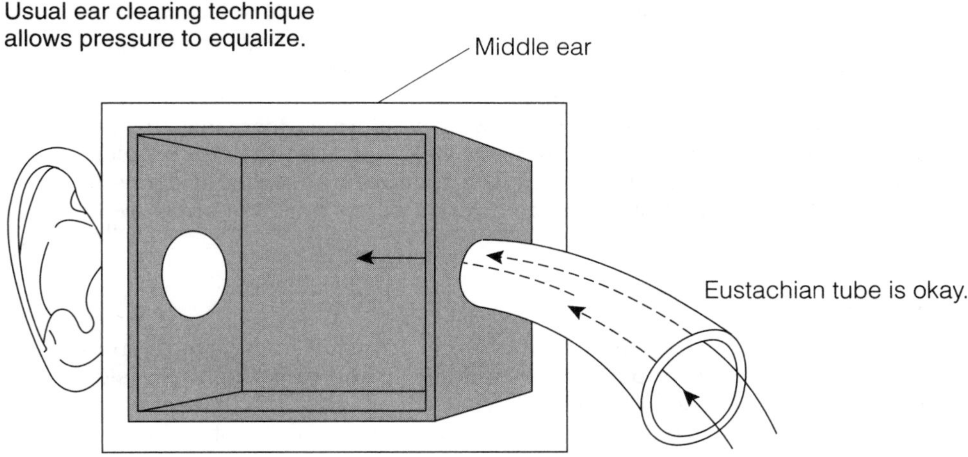
Usual ear clearing technique
allows pressure to equalize.
Middle ear
Eustachian tube is okay.

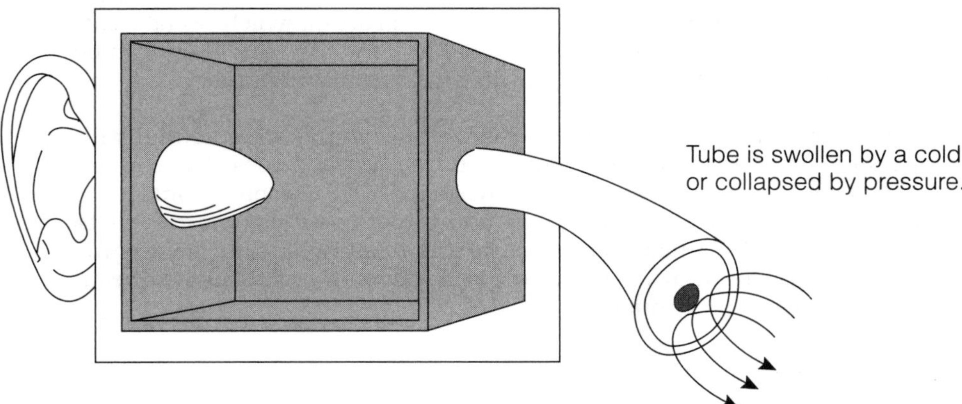
With ear blocked, pressure cannot be equalized.
Tube is swollen by a cold
or collapsed by pressure.

5-7 Common methods of clearing an ear block will not work if the block is the result of the mucous membranes being swollen from a cold or hay fever.

dle ear, a small bubble will form that escapes through the eustachian tube after it reaches a certain size. One can often feel this equalization of pressure as the air bubble "pops" out.

This pressure equalization should be relatively automatic without any conscious effort, especially under ideal conditions. Normal swallowing usually clears the ears. Problems arise when the eustachian tube does not allow passage of air, usually a result of the mucous membranes being swollen from a cold or hay fever. Furthermore, the nasal end of the eustachian is narrower. Often, the passage of air is only outward but not back into the middle ear, a form of a one-way valve. This is where problems develop.

An unexpected delayed blockage occurs when breathing oxygen, as in unpressurized cabins at altitudes requiring oxygen breathing masks. Here, pure undiluted oxygen enters the cavity. If a block occurs and it is unable to be cleared, the cavity now is pressurized with oxygen which, in turn, is absorbed by the mucous membranes. Now, instead of high pressure, a lower pressure develops, with equally disabling symptoms. Another phenomenon that is fairly frequent is passengers who sleep through the flight, are not actively clearing their ears (swallowing rate decreases during sleep), and during descent develop a mild block; however, the real problem occurs a few hours later, after the oxygen has been absorbed in the still blocked ear. The pain is now worse as a result of the negative pressure in the middle ear.

Symptoms

Inability to clear your ears (equalize pressure) can result in everything from a mild discomfort to severe pain in the middle ear and radiating to the side of the head—a true disabling earache. Anyone who has ever had an ear block never wants to experience it again. It is worse than a severe toothache. The pain becomes worse as the pressure changes with the continuing descent and will not always go away even when you have landed. Sometimes the trapped gas within the middle ear reaches a pressure that is extreme enough that the eardrum ruptures. The good news is that the pain will almost instantly go away but now you have a tear in the eardrum, which needs treatment by a physician. If there is an unequal block in both ears, there can be a disruption of the vestibular system with resulting disorientation.

Whatever the cause of the ear block, you are now impaired, potentially incapacitated, or at the very least, distracted from flying duties. The ear can be physically damaged, sometimes permanently. Despite continuing awareness of the causes of ear blocks and the dangers associated, pilots still get into trouble. A major cause of the pilot getting to that point is his delay in recognizing that an ear block is developing and failing to begin clearing the ears soon enough.

Treatment and prevention

As with so many things in flying, the best treatment for a medical problem is to prevent it from happening. Every pilot has heard that flying with a cold or congestion is asking for trouble, yet many pilots will fly under these conditions. Admittedly, only a few develop a serious block, but that infrequent event is a major risk to safe flight and not worth playing the odds.

In reality, if every pilot grounded himself or herself for every sniffle, there would be few flying; therefore, if you have a mild congestion (and there is no good definition of *mild*), then taking some decongestant medication before you begin the flight might be appropriate. *This medication must not contain antihistamines.* Selected long-acting nose sprays might also allow you to fly. The dilemma is that, technically, no medications are legal or safe if there is any possibility that side effects or the reason for the treatment could compromise your performance. Most AMEs should be able to help you decide, but the ultimate decision and responsibility is yours.

If you are prone to early ear blocks, it might be wise to carry a small bottle of nose spray in your flight bag. This should be used only when you sense a block developing—early enough that the spray has a chance to take effect, but only while all other measures are used in relieving the block.

The safest and most common method of clearing an ear block is to move the jaw and swallow at the same time. Chewing gum is a great way to do this, and airlines used to routinely give out gum to passengers just prior to descent for that reason. Yawning, moving your head around, or a combination of all will often clear the ear.

Another technique, and probably the most effective, is the Valsalva maneuver. Again, as soon as you feel a block beginning develop, hold your nose tight (as if you were going to jump into a swimming pool). Then blow hard against your nose (as if you wanted to blow your nose). This will often increase the pressure within the nasal area and force air back into the middle ear through the eustachian tube. It is not likely that you can blow hard enough to injure the eardrum, so it's okay to blow fairly hard. This is the same procedure that scuba divers always use when descending.

If all this fails, consider taking the aircraft (or cabin altitude) higher, to a level above where the block began. Then begin a very slow descent, using the techniques already mentioned but now under controlled conditions.

If a block persists after landing, see your doctor. Unresolved blocks can lead to inflammation and sometimes an infection. A middle-ear infection will temporarily ground you.

Sinus block (barosinusitis)

Basically, the same events occur with the sinuses except the openings to the sinus cavities are much smaller and the Valsalva technique is not as effective. The pain from a sinus block is as incapacitating as an ear block, and often more so. It's said that a sinus block feels like "an ice pick in the face." All the more reason to be cautious in flying with a cold or hay fever. Additionally, the sinuses are affected by both ascent and descent, unlike the middle ear.

Teeth (barodontalgia)

Tooth problems are caused by cavities, abscesses, and inadequate fillings (Fig. 5-8). Pain often increases with altitude and can become disabling. Descent will usually bring some relief. Dental intervention is essential once such a block is recognized. Explain to the dentist what the circumstances are relating to the tooth pain (while flying). If it feels as if the entire top row of teeth is hurting, or if the pain occurs on descent, it is probably a sinus block rather than a tooth problem.

Gastrointestinal (GI) tract

The gastrointestinal (GI) tract consists of everything from the mouth to the rectum/anus and includes the stomach and the small and large intestines. These organs already contain a variable amount of gas from swallowing and from formation by the digestive process (Fig. 5-9). The pain, which is similar to the cramps felt when constipated, can be quite severe. The symptoms vary from pain to a mild discomfort in some individuals, with reluctance to allow equalization through burping or passing flatus.

Prevention

It is common knowledge that certain foods can produce gas (flatus). Beans are the most common, and other foods include cabbage, onions, raw apples, radishes, cucumbers, and some melons. Fluids are very important to ingest in flight to prevent dehydration, but carbonated beverages will increase the gas in the gastrointestinal tract. Gum chewers also swallow a lot of air. A pi-

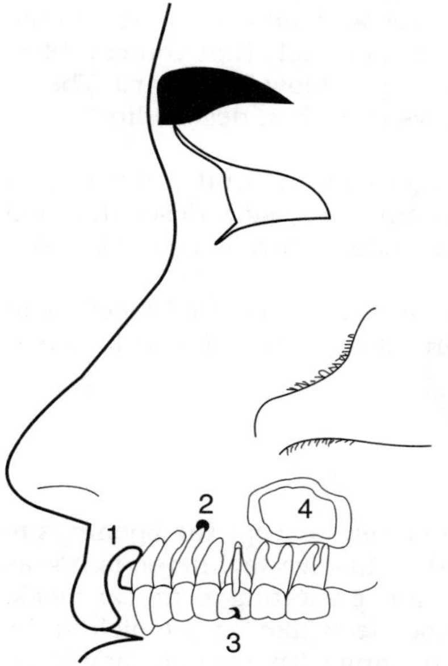

1. Gum abscess—dull pain during ascent
2. Root abscess—dull pain during descent
3. Inflamed pulp—sharp pain during ascent (from decay or recent dental treatment)
4. Inflamed maxillary sinus—associated dental pain during ascent and descent

5-8 See your dentist as soon as possible if you experience unusual tooth pain while flying.

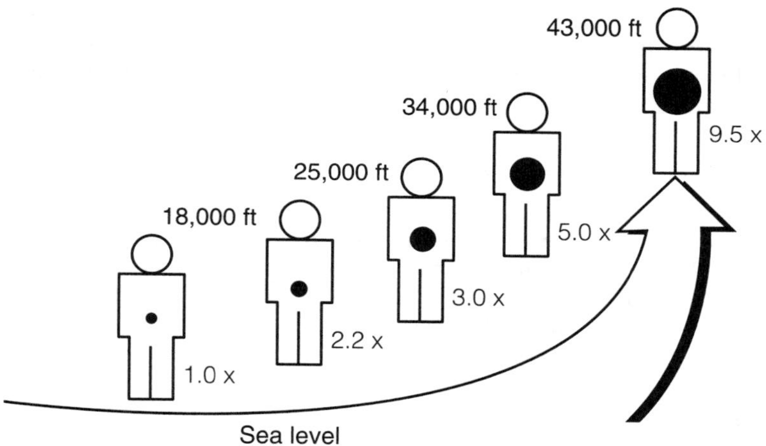

5-9 Trapped gas will expand dramatically as altitude increases.

lot who develops gastrointestinal gas as a result of any of the above cannot be modest. Getting rid of the pressure by allowing the gas to escape at either end of the gastrointestinal tract will help relieve the discomfort. Failure to do so could be more embarrassing. As the old saying goes, it's better to burp and bear the shame than not to burp and bear the pain.

EVOLVED GAS DISORDERS

> Richard frequently flew his turbocharged but unpressurized Bonanza from his home in Nebraska during the winter, over the mountains, and on to Santa Barbara and his beach home. This trip he slipped on some ice as he was getting into the airplane and bumped his elbow. He thought nothing of this and pressed on to California. He cruised at 21,000 feet over the mountains, sucking on oxygen through a loose-fitting mask. He had stayed for only a day when he got called back to Nebraska. He began his return flight uneventfully, except he began to notice pain in the injured elbow after he had reached his cruise altitude. It was bad enough that he thought he'd better land in Denver and have it checked; however, as he began his descent the pain began to go away, so he canceled his approach and kept going for home. Now the pain was severe and he landed at his home base to see the doctor. The X-rays were negative as was the exam. His doctor, who was not familiar with flight physiology, figured it was still a sprained elbow and told him to wait a few days. In a few days the pain did go away. It wasn't a sprain.

Evolved gas disorders, the physiological/physical event that occurs in varying ambient pressures, is known by different names, and the entire issue is often minimized by the aviation community. Scuba divers are very much aware of what can happen to gases dissolved in the blood and released under lower ambient pressures. The physiology and anatomy is the same for flyers, and decompression sickness can and does occur in the flight environment, although slower and at more extremes in altitudes. The following are some of the terminology used when discussing evolved gas disorders.

Decompression sickness (DCS)

This is the generic medical term used to identify all degrees and variations of evolved gas disorders. It does not include trapped-gas disorders. DCS means only those conditions that result from gases dissolved within the blood and tissues that are released when the surrounding pressure decreases, resulting in small bubbles now being transported by the blood to various organs and parts of the body.

Dysbarism

This is an old term no longer used, meant to identify any problems (*dys-*) that were related to pressure changes (*-barism*). This term was often confused by many people because most felt it meant trapped-gas problems; others related dysbarism to the bends. Consequently, these two basic issues have specific definitions: DCS and trapped gases. Do not confuse a conversation by using the term *dysbarism*.

Aeroembolism

By definition, this means air (as a bubble) acting as an embolism that can move throughout the body. An embolism often means a blood clot but generally can include anything that could plug or obstruct a blood vessel, such as a bubble of air.

Bends

This is the common term, especially in scuba. It is one kind of DCS, specifically involving the joints. It does not, however, refer to the other kinds of DCS that are actually more dangerous. DCS usually refers to involvement of the nervous system. In actuality, bends is by far the most common type (over 85 percent) of DCS problems.

Basic physiology

Henry's Gas Law states that the amount of gas dissolved in a solution is directly proportional to the pressure of the gas over the solution. This can be demonstrated by what happens when you remove the top or pull the tab on a bottle or can of pop or beer. The contents are under pressure. When the pressure over the surface of the liquid is released, so is the gas (in this case, carbon dioxide), and bubbles immediately form and come to the surface. Nitrogen dissolved in the blood responds the same way.

Inert gases in body tissues and blood (principally nitrogen) are in equilibrium with the partial pressure of that gas in the atmosphere (such as with nitrogen, which is 80 percent of our atmosphere). When barometric pressure decreases, as ascending in an aircraft cabin, these dissolved gases are released as small bubbles. Medical technology has reached the state where there are instruments that can actually hear the bubbles moving in the blood through the body, confirming that gas bubbles are present. Of additional interest is the fact that fat tissue dissolves five to six times more nitrogen than blood can absorb. That extra fat that some pilots have covering the body and internal organs now becomes an additional source of nitrogen and puts that pilot at increased risk of developing DCS problems at higher altitudes.

Physiologically, these gases, especially nitrogen, have the potential of being released into the bloodstream as pressures change over time. For example, inhaled air at sea level is absorbed into tissues and blood. After ascending to higher cabin altitudes (especially when greater than 18,000 feet) and staying there for a few hours, more of those gases are released into the blood. If this continues, additional bubbles can form until the cabin altitude returns to sea level. The symptoms that result along with the effects on the body are what concern the pilot who flies at these altitudes in unpressurized aircraft, even helicopters.

Effects on the body

Circulatory system

These gas bubbles can act as a blockage within smaller vessels and are called *aeroembolisms*. Such blockage is of minimal consequence until the bubbles get larger or bigger and important vessels become blocked. The ultimate problem exists in the body's organs.

Musculoskeletal system

Pain will be noticed in the larger joints, such as the knees, shoulders, elbows, and the like. It is often described as a "gnawing" ache, ranging from mild to severe, and obviously either distracting or incapacitating in the cockpit. This is also called *the bends*. The pain tends to be progressive, especially as ascent is continued. There is also the tendency for pain to recur in the same location during subsequent flights. It can involve several joints at the same time.

Paresthesias

By definition, this is involvement of nerves, causing abnormal sensations such as tingling, itching, and cold and warm feelings. It is felt this reaction is from the bubbles formed locally around nerves in the skin. Such symptoms are also found in hypoxia and hyperventilation. The pilot must be aware of the environment that she is in and consider which has the greatest chance of occurring before deciding if she is suffering from DCS or some other condition. Another term is *skin manifestations of DCS.*

Chokes

These symptoms are related to the chest (thorax) and lungs. Here the bubbles intrude on the pulmonary vessels (blood vessels carrying blood into the lung tissue). The first signs are a burning sensation in the middle of the chest (sternum), which progresses to a "stabbing" pain accentuated by deep breathing. It is similar to the feeling after a sudden burst of exercise where it is difficult to catch your breath. There is an almost uncontrollable need to cough, but the cough is ineffective and doesn't come up with any material. Finally, there is a feeling of suffocation, with typical signs of cyanosis (decreased oxygen supply). Immediate descent is necessary because it is a potential medical emergency. If no intervention takes place, this leads to collapse and unconsciousness. In fact, after getting to the ground, there will still be fatigue and soreness about the chest for several hours.

Disorders of the brain (central nervous system)

Although uncommon, these disorders are very serious. What is happening is that the brain and spinal cord are now involved with these nitrogen bubbles. The most common symptoms are visual, such as flashing or flickering lights. Other symptoms include headaches, paralysis, inability to hear or speak, and loss of orientation. Any of these symptoms (in the appropriate conditions of flight) should be highly suspect and are considered a medical emergency. As with other symptoms, these are easily confused with hypoxia and pulling Gs. Keeping a high index of suspicion is important if you are flying at a high altitude over a period of time.

Prevention

DCS is quite uncommon in most civilian flying. The bends do occur, however, in unpressurized smaller aircraft flying at excessive altitudes, even though oxygen is being used; therefore, if you are in conditions of high cabin altitudes or know you will be going there (even in a helicopter going from sea level to mountaintop rescues), have a high index of suspicion. If you anticipate being subjected to these conditions, the following should help prevent problems.

Denitrogenation

Breathing 100-percent oxygen prior to takeoff reduces the partial pressure of nitrogen in the body over a period of time. Thirty minutes will usually cause sufficient, although not total, reduction of nitrogen in the body to prevent decompression sickness. This reduced partial pressure of nitrogen over a period of time helps prevent bubble formation (Fig. 5-10). Denitrogenation will take place over several hours, but a large percentage is in the first half hour.

Pressurized cabin

This is common sense because most cabins are pressurized to less than 10,000 feet. If there is no pressurization or it is lost through mechanical means (decompression), then one must be aware of the possibility of DCS.

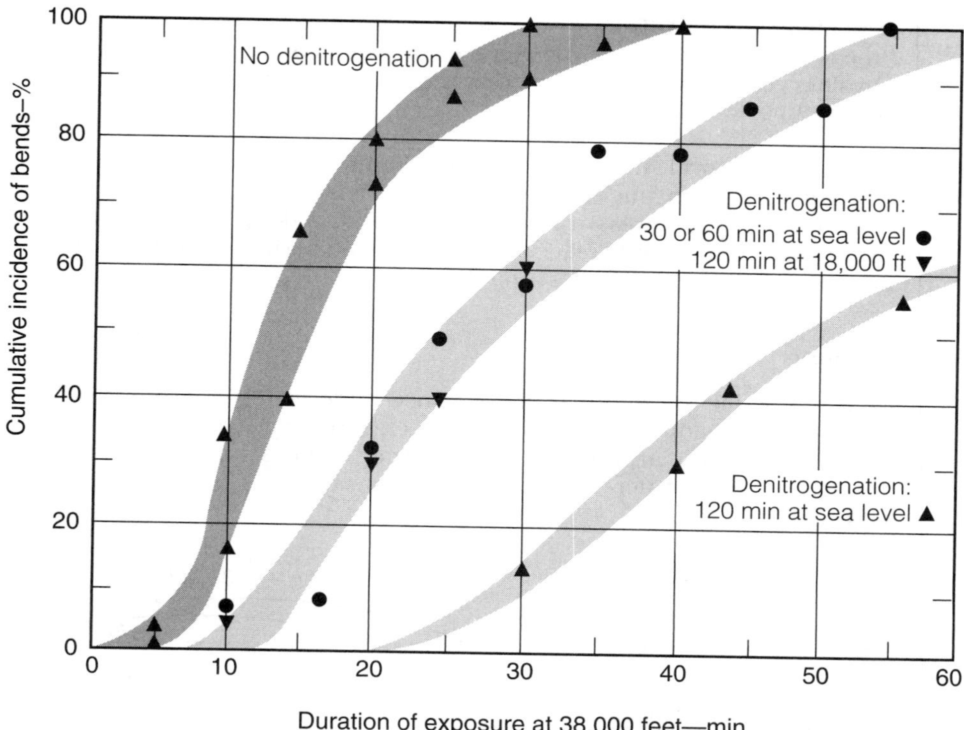

5-10 Ridding your body of excess nitrogen prior to flying will help prevent the bends.

Factors affecting severity

Six contributing factors can determine the severity of DCS:

- Time at altitude. The longer one is at an excessive altitude, the greater the risk for developing DCS because more bubbles can form and be transported throughout the body.
- Rate of ascent. The more rapid the ascent, the greater the chance of bubble formation because the body has little time to equalize the pressure changes.
- Altitude. The threshold for developing DCS problems is usually above 18,000 feet; however, DCS cases below 18,000 feet are now being identified. It increases more rapidly above 26,000 feet.
- Age and obesity. Older people tend to be more susceptible to DCS under minimal conditions. The same is true for obese people whose fat will store more gas and nitrogen. It is also suspected that recent alcohol use can increase the risk.
- Exercise. Exercise is one of the most important factors influencing susceptibility. It is wise not to exert oneself prior to a flight or immediately after flying at high altitude.

- Repeated exposure. Taking several flights over a short period of time (48 hours) increases the chances of problems. Repeated physiological "insults" to affected organs make them more susceptible.

In the real civilian world, the chances of developing DCS are small; however, the right conditions, such as having an injured knee, being caught at altitude in IFR in a nonpressurized airplane, and making several trips within a two-day period, could lead to a situation that is ripe for DCS. Keep it in mind, especially if you are a scuba diver.

Scuba diving

DCS is a common term to scuba (self-contained underwater breathing apparatus) divers and is drilled into each diver continuously. The physiology is the same as for pilots at altitude, but the diver is more susceptible because there is a greater differential in pressure in water and in shorter distances (Fig. 5-11).

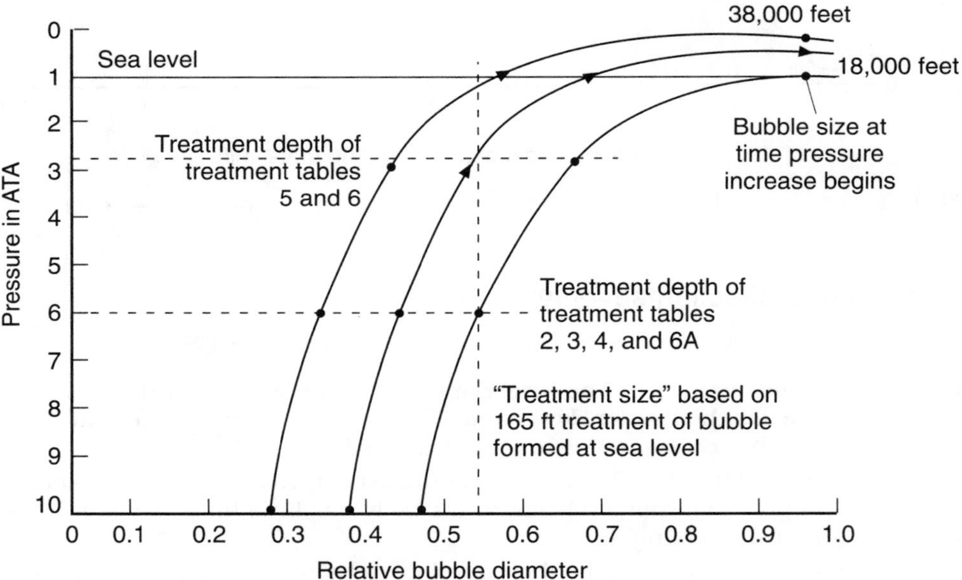

5-11 Pressure in water doubles at 30 feet, whereas in flight you have to go down from 18,000 feet to sea level to double the pressure. R.L. DeHart. *Fundamentals of Aerospace Medicine.* Philadelphia, Lee & Febiger, 1985. Reproduced with permission.

What happens is that at only 33 feet in water (2 atmospheres of pressure), the body will absorb twice as much nitrogen from the scuba air tank (the tank does not carry oxygen; it's *pressurized* air). Every diver knows this and anticipates problems when returning to the surface. Prevention of problems is understanding the dive tables, known to all divers, which determines the steps to take to prevent DCS. These steps do not include climbing into an airplane and essentially further decreasing the pressure surrounding the body. Furthermore, dehydration, common with pilots, will make the risk higher for DCS.

Bubble size is another concern, found mostly with divers and to a lesser extent at altitude; however, the problems that can occur with the combination of flying and diving can be severe. Much of this is a result of varying bubble size at different atmospheres of pressure (Fig. 5-12).

Depth in feet	Pressure in ATA	Relative volume (percent)	Relative diameter (percent)
0	1	100	100
33	2	50	79.3
66	3	33.3	69.3
99	4	25	63
132	5	20	58.5
165	6	16.6	55
297	10	10	46.2

5-12 The volume of a nitrogen bubble increases with altitude.

A rule of thumb: Don't take one more dive the same day before leaving for home in a plane with a cabin altitude that is probably around 8,000 feet. The same situation is present as if you were flying at 40,000 feet unpressurized. *The recommended waiting time is 12 hours after the last dive and at least 24 hours if you were a part of a dive that required a controlled ascent (decompression stop dive). The FAA recommends 24 hours after any type of diving. Some studies indicate that you should not dive for about 12 hours after arriving by plane.* There is continuing research as to what limits of times are safe for different types of dives. Keep in touch with your diving instructors as to what the current recommendations state.

If DCS is suspected in a scuba diver, then that diver must get to a hyperbaric chamber (recompression) and be monitored by medical personnel. The dive shop should have the number of your closest chamber for treatment through an organization called Divers Alert Network (DAN).

Altitude-chamber ride

Training in an altitude chamber is a common experience for military aircrew; the training is offered to the civilian pilot but is not required by the FAA. It is highly recommended that a pilot have at least one chamber ride for a number of reasons:

- To experience basic symptoms of hypoxia and decompression.
- To recognize personal symptoms before they become disabling. Because everyone responds differently to the same conditions, it is important to know what your unique symptoms are. Only through the chamber ride can you determine your signs and symptoms.
- To experience rapid decompression (R/D) in controlled conditions and realize that it is not the terrifying situation you might think it is. Also, it helps you realize the shock of R/D and how it can immobilize you and hinder your reactions.
- To gain confidence in and respect for high-altitude flight, especially for the insidious effects of hypoxia.

The ride profile (Fig. 5-13) might vary, but basically the pilot is taken to altitude, experiences hypoxia and trapped gas, is shown the effects on other participants so that he or she can detect these problems in others, and then is brought down. The most useful part is to be able to recognize his or her own symptoms of early hypoxia so that immediate action (getting on oxygen) can be taken before he or she is unaware of a disabling or incapacitation beyond EPT. The pilot is also shown what it is like to pressure breathe and the effect of hypoxia on night vision. The FAA has access to several chambers and offers a short course in altitude physiology. Some nonmilitary organizations now have chambers for such training.

The FAA has a chamber in Oke City (the Airman Education Division of the Civil Aeromedical Institute) for no charge but will work with nearby military chambers for a nominal fee. A third class medical issued within the previous three years is required.

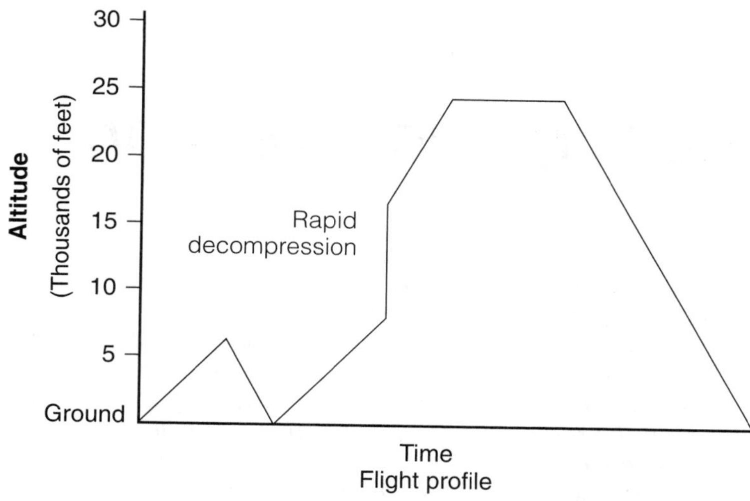

Altitude chamber flight

1. Preflight briefing
2. Mask fitting and operation of oxygen regulator and intercom system in chamber.
3. Ascend to 6,000 feet and descend to ground level to determine proper clearing of ears and sinus cavities. Rate of ascent and descent is 3,000 feet per minute.
4. Reascend to 8,000 feet at a rate of 3,000 feet per minute.
5. Rapid decompression (8–10 seconds) from 8,000 feet to 18,000 feet. Masks will be donned during the decompression.
6. Ascend to 25,000 feet at the rate of 3,500 feet per minute.
7. Removal of masks at 25,000 feet to enable trainees to experience symptoms of hypoxia. Five minutes time limit off supplemental oxygen.
8. Descend to ground level at the rate of 3,000 feet per minute. Trainees will practice pressure breathing and discuss their hypoxia symptoms on descent.

5-13 An example of an altitude-chamber-ride profile.

6
Hearing and vibration

The young flight instructor had a busy day with three students doing work in the pattern and a 4-hour cross-country. The Cessna 152 performed as expected, and everything worked well. After the long trip, she got into her car and headed home. She felt tired, more so than usual after a long day, but thought nothing of it. In the car, she had to turn the radio up to hear the news. The following morning, she overslept, even though her clock radio came on at the usual time. After a good night's sleep she felt more refreshed, but now noticed some ringing in her ears. She called the FBO to see when her first student was and couldn't understand the response: "Did you say 9:15 or 9:50?" The dispatcher had to repeat the time twice before she understood. Now she was getting concerned that something was wrong with her hearing, and so she went to her personal doctor. He found a high frequency hearing loss and informed her that it was permanent.

HEARING IS SECOND ONLY TO VISION as an important physiological sensory source of information in flight. What and how we hear is how we communicate. It lets the pilot detect audible changes in engine sounds, the flow of air over the airframe, and subtle changes in the sounds of gyros. Unexpected sounds alert the pilot to possible problems. The absence of sound in crucial instruments and engines tells the pilot of potential failure.

The well-trained and experienced pilot learns to depend on sounds and, of course, communication to make her flight safe. She is able to pick out significant signals from the myriad sounds that surround her during flight and is therefore able to respond appropriately. For example, in the maze of aircraft in a Class B terminal area with ATC communicating with each aircraft, a pilot can single out her call sign without consciously listening. Many seasoned pilots are able to accurately judge speed simply by listening to the sound of airflow.

Loss of hearing can be disabling, depending on the severity of the loss. Helen Keller, famous for her ability to cope with both blindness and deafness, stated that there is nothing more isolating than the affliction of complete loss of hearing. The irony is that for most pilots, virtually all high-frequency loss of hearing is secondary to prolonged noise exposure. Because the pilot can be protected from noise, there is no reason to suffer hearing loss.

A derivative of noise is vibration, which is sound at a lower frequency (10–60 cycles per second or less). There is a physical energy involved, which is felt more than heard. Its effects on the body are also capable of impairing a pilot.

ANATOMY AND PHYSIOLOGY OF THE EAR

The ear is meant to take sound waves (a form of energy) from outside the body and concentrate them like a megaphone onto the eardrum, which translates the energy into electrical currents into the brain via the acoustic nerve for interpretation (Fig. 6-1). Hearing is the mental perception of that sound energy. Discrimination by the brain of the various kinds of sounds determines how the pilot interprets the significance of the sound signals. Are the signals voices, music, engine sounds, or noise?

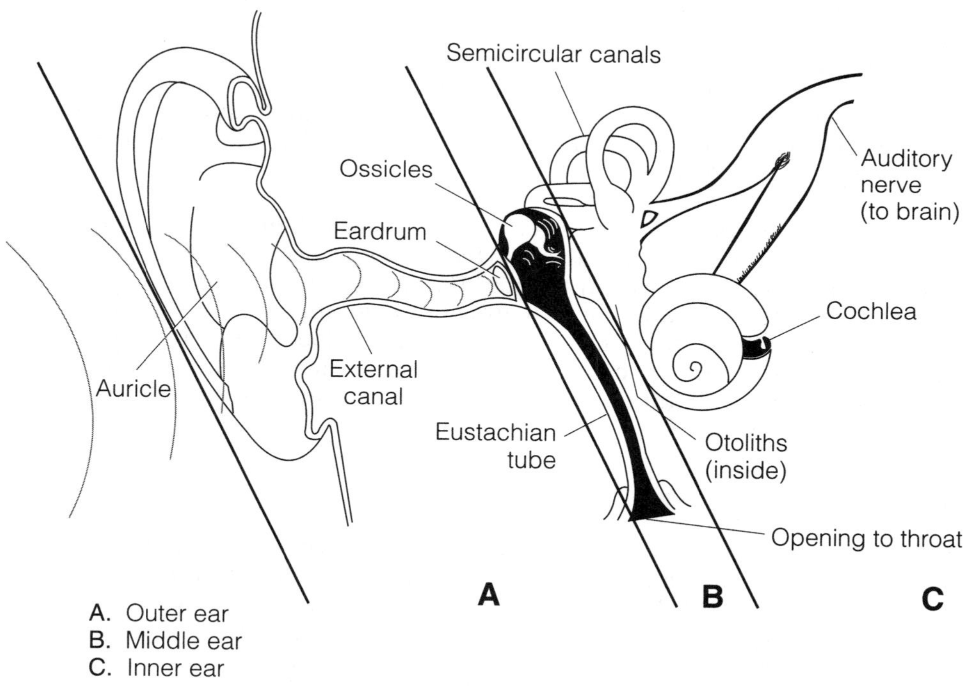

6-1 The ear is divided into the inner, middle, and outer parts.

The outer ear

The outer ear includes the *auricle*, the *external canal*, and the *eardrum*. Sound waves are picked up by the auricle and directed and focused down the canal to the eardrum. The eardrum is set into motion by the force of the sound waves, the drum moving back and forth with each wave. It's somewhat like the antique needle-type phonographs where a diaphragm is moved by the energy transported by a needle in the groove of the record.

Anything that obstructs this transmission of wave energy through the air diminishes or distorts the energy necessary to move the eardrum. Earwax, which everyone makes in varying amounts, is the most common obstruction. Those who make large amounts need to have this wax removed periodically. It cannot be removed with cotton-tipped sticks because the wax is only pushed further into the canal. The wax can be removed by softening the wax with over-the-counter solutions and then flushing with water, or seeing your doctor for his or her assistance. The eardrum is also affected by infection, swelling from an ear block, scarring from an old infection, or even perforation, all of which can impair the transfer of sound energy into mechanical energy.

The middle ear

The middle ear is also an air-filled cavity under equal pressure to the ambient air. Within this cavity is a chain of three tiny bones (*ossicles*) that are set into motion like a series of connected levers by the vibration of the eardrum. This transforms airwave energy into mechanical energy, which in turn is connected to the cochlea of the inner ear as the signal continues its journey to the brain. The cochlea has a "window" much like the eardrum and moves in amplified concert with the eardrum if the ossicles are functioning properly.

Interference of motion of these ossicles, especially at their joints, is common, leading to diminished and distorted transmission of a signal. Interference includes fluid within the middle ear from an infection, which is an *ear block*, inflammation or stiffening of the joints between the three bones, calcification of the joints, and physical dislocation of the three bones. The most common problems occur with a middle ear infection (often associated with fluid) and an ear block from trapped gas (see chapter 5).

The inner ear

The inner ear contains the *cochlea, utricle, succula,* and the *semicircular canals*. The organs of balance—utricle, saccula, and semicircular canals—are further discussed in the chapter on orientation. The cochlea resembles a sea shell in appearance, and it is the organ of hearing that converts mechanical energy from the movements of the ossicles into nerve impulses through their attachment to a movable window in the cochlea, which then moves the fluid behind the window.

The cochlea is filled with endolymphatic fluid. Lining the inner walls are thousands of hairlike receptors, which move when the fluid moves. This is similar to tall grass moving when wind passes overhead or sea plants moving with the water currents. This movement of the hairs stimulates the nerve ending at the base of each hair cell and produces an electrical impulse. This impulse is then transmitted through the cochlear nerve (also called the acoustic or auditory nerve) to the brain.

Hearing loss secondary to nerve damage (infection or injury) involves this nerve. Hearing loss secondary to noise involves the hairs in the cochlea. It's like walking on wet grass and looking at your footsteps. Damage to the hairs is the same. Like grass, most will repair themselves over time if there is no further damage; however, with enough insult, the hairs, like the grass, will remain damaged.

Tinnitus, or ringing in the ears, is very common, often the result of some degree of insult or damage to the cochlear nerve. Tinnitus is also often associated with noise exposure. This hearing loss from nerve damage, or *ossicular pathology*, is different than that associated with noise. Both will be discussed later in this chapter.

The sound of a voice, therefore, begins as an airwave from our vocal cords and is focused through someone's ear canal to the eardrum, which moves back and forth in concert with the sound wave. This movement is transmitted to the cochlea via the ossicles, which accentuates (through leverage of their joints) the energy as it reaches the window of the cochlea. With the resulting movement of the fluid within the cochlea, the hair cells within that fluid move and activate their associated nerve endings, which in turn is transmitted to the brain for final interpretation.

DEFINITION OF SOUND

Sound is defined in the dictionary as energy that is heard, resulting from stimulation of the auditory nerve by vibrations of the ear parts as just described. Sound is a form of wave energy, like electricity, light, and microwaves. Sound waves follow the same physical laws as any other energy wave. When they reach the ear, variations in sound waves describe and define a myriad of signals and cues, processed by the brain and which we use to communicate and interpret meaning and source. Like vision, hearing is an important physiological sense that plays a significant role in safe flight.

Physical properties

There are three measurable objective properties with all sound, whether characterized as useful sound or noise (or even vibration). How the pilot perceives and interprets the meaning of these physical signals becomes very subjective and ranges from indifference through inability to detect the sound efficiently to being keenly aware of the cues and their role in flight.

Frequency

Like other wave energy, frequency refers to the number of wave cycles (or oscillations) per second (CPS), also called hertz or Hz. The frequencies of various sounds as perceived by humans usually are in the range of 20 to 20,000 hertz. Humans can detect a fairly wide frequency range. Other animals have wider ranges; bats, for instance, can detect frequencies in excess of 80,000 Hz.

Intensity

This measurable property describes the loudness of the sound. Because it defines a physical property of energy, this measurement is directly related to how well the pilot hears or interprets a certain frequency or combination of frequencies. It is this intensity at individual frequencies that is tested by the audiometer. The unit of measurement of intensity is called the *decibel* (dB), which is a logarithmic function of pressure (Fig. 6-2) and is used as a comparison to another sound. A sound 10 times as powerful as another is called a decibel (deci=10), and each tenfold increase in sound pressure produces a sound that is 20 dB greater, *by comparison*, than the preceding sound. A graph of dB versus sound intensity is a straight line. A sound that produces an energy level that is 1,000 times as intense as the reference noise has a value of 30 dB (10^3). The decibel unit of measurement allows us to express the large magnitudes of sound differences as a small number rather than a large one.

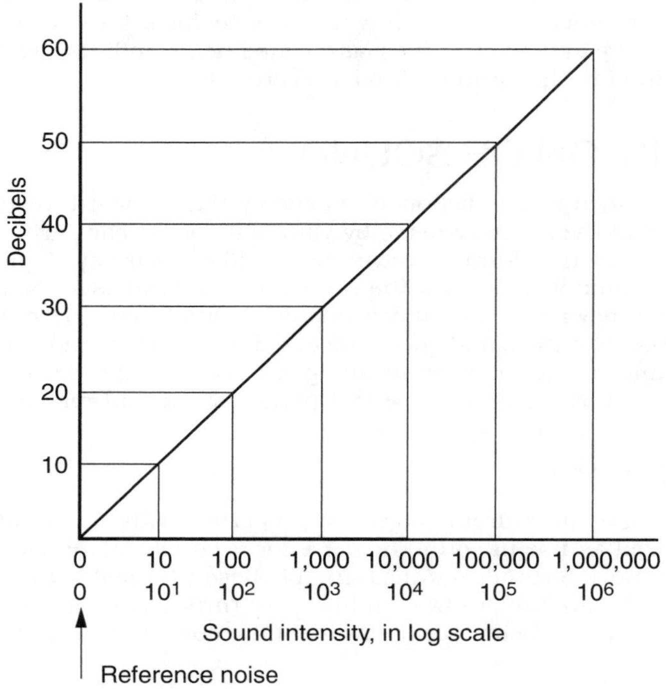

6-2 The decibel scale versus sound intensity.

For example, at 60 dB, a sound (or combination of separate frequencies) is *1 million* times greater than at 0 dB. The higher the dB level, the greater the energy level, exponentially. Intensities greater than 130 dB, therefore, are physically painful and damaging.

Duration

How lcng the ear is exposed to a sound determines how well the sound is heard and discriminated or interpreted as being important. It also determines the level of potential risk of injury to the ear. Because sound is energy, it follows that short periods of intense sound are potentially physically detrimental to the organs of hearing. Likewise, the same degree of damage can occur with less intense sounds that occur over longer periods of time; therefore, the potential damage is time related as well as intensity related.

Conversational range

Interpretation of sounds in flight is important for two major reasons. One is to determine audible cues or changes to these cues for purposes of detecting the performance of the aircraft, its power source, and other crucial parameters of safe flight. The other, of course, is to communicate. Although the ear can detect a wide range of frequencies (which enhances the first need), the only frequencies necessary to communicate are in the conversational range, or 500 to 3,000 Hz. Above and below this range, the sensitivity to the ear decreases markedly.

Most communication systems in aircraft are not high fidelity because they only operate in the conversational range. Consequently, this range is the one most carefully monitored by the FAA in its medical evaluations. Specific intensity levels at 500, 1,000, and 2,000 Hz are among the few stated specific FAA standards, and then only for first-class medical certification.

DEFINITION OF NOISE

A sound perceived as being too loud, disagreeable, or distracting is considered noise. This is a subjective cue to each pilot because sound that is noise to one pilot might be an important cue to another. In any case, noise still implies sounds that need to be filtered out or become a background level over which more important sounds, such as conversations, can be heard and understood.

Types of noise

Two types of noise are steady-state and impulse. *Steady-state* is a type encountered around and in an operating aircraft, often during flight. Here is where duration of exposure becomes important because there is a direct correlation to intensity, duration, and amount of damage to the ear (Fig. 6-3). Sources include prop noise, which ranges from 90 to 113 dB, ventilation sounds in small aircraft and even in the new glass cockpits, and the turbulence of air around the fuselage.

Impulse noise is the short burst type that can be produced by a backfiring engine, switching to a communication channel already turned up to full

Exposure duration per day (hours)	Maximum exposure level (dBA)
8	85
6	87
4	90
3	92
2	95
1½	97
1	100
½	105
¼ (ceiling)	110 or less

6-3 Allowable noise exposure.

volume, or other short-term (often lasting only a few milliseconds) but loud sounds. Often these sounds come as a surprise, and there is no way to anticipate the need for added protection; however, within the middle ear, where the ossicles are located, there is a small muscle attached to one of the bones that functions like a "bungee" protection against sudden and extreme movement of the joints.

At the very least, noise is annoying and distracting in flight and can seriously interfere with concentration, communication, and performance.

Noise and fatigue

An often overlooked physiological effect of noise on the entire body is fatigue. Noise, as a form of energy acting directly on the body as well as on the hearing mechanism, therefore, is more than just an annoyance. Pilots returning from a long flight in a noisy aircraft are often more fatigued than those who are wearing noise protection. Some have recognized that fatigue is more noticeable in the pilot who is mentally active than in someone who is resting in a noisy environment.

Other nonauditory effects of noise include interference with sleep, rest, and the proprioceptive system, which will be discussed in chapter 8. Other subjective symptoms include nausea, disorientation, headaches, and general irritability.

Another result of noise exposure is *tinnitus*, or ringing in the ears. Everyone has some degree of ringing, and this is very noticeable to an individual when he or she is in a superquiet room with no background noise. There are very few places without some noise: circulating air, vehicle traffic, wind, voices, and the like. Tinnitus is especially noticeable at night, when one wakes up and becomes aware of the ringing.

After exposure to noise, the ringing is more noticeable but tends to diminish over time; however, with significant noise-induced hearing loss through prolonged exposure, increased tinnitus is common. There is no cure for tinnitus. Trying to ignore the ringing is the best solution. Maintaining a heightened awareness of tinnitus seems to increase the subjective perception of this sound to the point that a pilot is overly distracted or annoyed. If tinnitus is severe and more than just distracting, an ear specialist should be seen.

There also appears to be an individual variation of tolerance to noise. Some are able to withstand periods of intense noise without detrimental effects. Some recover more efficiently from noise than others, even without protection; however, this is the exception, and only through adequate monitoring of effects to noise exposure by audiometric testing can one minimize unexpected damage or impairment.

PERCEPTION OF SOUND

The hearing apparatus within our body provides the pilot with a multitude of valuable signals and cues. The properly functioning ear can tell where a sound is coming from by providing variations of sound intensity from two sources and then processed by the brain for location. This ability can be demonstrated by having someone walk around you while your eyes are closed. As long as some sound is heard, either footsteps or conversation, you can always tell where that person is, without looking.

The ear is always "on"; that is, loud or abnormal sounds will awaken a sleeping person. And some individuals have the ability to tune in and screen out the important sounds. Mothers hear their babies cry in the middle of the night in a house with many other sounds. A pilot hears her call sign amongst many others in conversation with ATC, even if that pilot is concentrating on some other activity.

If the ear can detect its full frequency range of sounds at low levels of intensity, there now exist many cues that the pilot uses for situation awareness, often unconsciously. The proficient pilot, for example, can detect a change in airspeed without looking at the gauges but listening to the various sounds of flight. This perception comes only from training, experience, and a well-functioning pair of ears.

Even surrounded by noise, the ear and brain can discriminate important sounds from unimportant noise, providing the noise isn't overpowering. The healthy ear can detect faint sounds, especially spoken communications, under a variety of interfering conditions: noise, thermal extremes, distractions, and multiple activities occurring at the same time. By the same token, many situations interfere, including fatigue, hypoxia, stress, and motivational factors, to name a few. One of the most important constants in efficient and meaningful hearing is proficiency of flight: becoming accustomed to familiar sounds, ATC communications, and expectations in aircraft performance.

A common problem, however, affecting hearing is noise and hearing loss, especially loss that goes undetected. This combination is a major cause of missed communications, misunderstood ATC clearances, and misinterpreted audible cues.

MEASUREMENT OF HEARING

Because sound is a form of energy, it can be quantified. By sending a pure tone of a specific frequency and at a specific intensity to the ear, the perception of that sound can be identified and measured for all frequencies. The testing for this is from an audiometer. The ideal setting is listening to each pure tone in a soundproof booth using a headset. For screening purposes, a good headset in a quiet room will do the job.

All FAA exams require testing of hearing. A common method is the *whispered-voice test*, now called a *voice test*. This is a poor substitute for the audiometer because it cannot quantify or identify the frequency at which a hearing loss might be present. Only with an audiometer can a reliable result be achieved.

The test results from an audiometer are recorded as an audiogram, which describes a hearing level curve for each ear. Normal hearing should be a straight line in the 5-dB range. Variations from normal are acceptable, but any frequency with a loss of greater than 20 dB is considered potentially significant and needs to be further evaluated to rule out a progressive loss of hearing.

Over time, by comparing audiograms, one can see shifts that are occurring. If high frequencies are being affected, then better hearing protection is necessary. A baseline audiogram is very important so as to compare subsequent curves of changes. These changes are called *threshold shifts*, meaning there is a shift from the baseline of years past and a strong indication of a noise-induced hearing loss.

NOISE AND HEARING LOSS

Hearing loss is not inevitable. Several preventive measures can be taken. The first is to recognize that noise is present everywhere and is potentially detrimental, both in performance and in the health of the ear and hearing system. The next is to protect the ears from noise by wearing appropriate hearing defenders, or ear protection. The third is to be part of an ongoing hearing conservation program with annual monitoring of hearing abilities and retraining in noise avoidance.

Sources of noise

Airplanes and airport ramps are some of the noisiest sound sources to work around. Pilots are exposed to sounds varying from continuous low-level noise on a cross-country flight to short periods of engine runups on the ramp. In addition, pilots encounter noise associated with many activities not related to aviation, such as hobbies, mowing lawns, downtown traffic, and the like.

The Occupational Safety and Health Administration (OSHA) has determined certain levels of noise for many different events and activities (Fig. 6-4). Note that each is "time weighed," which means that damage is done with lower levels for longer periods of time. For example, eight hours of exposure to noise levels of 90 dB is as harmful as two hours at 100 dB. Then note the sources. It is surprising how much energy there is from some sources. Rock concerts near 120-dB levels and walking around with headsets at full volume virtually focus that energy right into the ear, much like a megaphone. Studies are now proving that such exposure does cause permanent hearing loss.

Decibels (in dB/A)	**Sounds**
0	Threshold of hearing for young ears at 1,000 Hz
20	Whisper at 5 feet
30	Broadcasting studio speech
30–40	Country residence
40–60	Noisy home or typical office
50–70	Normal conversation
60–80	Noisy office, average street noise, average radio
70–80	Automobiles at 20 feet
80–100	Loud street noise
90	OSHA standard for 8 hours work
90–100	Public address system
100	OSHA standard for two hours work
100–110	Power lawn mower
100–120	Thunder, snowmobile
105	OSHA standard for one hour work
110	OSHA standard for one-half hour work
115	Rock concert
130	Pain threshold for humans
140–160	Jet engine
167	Saturn V rocket

6-4 OSHA standards for noise exposure.

Airplanes are not quiet. Many small reciprocating airplanes are very noisy. Prop noise ranges from 90 to 115 dB. Even in the passenger cabin of airliners, certain noise levels are potentially harmful over periods of time. As a general rule, whenever you must raise your voice to be heard by someone next to you, you are in a noisy area. Furthermore, for every 4–5 dB increase in sound above 85 dB, *the safe time limit for exposure is cut in half.*

Noise-induced hearing loss

Equally important as the interference with communication, noise also causes permanent high-frequency hearing loss. It is an insidious change, taking years to develop significant loss. Hearing aids do not help noise-induced hearing loss. In other words, noise alone can reduce hearing over the years and there is nothing that can be done to reverse the process. You won't know there is a change occurring over time unless you check for it. Like so many other issues in medicine and as frequently stated throughout this text, the best and only treatment is recognition, respect, and prevention.

The loss can be rapid in uncommon occurrences of extremely intense noise. Sudden impulse noise is physically damaging to the eardrum as well as the ossicles. This usually occurs with noise in excess of 140 dB, such as gunshots, explosions, and jet noise.

Many pilots have experienced the temporary loss of hearing after a long trip in a noisy airplane. There are stories about people who start their cars for the ride home and then try to restart the engine, which is already running. This sort of loss will disappear in a few days but is also an indication that there was a significant noise exposure that eventually could lead to a permanent loss.

Threshold shift

As unprotected exposure to noise persists, there is an increasing risk of hearing being impaired, especially in the higher frequencies. This threshold shift is not noticed by the pilot because there is no change in the pilot's perception of sound, plus it occurs over several years. There is no pain to warn the pilot. Short of undergoing testing, there is no way for someone with a threshold shift to compare his hearing to normal or what is expected from previous perceptions of sounds.

In other words, sounds are different but go unnoticed over time. With even partial loss of high frequencies, such sounds as brass and string instruments sound bland and without clarity. Voices are still heard but are not sharp or clear. As the loss increases, sound becomes muffled. Because this change occurs over several years, the pilot does not recognize it until extensive loss is noticed. As the loss becomes even more severe, speech recognition becomes difficult if not impossible, especially when background noise is also present, which is at a dB level equal to the hearing loss.

There is also a subtle reduction in the interpretation of certain words. Over a period of time, some consonant sounds of words are diminished and only some of the vowel sounds are distinguishable because the higher frequencies are not heard. A word, therefore, that depends on a full range of frequencies will not be understood. It might sound like mumbling. Communication then becomes more difficult.

Threshold shifts can be temporary, as when you are subjected to short periods of sound infrequently; however, with continuous (over months) exposure to even moderate noise, this threshold shift becomes more and more permanent. Only monitoring and comparing results of audiograms will alert you to any shifts (Fig. 6-5).

The audiometer will detect even subtle changes before more permanent damage is done. Even without such documentation of loss, the pilot should expect loss if he or she continues to subject himself or herself to noise without protection. The primary concern is a permanent threshold shift.

Normal hearing: Has no difficulty with faint speech and little difficulty with speech in a noise background.

Moderate hearing loss: Has difficulty with normal speech, ringing sensation (tinnitus), many combat sounds not detected, possible hearing aid candidate.

Mild hearing loss: Has some difficulty with speech in a noise background, ringing sensation (tinnitus) might be present, some combat sounds might not be detected, e.g., clipping barbed wire, dog tags rattling, loose cartridges in pocket, man walking through grass.

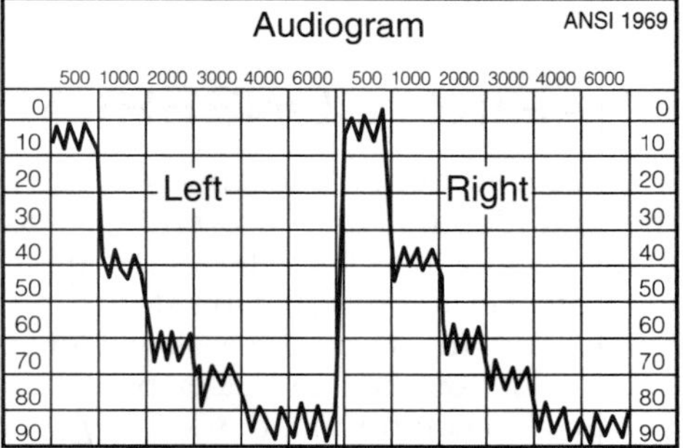

Severe hearing loss: Has difficulty even with loud speech, ringing sensation (tinnitus), needs hearing aid, auditory training, lip reading, speech conservation and speech correction.

6-5 Audiograms show progressive hearing loss.

OTHER TYPES OF HEARING LOSS

Although noise is the greatest and most common cause of loss, especially in the pilot group, there are other causes. Age, health, and individual sensitivity increase hearing loss. In addition, the following also are found when tested with proper equipment and techniques.

Conductive

When there is some defect or impediment of sound transmission from the external ear to the inner ear, *conductive* loss is identified. Such a loss involves all frequencies, not just high frequencies. Such situations develop from wax buildup, fluid in the middle ear from infections, ear block, or allergies and colds. Also, the ossicles can become calcified around their joints similar to what can happen to any joint in the body. Most of these situations are treatable and hearing can be restored.

Sensorineural

Sensorineural loss occurs when the hair cells of the cochlea are damaged or inflamed. This can occur from noise exposure, age, or infection. Tumors have been known to cause hearing loss in this category. Because the lower frequencies are also affected, this is a clue that this form of loss is more than just noise related. This type of loss is generally not curable.

Presbycucis

Presbycucis loss is associated with aging and can be conductive and/or sensorineural in nature.

HEARING CONSERVATION

It bears repeating. The only way to cope with noise and hearing loss is through identification, prevention, and conservation of the hearing one already has (or has left after suffering a loss). This program, required in the military of everyone on flying status, must be adopted by the individual pilot. Some companies and the FAA will let you damage your ears as much as you want until you can no longer pass the medical exam or you are noticeably impaired in communications. *The pilot is responsible for protection of hearing and is ultimately the one to blame if preventable hearing loss develops.*

A successful program consists of three phases. The first is to define a baseline audiogram. If you have never had an audiogram, then that needs to be done as soon as possible. If your AME doesn't do audiograms, seek out one who does or see an audiologist (usually listed in the yellow pages of the phone book).

The next phase is to structure a monitoring program. The ideal is to have your FAA exams done by an AME who already does audiograms as part of his or her FAA medical exam. If this isn't the case, then an annual audiogram from the audiologist is important; you would keep a copy of the audiogram in your personal file to track any shifts.

The final phase is to wear hearing protection in all noisy areas, especially at work. Noise-induced hearing loss is easy to prevent, if you wear protection. Many of the newer headsets have adequate noise attenuation. A critical condition is the seal between the headset and around your ears. If that seal is broken or interrupted (as with glasses or old and cracked seals on the headset), then noise can enter the ear. This is especially true with the electronic attenuating headsets.

The simplest and least expensive is the common moldable foam-type plug, which you compress by rolling into a tube, inserting into the ear canal, and holding it there until it expands and stays in place. The plugs are very effective. The more expensive ear "pieces" offer little or no better protection. The inexpensive foam plugs lend themselves to being purchased by the dozen so as to always have them available. Keep them in your aircraft, your flight bag, at home, in the passenger section, and in your flight suit or map case. Plain cotton or tissue is useless as it does not block the sensitive frequencies.

Custom molded earplugs are somewhat controversial. Many pilots swear by them; others find them more uncomfortable and ineffective. Be sure to have the manufacturer quantify the amount of attenuation, if any, before you purchase them, and compare their performance and attenuation standards to a traditional headset. These ear pieces are not meant to protect from noise and they do have their place in low-noise cockpits; however, consider the situation where one communication channel is set low and the other is set high by someone else and you switch to that channel. The ear piece funnels all the sound energy into the ear canal and against the eardrum with potential physically damaging effects.

In extremely noisy environments, both a headset and earplugs might be necessary. More important in this situation is to be dedicated to an ongoing monitoring program.

A common question concerning the wearing of earplugs is their effect on hearing what you want to hear. Is there a reduction in communication? Ear defenders, hearing protection, and earplugs all do the same thing. Like wearing sunglasses, earplugs have a place in the right situations. Sunglasses are not needed in low-light conditions. In bright light, they are necessary. In quiet situations, there is no need for earplugs; however, if the ambient noise level is such that you can't hear or communicate without shouting or even raising your voice, then protection is necessary to block out the noise.

The intent is to be able to maintain clarity of speech and to detect changes in the sounds of flight. Noise is an interference and can be blocked out with earplugs without sacrificing the ability to hear meaningful sounds. Masking unnecessary sounds, or noise, allows for more clarity in otherwise difficult conditions of flight communications and hearing.

The other equally important reason is to prevent permanent hearing loss and other associated symptoms such as fatigue. Once a pilot is in the habit of wearing earplugs or attenuating headsets, there is no compromise of communication.

VIBRATION

Like sound and noise, vibration's effect on the body is related to frequency, intensity, and duration. Unlike noise, vibration exerts very little physical effect on humans at frequencies above 4,000 Hz. Furthermore, vibration is related to the physical displacement of parts of the body, some parts moving more and easier than others (bone, soft tissues, internal organs, etc.).

Vibration is noticed more in helicopters, less in reciprocating engines, a little less in turboprops, and minimally in jet aircraft. A common component is the fact that vibrations are communicated directly to the subject via a solid medium, or through direct contact with the vibrating aircraft.

These waves of vibration are transmitted to various body parts, each of which reacts differently. Only very intense vibrations disrupt normal cellular and organ function; however, generally speaking, all vibrations can cause body discomfort, chest pain, diminished visual acuity, and distractions. Vibration is also an important cause of fatigue.

The most important frequency range associated with adverse effects lies between 1 to 40 Hz. Different parts of the body oscillate with the vibratory frequency but at different levels.

There is little one can do to change the causes of vibration because once the aircraft is designed, little can be changed to reduce vibration. The most important modification is in the seats; the more the body is cushioned, the more the vibration is dampened. Shoulder and lap belts tend to transmit some vibration.

The healthy body is more tolerant of the effects of vibration; however, because vibration is fatiguing, this fatigue now allows the vibration to be more of an impairment, a vicious circle. This fatigue is probably the most noteworthy result of vibration. Awareness of this source of fatigue allows the pilot to be more suspicious of substandard performance.

7
Vision

Fred was more than 50 years old, but he strove to maintain that image of the professional pilot, which meant that wearing glasses for reading was necessary only when they were needed. It was night, and the last takeoff of the trip was after a quick turnaround. They had been flying for 3 hours (cabin altitude 6,500 feet), which was bad enough, but being a smoker, he longed to have a cigarette. He was tired, and it was his leg to fly. The pretakeoff checklist was completed, and they had clearance for immediate takeoff. He pushed throttles forward in the darkened cockpit and the plane began to roll. As they approached V_1 Fred began to rotate. His copilot immediately pushed the yoke forward. They were too slow. What happened? Fred's reading glasses were in his shirt pocket, and he misread the instruments, thinking he was at rotation speed. His uncorrected vision was severely compromised by a multitude of causes: hypoxia, low light, age, fatigue, and not wearing his glasses. His copilot saved his day.

OF ALL THE SENSES USED FOR FLYING by the human body and mind, vision is by far the most important. In addition to supplying nearly 90 percent of the cues for orientation (vestibular and proprioceptive are about 10 percent), vision supplies data for monitoring instruments, traffic, and written materials, and it also determines visual references for taxiing, takeoffs, and approaches to landing. Put another way, the disruption of vision virtually incapacitates a pilot, and impaired vision increases the likelihood of an unsafe flight.

Most people are born with healthy vision, even if some have to wear glasses for distant vision at an early age. Faulty vision is usually easily corrected. In most situations, one has the flexibility to control the environment in order to improve acuity (clarity of vision), such as improving lighting, ignoring unclear characters or objects, changing position to see better, and the like. In flight, however, this flexibility is not present and the pilot has to make use of available information, often in dim light, often under restricted and harried situations. Furthermore, the effects of fatigue, hypoxia, minor illness, dehydration, and other physiological decrements add to the loss of visual acuity.

Most people take vision for granted, as long as they can read and see distant objects. But without testing for visual acuity, the pilot has nothing to compare his vision to and might not be aware that what he is seeing is not as clear or sharp as it could be. A common comment by children who begin wearing glasses for the first time is that they now realize that brick walls have lines. Until they began to wear corrective lenses, the bricks were a blur, but with no basis for comparison of this image, it was assumed that there were no lines.

The same holds true for older pilots (over age 45) who begin to wear reading glasses for the first time. It's almost startling to them to see how clear their approach plates and cathode-ray tube and liquid crystal display (CRT and LCD) characters really are. They might have been able to pass the FAA vision test, but that didn't mean they saw the near-vision characters clearly.

Furthermore, pilots with 20/20 day vision might not possess adequate night vision as a result of the physiology of how the retina converts light energy into electrical energy for the brain. There also appears to be an individual variation in pilots of how light energy is perceived by the brain and translated into data necessary for flight. In other words, with the same visual acuity, one pilot sees better and uses vision more effectively than another. And we all have good and bad days with vision; one day we are able to read the 20/20 line and the next day only the 20/30.

THE LIGHT SPECTRUM

Visible light is defined as radiant energy that excites cells of the retina of the eye and produces an electrical current that is transmitted to the visual cortex of the brain for interpretation. Visible light is only a small part of the radiation spectrum, ranging from violet at 380 millimicrons to red at 760 millimicrons (Fig. 7-1).

Although not visible, shorter (ultraviolet) wavelengths often cause damage to the ocular tissue (especially the lens) if the energy is sufficient. The surface tissue of the eye (the cornea) can be burned by the ultraviolet rays in a tanning booth or at high altitudes. Chronic excessive exposure to ultraviolet rays is now thought to be a cause of *cataracts*, a cloudiness of the eye lens. Most of these harmful rays are absorbed by the atmosphere; however, there is growing concern about flying at high altitudes for long periods of time.

Longer wavelengths of light (*infrared*) cause heat and are a source of increased energy. Damage to the eye, especially the retina, is dependent upon the degree of energy and duration.

Light rays (or energy) reach the eye either directly from a light source or indirectly, as reflected from a source (such as water or snow) back to the eye. Diffusion and refraction of light follows basic physical laws and can be quantified depending upon strength and distance. For our purposes, we will concentrate on the light that enters the eye, not the source of the light; however, it is necessary to recognize that there are individual light *rays* reaching the eye from the light source or reflected from an object. These rays are virtually parallel to each other from a source beyond 20 feet from the eye. Light rays originating closer than 20 feet come to the eye at an angle.

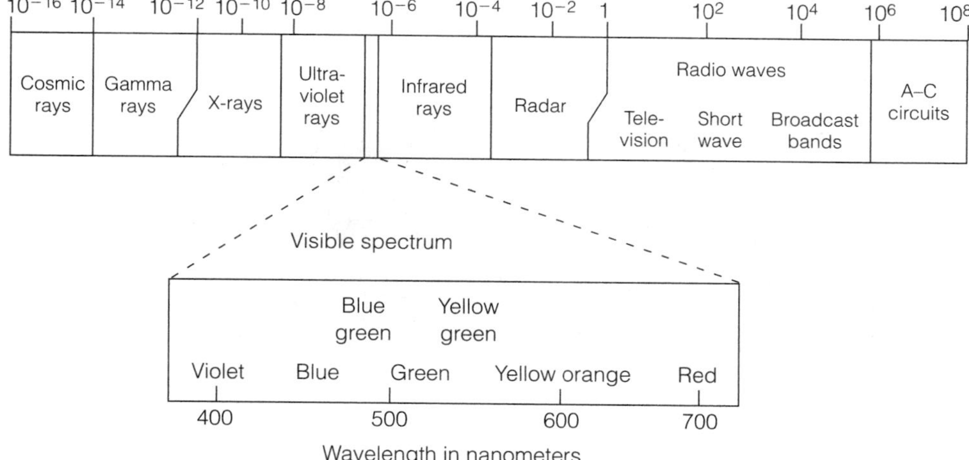

7-1 The electromagnetic spectrum.

ANATOMY OF THE EYE

Light is focused onto the retina after it enters the eye through the cornea and passes through the lens (Fig. 7-2). The retina translates this radiant energy into electrical energy by photosensitive cells called *rods* and *cones*. This current is transmitted via the optic nerve to the brain where it is deciphered into a meaningful signal.

Movement of the eyeball is by the extraocular muscles. Each eye has six independent muscles that can rotate the eyeball to track or fixate on any object within the field of vision. Imbalance of the strength of these muscles leads to tracking as if with two separate eyes, resulting in double vision, especially when tired; therefore, the coordination of muscles and the brain's signal to concentrate on an object is essential for *binocular vision*, as opposed to *monocular vision*, or one eye.

The cornea

The outer layer of cells of the eyeball is the cornea. This is a transparent "lens cap" that has considerable *refractive* (focusing) properties. Also, the transparency of the cornea can be affected with age and temporarily with overexposure to ultraviolet radiation.

The lens

Like in a camera, the lens refracts (bends) light rays as they enter the eye and focuses each ray onto the retina, which is analogous to the film in a camera. If the focus is in front of or behind the retina, the perception will be a blurred object. A camera changes this focus by shortening or lengthening the distance of the lens from the film because its lens is hard.

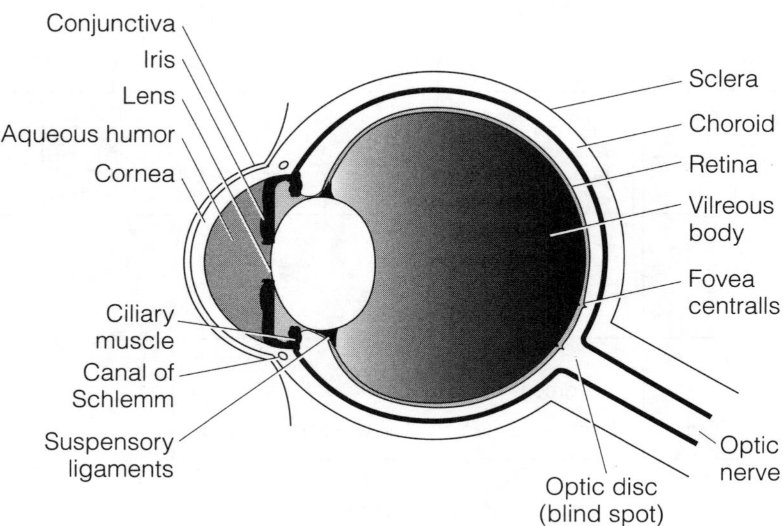

7-2 Anatomically, the eye is an extension of the brain, and as such, the cells of the eye are adversely affected like other brain cells by fatigue, hypoxia, or alcohol.

The lens of the eye is soft and moldable, and its shape is changed by the ciliary muscles that surround the lens. When these muscles contract or relax, the lens changes from fat to thin, allowing for more or less refraction of the light rays; therefore, focusing is accomplished by changing the shape of the lens rather than changing the distance. How well this lens can be altered determines how well the eye can clearly focus an object onto the retina.

The lens can become opaque in some people as they age, resulting in cataracts. Also, as we age, everyone's lens begins to stiffen, which prevents adequate sharp focusing and slows down the rapid change of focusing from distant to near objects, which is *accommodation* (defined in the subsequent section on optics in this chapter).

The iris

The iris is the colored membrane seen in the front of the eye that determines the eye color. The center portion is called the pupil, and like a camera's aperture, the size of the opening can be changed to allow more or less light to enter the eye. This gives the eye more flexibility to control the amount of light reaching the retina: a larger diameter in low light and a smaller diameter in bright light.

The brighter the light, the smaller the pupil, which increases the depth of field, like a camera. *Depth of field* is the nearest and farthest distances from the lens within which everything is in focus; therefore, in low-light conditions, the depth of field is very short because the pupil is larger. Squinting duplicates this action to a small degree and allows for some focusing.

The retina

The retina is the "photographic film" of the eye, using photosensitive cells to detect light energy in place of the film's light-sensitive silver nitrate crystals (Fig. 7-3). A big difference is the distribution of two different kinds of cells within the retina. Each of these cells connects to a nerve, which in turn ends up in a bundle of nerves called the *optic nerve*. An electrical impulse is generated from each cell and transmitted through the optic nerve.

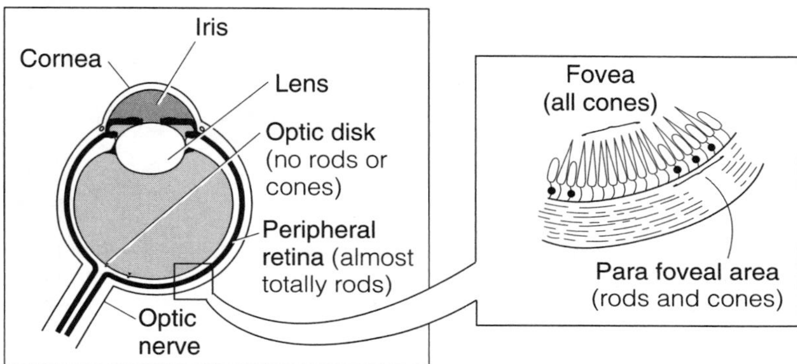

7-3 Anatomy of the eye.

One cell is the *cone*, which is sensitive to bright light and colors, as well as clarity. Cones are thus used primarily in daytime and for most of our accurate visual sensing (*photopic vision*). Cone cells are concentrated in a small area called the *fovea* (part of the macula). This is the spot on the retina, called *central vision*, on which most visualized objects are focused.

Rod cells are reactive to low light but not to colors and thus are more suitable for night vision (*scotopic vision*). Surrounding the fovea is an increasing number of rods and a decreasing number of cones until on the periphery there are no cones. This peripheral vision is more suited to orientation and supplementary in night-vision techniques. It is estimated that although images are poorly focused, the concentrated rods are about 10,000 times more sensitive to light than the cones in the fovea (Fig. 7-4). There are no cones or rods where the optic nerve is formed on the retina. This becomes a blind spot in vision but is compensated for by using our two eyes, each one overcoming the other's blind spot.

Because the eye, and especially the retina, are direct extensions of the brain, the rods and cones are particularly susceptible to hypoxia, medications, alcohol, and other factors known to affect the brain. The other senses are not as susceptible to these abuses.

OPTICS AND PHYSICS OF LIGHT

A single light ray has a variety of physical properties that are associated with vision. *Refraction* is the property of a lens to bend the light ray so as

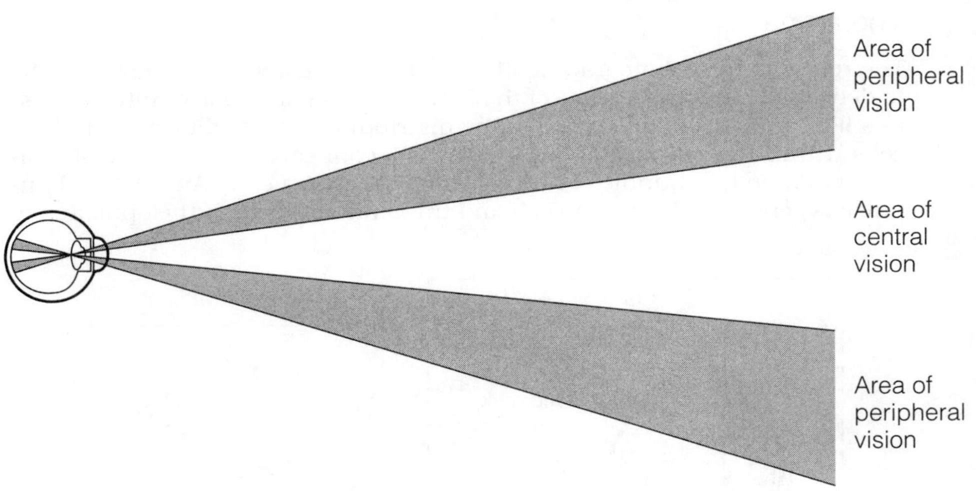

7-4 The central area gives the most acute vision in light down to the intensity of moonlight. Below this level of illumination, central vision cannot function as well, and the central area is nearly blind. Any object that an individual looks at directly will not be seen clearly.

to reach a given point (on the retina). Several light rays must also remain in focus through the same lens. Furthermore, the lens must be able to focus on a near object as well as on a distant one. This is called *accommodation*, where the refractive powers of the lens are continuously changing, allowing for uninterrupted focus, as opposed to a camera, which has to change focal length to focus.

Several other properties determine the clarity of the perception of these images even when focused. These include brightness, contrast, and color. Reaction time, or the time from seeing or identifying an object and recognition by the brain to some physical response, must also be considered, and such properties will be discussed in the section on vision in flight. In addition, atmospheric conditions such as fog, haze, rain, and the like, can also interfere with light-ray transmission and its diffusion through the atmosphere.

The brighter the light, the easier it is for the eye to gather light rays onto the fovea. As with sound, low-energy sources result in a poor signal, or lack of clarity. The opposite extreme is too much light and glare, which interferes with proper acuity, plus being uncomfortable.

The greater the visual difference between an image of an object and its background (contrast), the easier it is to distinguish what the object (or character) is. If the edge is fuzzy for any reason, the object appears out of focus. Look at a painting, and you will notice how the artist has clarified edges by outlining them with a darker color. Fog and haze interfere with this contrast and lead to poor acuity and clarity as well as the perception of visual illusions.

Color also affects acuity. Each color is refracted differently through a lens, as evidenced by the rainbow effect from a prism; therefore, some colors (other than white light, which contains all colors) will not be in focus relative to other colors, and their individual rays will land outside the focused area. Red, for example, is refracted much less than others, which leads to unfocused perception by the pilot. It will not be as clear as an image illuminated by other colors or with white light; therefore, using red light in the cockpit is not recommended because small characters on maps and checklists will not be clear or sharply focused and might be misread.

Another potential problem is present in new cockpit instruments and CRT/LCD displays that use not only all the primary colors of the visual spectrum, but up to 250 varying hues of these colors. Someone who is already color deficient will have problems distinguishing the differences between some characters.

Two objects of the same brightness can be distinguished easily if there is a recognizable color contrast, even in low light. The clearest is black against white; however, few situations have such a clear definition. If the pilot is not sure what is being visualized, she will take more time than necessary to discern what the object or character really is.

It is obvious that such factors must be considered when designing and using instruments, gauges, maps, and checklists; however, since such design is not always the best or has already been constructed into the system, one must be aware that vision will be compromised under conditions of low light, poor contrast, and inefficient colors.

CORRECTION OF ACUITY

Several variations of the lens affect acuity; *acuity* means the sharpness, clarity, or "keenness" of vision, with perfect acuity meaning that the eye sees or perceives exactly what the object is, regardless of its distance from the eye. Acuity also refers to the ability to discriminate one small character from another when very close to each other. Poor vision will see these characters as one instead of several distinct ones. Because focusing and accommodation are relative to the status of the curvature of the lens, we can modify the curvature with artificial lenses (glasses or contacts).

Variations of acuity

Testing for acuity is accomplished by comparing what you can see clearly in specified conditions compared to what a normal eye with perfect visual acuity can see in the same conditions. The eye doctor will have you distinguish various characters of different sizes. The bottom line of the eye chart (20/20 or 20/15, depending on the chart) can be read without difficulty by a normal eye; however, as the ability of the eye to see clearly diminishes, the characters must be larger, as in the second line (20/30).

This means that the poorer eye can only see a character size at 20 feet that a normal eye could see at 30 feet. These tests are also done under ideal conditions with bright light and black-and-white contrast. This doesn't imply that an eye that can't see 20/20 without correction is abnormal; it's just unable to focus as well.

Myopia (nearsightedness) is the most common acuity problem (Fig. 7-5). This means distant objects are out of focus without correction. Optically, the focal point of the lens at rest is in front of the retina.

Astigmatism is the unequal and variable curvature of the lens and the cornea that prevents an equal focus of varying distances; that is, light rays are not refractive equally through the lens. For example, a vertical light pole might be in focus, but the horizontal power lines might be out of focus. If one has a evenly spherical cornea and lens, there is no astigmatism.

Presbyopia, a form of farsightedness, is a predictable change to the lens, occurring with age. The lens stiffens and is unable to accommodate effectively. The focal point at rest is behind the retina (Fig. 7-6). To be in focus, the object being visualized must be farther from the eye, which, in turn, makes the object smaller. It affects near vision and necessitates "reading glasses" while distant vision remains unchanged. Initial reading glasses are really just magnifiers. There is also less light passing through the older lens because of age related yellowing, which further interferes with depth of field. It is common for the older pilot to have uncorrected distant vision remain at 20/20 but to need glasses to read.

Refractive error

The refractive powers of the lens can be measured to determine how well it can focus. This is accomplished by using a variety of different-strength artificial lenses placed in front of your eye. You then tell the doctor which "supplemental" lens makes a character clearer on a chart that you are visualizing. Astigmatism can also be measured.

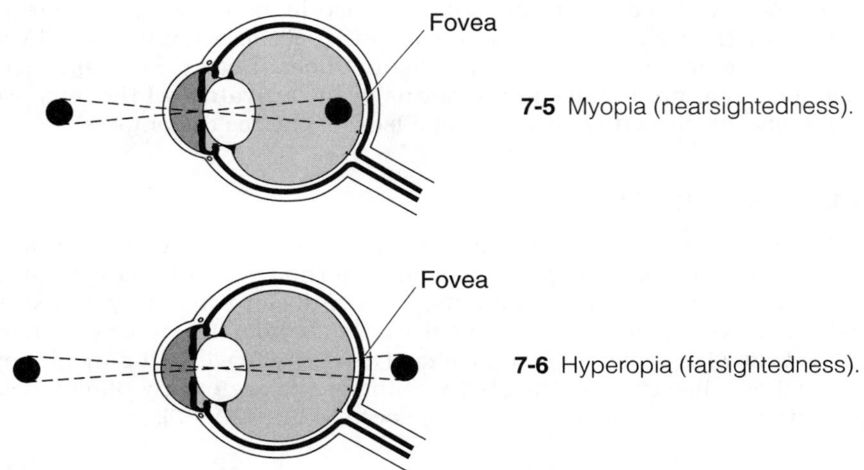

7-5 Myopia (nearsightedness).

7-6 Hyperopia (farsightedness).

The measurement of these optical properties (measurements) is stated in *diopters*, which is the curvature of the lens in different planes. The entire definition of the lens' ability to focus is called the *refractive error*. *Plano* is essentially perfect refraction. Corrective lenses can be ground using this refractive error when stated as a prescription.

It is a good idea to get periodic checks of your vision by an eye doctor and not rely solely on FAA exams. Certainly the key ages where there are changes are in the early 20s and then progressively after age 40, when near vision will change. When it comes to vision in flight, it is crucial that you have strong 20/20, uncorrected or corrected, for both near and distant. The conditions of flight, as subsequently explained in this chapter, will impair vision and necessitate the best vision you can achieve before you begin the flight.

Corrective lenses

Once the refractive error is defined, lenses can be ground to make up for the error of refraction, allowing the focal point to be on the retina and in focus. This correction can be in the form of frame glasses or contact lenses. There are trade-offs. Frame glasses are generally more easily tolerated whereas contact lenses allow for better peripheral vision. Also, as one gets older and near vision changes, there will be the need to wear glasses for near vision. If one already has "distant" glasses, then bifocals are needed.

Bifocal glasses (one part of the lens ground for distant vision and the other lower part for near vision) are common, and one needs to get used to them. Many pilots are reluctant to wear glasses, especially for reading; however, the sooner such glasses are worn as a habit, the easier it will be to adapt to the correction.

A common misconception is that not wearing prescription glasses strengthens vision. The pilot therefore wears her glasses only when she feels she needs them. This is a poor policy. The eyes are not made stronger by not wearing glasses with the belief that the eyes are forced to maintain acuity. Although this might be compared to building up muscles by using them more often, this is not the case with the eyes. In fact, there is eye strain and fatigue resulting from the eyes and the brain trying to focus unsuccessfully on an image without the help from correction; therefore, if you need glasses for correction, wear them.

There is also the technique of having one eye already 20/20 uncorrected for distant vision and then correcting the other eye just for near vision. The brain, in time, will adjust to these two separate sources of information; however, the FAA requirements are for both eyes to be corrected equally.

NIGHT VISION

Vision under low light is a different physiological activity than vision under bright light. A chemical change within the cells of the retina processes light energy. During the day and in well-lighted conditions, most light rays are focused on the *fovea*, which is packed with cone cells along with some rods. Such cells have immediately available a chemical called *iodopsin*, which allows the eye to be immediately stimulated with the high-intensity light.

Another chemical (rhodopsin, or visual purple) is found in the rods and needs less light; however, the chemical must be made by these cells at the time light levels are low. This requires time for this chemical to be formed, which means it takes longer (up to 45 minutes) to adapt to low light than to bright light (about 10 seconds); therefore, the retina has the ability to adapt to different light conditions by changing the amount of low-light-sensitive chemicals found in the rods.

Types of vision

Because of the varying resources, the eye and retina have to adapt to changing light conditions, our vision is divided into three types, depending on the amount of light available (Fig. 7-7).

Photopic vision

This is daylight with relatively sufficient light to activate the foveal cells (cones), either from sunlight or from artificial illumination. Here color is easily discerned and images are sharp because of the use of the fovea (central vision). Only cones are found in the fovea, but the visual purple chemical already present is bleached out by the light; thus, photopic vision is primarily through use of cones and central vision.

Scotopic vision

This occurs in low-light conditions, either at night or in darkened environments; cone cells are ineffective, resulting in poor resolution of detail. Visual acuity decreases considerably, to as much as 20/200. Color vision is totally lost. Because central vision depends almost solely on cones and not the now ineffectual rod cells, there is another blind spot in the fovea. This is in addition to the other blind spot, always present, and a result of the entrance of the optic nerve into the retina. This is why one must look off center to see specific stars (or air traffic) in the night sky. As a result, peripheral vision is the only real source of visual input in darkened conditions and is often based, in part, on seeing silhouettes against a contrasting background.

Mesopic

This is a situation of light availability being between adequate and inadequate, such as during dawn or dusk or at full-moon levels. As expected, visual acuity and color discrimination diminish as light levels decrease and cone cells become less effective. One might think he is seeing clearly because of the gradual change. Awareness of impending light deficient visual impairment is crucial during these midrange light levels.

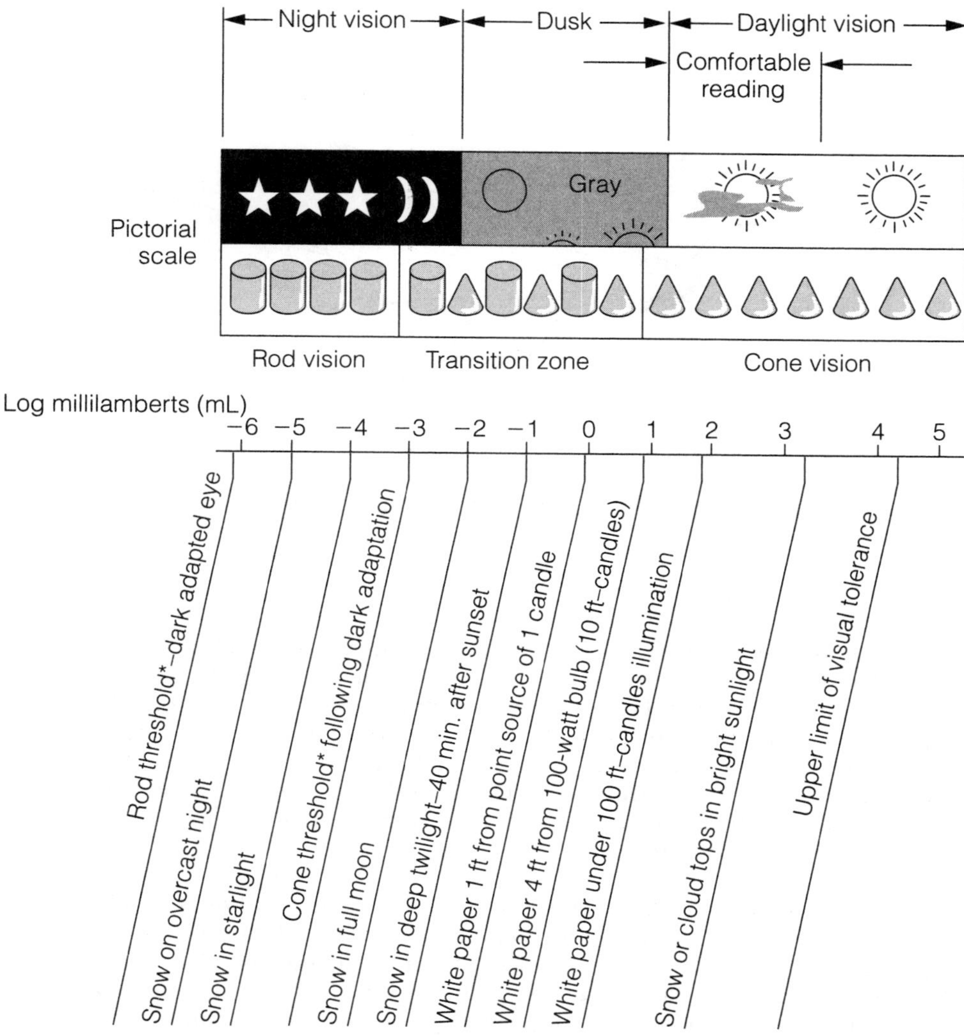

7-7 Characteristics of night and day vision.

Dark adaptation

Dark adaptation is the process by which the retinas increase their sensitivity to decreasing levels of available light. The chemical (rhodopsin, or visual purple) change occurs in the rods as the cones become less effective. The lower the starting level of available light, the more rapidly the adaptation will occur (Fig. 7-8).

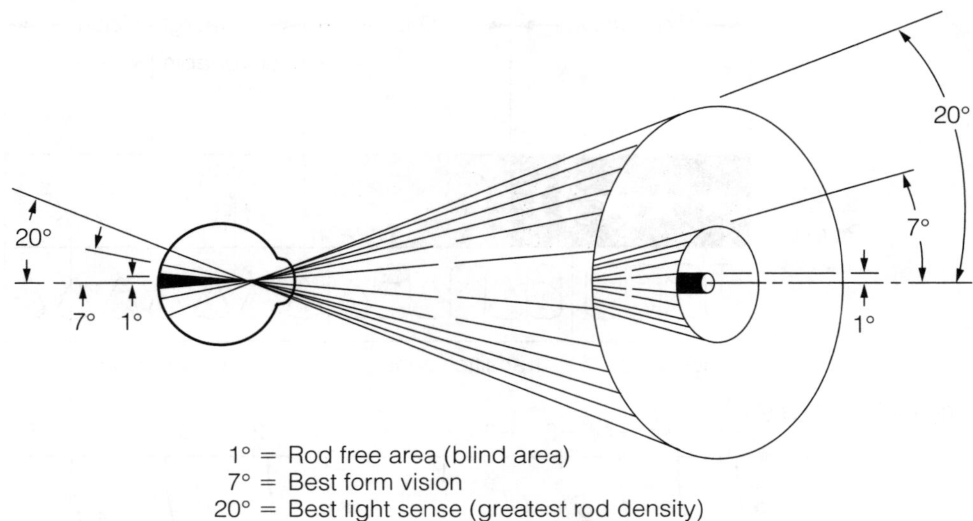

7-8 The eye in night vision.

It takes about 30–45 minutes for this adaptation in minimal light conditions. The lower the target's source of light in the dark, the longer it will take to adapt. This also is dependent on ideal conditions, during which there is no exposure to bright light, even for brief periods. This bleaches the changing chemical in the rods, which then has to be readapted. A very bright light, such as a flashlight directly into the eye for even a second or strobe lights on an aircraft, can seriously impair night vision. Recovery could take several minutes.

Prolonged exposure to very bright light for as little as 3–5 hours, such as sun glare off sand, water, or snow, can hinder dark adaptation later on. The more severe the intensity and the longer the exposure, the more diminished initial night vision can be. Some of these impairing effects are cumulative and can persist for days after the exposure. During wartime, pilots on alert at night would sit in dark rooms so their eyes would be already dark adapted for their flight.

The ability of the eye to see at night is enhanced by the technique of scanning to overcome the blind spots and poorer acuity. Under scotopic levels, the eye has a central blind spot about 2 degrees in diameter; thus, the eyes must continually scan to keep the object to be located or examined off this central area. Scanning also prevents the periphery, or rod area, from adapting to an image, which occurs if the eyes are kept stationary for more than a few seconds.

Preserving night vision

The obvious technique to preserve night vision is to avoid bright lights and to stay in a darkened environment for the time it takes for the rod cells to adapt. There is another method to protect your night vision, and that is by using red light for illumination.

Rod cells in the retina are least affected by the light wavelength in the red range (longer wavelengths). Red lenses block light at wavelengths of less than 620 millimicrons, effectively keeping the rods in the dark while minimal cone action, sufficient for vision, is still maintained.

This means that a red light source will not bleach the chemical rhodopsin as do other colors of shorter wavelengths. It follows that the use of red light, once vision is adapted to night, would not alter the night-adapted status of the retina and the rods. For this reason, wearing red goggles in certain situations prior to actual flight and in preparation for a trip at night is advisable as is using red as a light source for acuity (Fig. 7-9). This is also effective when preparing for night work by wearing these goggles in brighter light conditions, again prior to flight, allowing the rods to begin their adaptation. Dim white light works the same as red light as far as acuity is concerned.

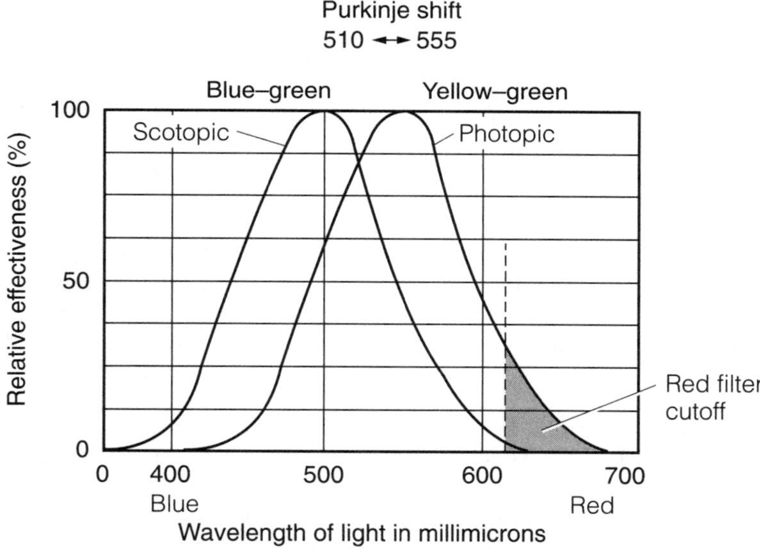

7-9 Luminosity curves for rod and cone vision.

There is a trade-off, however. Red light refracts less than the other colors of white light (remember that white light is a combination of all colors). Objects are no longer in focus using only red light, and it is more difficult to accommodate from near to distant vision. Red light also masks any red-colored objects, instruments, or map markings.

As more and more liquid crystal display (LCD) and cathode-ray tube (CRT) displays are used in "glass cockpits" with multicolored and multihued characters and backgrounds, red light (or any light source other than white) becomes counterproductive and should not be used. The other extreme is blue light (shorter wavelength), to which the retina is least sensitive. The same principle of color's effect on acuity is true with sunglasses with designer type colors (every color of the rainbow), as will be subsequently discussed.

A good compromise would be to use red light only when trying to enhance the adaptation process, but not during actual flight. Low-level white light on the instruments and within the cockpit is the best, provided the intensity can be controlled. As darkness outside increases and the eyes adapt more to low light, this white-light source should be reduced so as not to interfere with acuity outside the cockpit.

Techniques for improving night vision

In review, there are significant changes in how the eye perceives any level of light as darkness increases. Central vision is compromised because of the dependence of cone cells. Peripheral vision becomes the primary source of visual cues (Fig. 7-10); therefore, off-center vision must be used, which is looking at an object about 10 degrees away from the object. This allows the peripheral vision to be the source of light signals to the brain. The tendency is to look straight at the object, and it takes a conscious effort to look beyond the central-vision habit.

Another challenge is that if an object is visualized for more than 2–3 seconds, the retinal cells can become bleached out, which can result in the object not being seen. Continuous scanning or changing position of the eyes on an object can prevent this occurrence.

Color vision becomes less dependable. Colored lights from beacons, runways, and wingtips are discernible if the light source is strong enough to activate the cone cells; however, other colors outside (and often inside) a dimly lit cockpit will appear gray. Contrast with the background will be the primary way of defining one object from another in flight.

Reaction time of visual accommodation, recognition, and action is also impaired (Fig. 7-11). This reaction time is defined as that necessary to first recognize that an object is present. Additional time is necessary for this signal to get to the brain cells dealing with vision, and still more time is necessary for the eye and head to turn toward and focus on the unknown object. The brain must then determine the importance and significance of the object's presence, and further time is necessary for the body to react, such as moving the muscles necessary to take corrective action at controls. This whole process can take up to 4–6 seconds, depending on the condition of the environment and the pilot. A keen awareness of this discrepancy is necessary in night vision as well as daylight activities to allow for this greater risk of slower reaction time.

FACTORS AFFECTING VISUAL ACUITY

In either day or night flight, there are factors, many controllable, that can seriously affect the efficiency of vision. This impairment is more noticeable

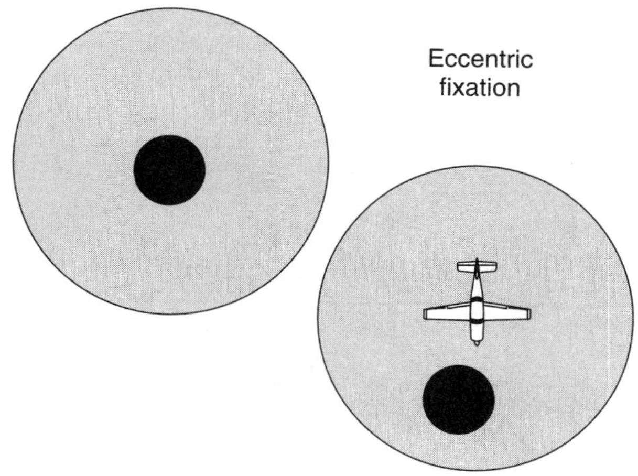

Eccentric fixation
Top left—The central blind spot present in very dim light makes it impossible to see the plane if it is looked at directly
Bottom right—The plane can be seen in the same amount of light by looking below (as is shown here), above, or to one side of it so that it is not obscured by the central blind area.

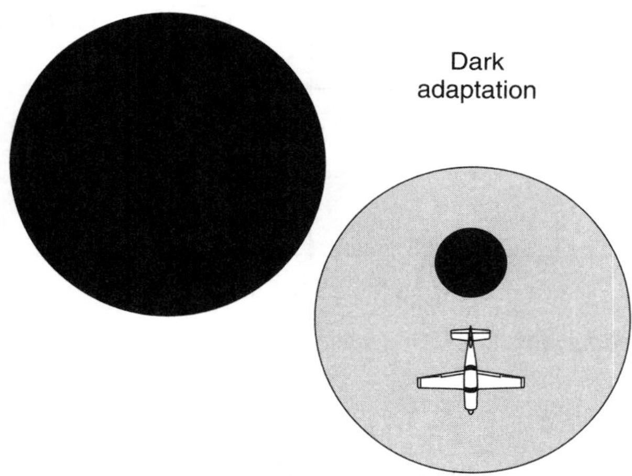

Top left—View seen by a person who is not dark-adapted.
Bottom right—The same view seen by a dark-adapted person who is looking at a point above the plane.

7-10 As darkness outside increases, off-center vision must be used.

during night-vision flying. The most important is hypoxia. Since the retina is an actual anatomical extension of the brain, anything that affects the brain immediately affects vision. Lack of oxygen first affects brain tissue; thus, vision is concurrently eroded by reduction of sensitivity of rod cells. Low-light conditions add to this erosion.

Hypoxia also increases the time necessary for dark adaptation. Such impairment is noticeable starting at about 5,000 feet above the ground level to which you are acclimated. This is easily demonstrated in the altitude chamber or even in flight. By using oxygen at night after several hours in a cabin at 5,000 feet or more, common in many commercial aircraft, there is a noticeable change in the colors due to being hypoxic. The oxygen will immediately bring the colors out and make characters clearer.

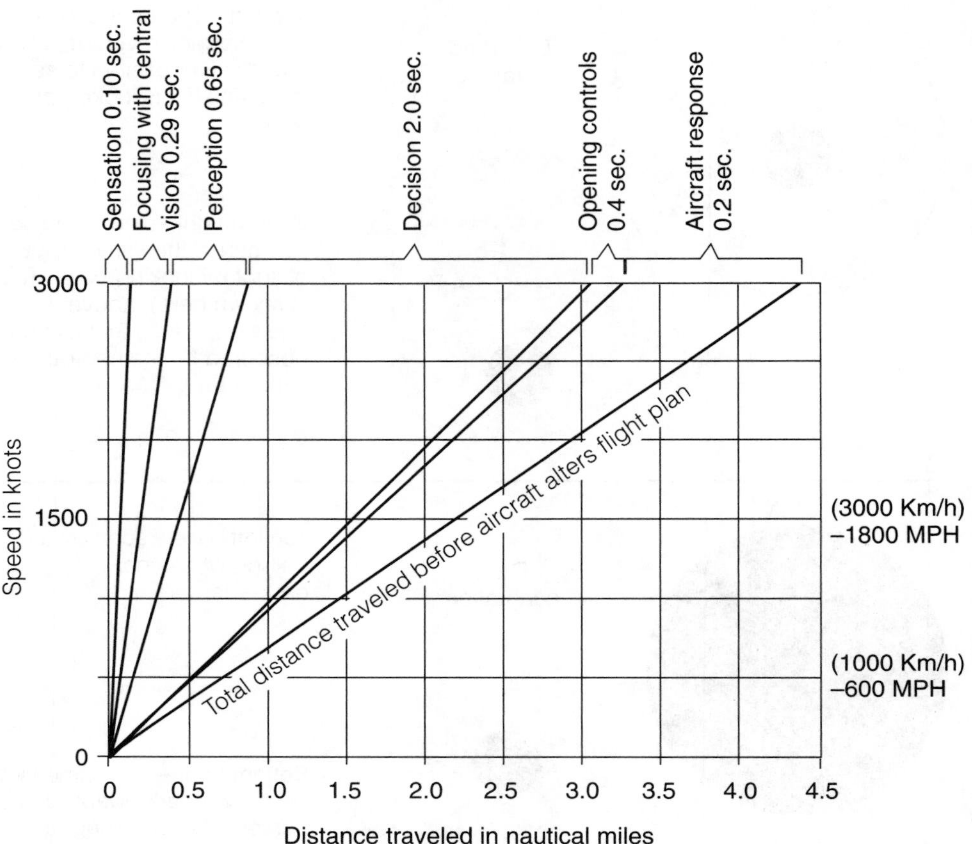

7-11 Visual reaction time.

Carbon monoxide, which builds up in the blood in all smokers, creates hypoxia with the same results. Some have observed that smoking just three cigarettes in a short period of time adds enough carbon monoxide to the blood cells to simulate to the body an additional 3,000–5,000 feet in altitude.

Fatigue is another factor that indirectly affects vision through reduction of mental alertness and visual recognition of critical situations. Accommodation is also compromised, impairing the pilot's ability to maintain a productive scan within the cockpit and observing activities outside. Other factors such as fixation, channelized thinking, and reduced judgment skills related to fatigue are covered in chapter 11.

Because alcohol can cause histotoxic hypoxia, the same reduction of acuity results even hours after the last drink.

Therefore, improving any visual techniques begins by requiring a high degree of suspicion for impairing conditions. Logically, the improvement comes from avoiding these known, controllable situations.

VISUAL SCANNING

One important difference of the eye compared to a camera is how we see a panorama of the surroundings. A motion-picture camera taking pictures of a large crowd at a football game, for example, will clearly see and record everything in focus as it moves from one end to the other. The human eye and our processing unit (the brain) cannot do that. The eye, if moved from side to side, will not see anything in focus. In fact, as we look over the crowd and try to see as much as possible, the eye automatically stops for a fraction of a second and then moves to the next "scene" (*saccatic movement*). The slower we move the eye, the more we perceive.

The ultimate technique is scanning, a deliberate pausing at each spot until we accurately visualize and recognize the object. This is true whether scanning the skies for traffic, scanning the instruments for changes in performance, or when reading this text. By necessity, you pause enough to read this or other text before moving on. Unfortunately, we do not practice this same technique in the airplane—it's more difficult than one would think—because we often do not know what we are searching for and what to focus on.

Furthermore, only our central (foveal) vision will see everything in focus, which means visualizing a target straight on before moving on to the next. While our peripheral vision detects motion well, it has compromised acuity. The message is clear. As with reading, a productive scan is essential. Take deliberate regularly spaced pauses between eye movements. Each scan might be different, depending on what you are trying to visualize. Try different techniques and keep using what suits you the best.

VISUAL ILLUSIONS AND MISPERCEPTIONS

Many of these illusions are covered in chapter 8 on orientation because they are often in association with our vestibular apparatus. From a purely visual perspective, most illusions are a result of light rays being distorted as they pass through air containing fog, dust, haze, and the like. These components tend to refract the light rays before they get to the eye, causing poorly focused objects, reduced light source, and distorted visual cues. Visual cues often refer to the comparison of objects of known size to unknown objects and are especially noticeable in judging relative size, speed, and distance.

Basic optics and our mind's interpretation of those signals distinguish that distant objects appear smaller. The distance of an object from a person is therefore a mental judgment comparing size to already known comparisons or related to nearby known objects. Distortion of these light rays further confuses the brain that is trying to interpret size and distance. Having two eyes is not as critical for depth perception because an object farther away than 20 feet (where light rays emanating from an object are nearly parallel) is judged by these relative sizes and not by binocular vision.

Depth perception and distance determination

Flying at altitude subjects the brain to interpreting visual images and cues from great distances, usually greater than 20 feet. Monocular cues, therefore, are just as important as binocular cues. Such cues include comparing unknown size to a known size (a 727 seems to move faster than a 747 on takeoff), texture or gradient (loss of detail at increasing distances), illumination perspective (light sources usually are assumed to be from above), and apparent foreshortening (a circle appears as an ellipse at an angle). Other conditions can confuse visual perception of any object or image.

One is *geometric perspective*, which includes linear or parallel lines that appear to converge in the distance. Also, the true shape of an object can be distorted when viewed from a distance, as with the circle appearing elliptical and when objects at a lower altitude relative to you appear higher on the horizon when at a distance from an observer.

Another perspective is *motion parallax*, which refers to the apparent relative motion of stationary objects as viewed by an observer moving across the landscape. Near objects seem to move in directions opposite the path of movement. Far objects appear to move in the direction of motion or don't move at all. The rate of this motion depends on the distance the observer is from the object. Consequently, objects that appear to be moving rapidly are judged to be near, while those moving slowly are sensed to be at greater distance. A good example is a car moving quickly by a picket fence. To the driver, the fence is moving quickly, whereas distant trees seem to pass more slowly.

A final perspective is the effect of haze, fog, or smoke when objects are viewed through them. In addition to being less clear simply because of the reduction of transmitted light, objects also seem to be at a greater distance. This is a result of a different refraction of light rays from objects seen in a fading color or shades in the distance.

Space myopia

Space myopia is also sometimes called *empty-field myopia* and means that at altitude there are no objects upon which to focus beyond 20 feet; therefore, unless the eye and lens are "stimulated" into focusing, the lens tends to go into a resting state, focusing somewhere between near and far.

As a result, small targets outside the cockpit are missed, such as distant approaching airplanes, and the reaction time to accommodate to cockpit objects is somewhat impaired. This is easily avoided and corrected by frequently picking out an object beyond 20 feet, such as a wingtip or a cloud edge, and consciously focusing on it. Staring into space during a boring flight is to be avoided for these reasons. This is just another of the many reasons to perform effective and continuous scanning.

BLACK HOLES AND WHITEOUTS

Two approaches to landing and departures are especially dangerous as a result of a visual illusion from our peripheral vision lacking adequate cues. The first is the black-hole effect, done at night and without reference to a horizon. This is common when flying over water or unlighted terrain and only the runway lights are visible. Without the visual cue of the horizon, the illusion is to be in a situation other than straight and level, especially if there is already the illusion of a sloping runway or the like (see chapter 8 on orientation). City lights located on rising terrain further confuse the visual perception.

A similar illusion is *whiteout*: atmospheric and blowing snow. *Atmospheric* whiteout occurs when the terrain is covered with snow underneath overcast skies. There are no shadows to give texture to the terrain nor can the horizon be distinguished from the ground. The only clue is the actual runway, and special care must be maintained to prevent disorientation or an inaccurate approach and landing.

Blowing snow gives the same illusion plus disrupts visual contact with the ground. Helicopter pilots have unique problems in this situation; they are disoriented because of vection illusion (see chapter 8), plus they are unable to see the ground. Flying through snow, as with rain, also gives a vection-type illusion and is especially troublesome when these conditions are entered inadvertently.

SUNGLASSES

Whether or not correction is needed for acuity, sunglasses are often necessary, as well as part of "pilot" image. Sunglasses are often worn for cosmetic reasons, even when they serve no other purpose, even at the expense of distorted colors reaching the retina. As a result, there are several considerations that must be recognized when discussing the use of sunglasses.

The primary reason to use sunglasses is to reduce the amount of light from every bright source—the sun is the most obvious source—to a comfortable level that doesn't affect the requirement for adequate visualization. Additionally, there should be no distortion of color. This translates into ensuring that all wavelengths of visual light rays enter the eye for comprehensive interpretation. At the same time they should filter out harmful ultraviolet and infrared radiation.

There is an individual variation in how one tolerates bright light and glare. Some are able to adjust without filtering through sunglasses; others cannot tolerate even moderate levels. If bright light does not seem to affect performance and acuity, sunglasses probably are not indicated. Intense glare, however, usually affects everyone to some degree, can be fatiguing, and can impair acuity and dark adaptation later in the day. Wearing sunglasses under these conditions is recommended for all.

Because sunglasses are meant to decrease the amount of light entering the eye, the eye automatically reacts by opening the iris (aperture) to allow more light onto the retina. Remembering the analogy to a camera, a larger opening, or aperture, decreases depth of field, which means that objects within a narrower area are in focus and objects within wider areas (fore and aft of the narrower in-focus area) are out of focus. (More objects are in focus when the iris is smaller and the depth of field is larger.) Observe professional tennis players, and you will rarely see them wearing sunglasses. They do not want to risk a shortened depth of field and impaired focus between each player. An expanded depth of field allows them to clearly see the tennis ball come from their opponent's racket to their side of the court without having to rely on the focusing mechanism. This same principle holds true in aviation, especially in the traffic pattern. There is a trade-off with sunglasses in that decreasing light into the eye initiates a reflex opening of the iris, which produces the narrower in-focus area.

One specification for sunglasses includes maintaining a reasonable filtering of the amount of light, probably no less than about 85 percent. More filtering than that compromises the light source necessary for safe flight. Wearing sunglasses in low light results in a definite loss of acuity. Eyeglasses that automatically darken in bright light should not be used in the cockpit because of the varying levels of light encountered when you look outside to scan for traffic, landmarks, or the runway environment then have to look inside to read instruments deep in the cockpit or charts spread out in your lap. The eyeglass tinting does not change fast enough to adapt to instantaneous light intensity changes. Also, these glasses, when darkened, do not filter out enough bright light (much less than 85 percent).

Sunglasses should be neutral in color, allowing most wavelengths to pass through. Any other colors besides neutral gray could distort the image as seen by the eye. For example, green-colored sunglasses absorb a high amount of every color except green. The same is true for other colors. With so many cockpit controls, CRT/LCD characters, displays, and maps in varying colors, it is not safe to wear any eyeglasses that filter out key colors.

A popular color is yellow, or amber, which is claimed to make it easier to see through fog or haze (blue blockers). There is a perception of clearer visibility and sharper focus, but this is more subjective than proven. Whatever the color of the lens, the other colors will be washed out when they reach the eye. With vision a key source of sensory input, anything that is not a true color can mislead the pilot's perception of objects and discrimination of essential characters in cockpit displays.

Neutral color filters out equal amounts of all colors: hence the gray appearance. It can reduce the total amount of light passing through the glasses without compromising color cues.

Polarizing glasses are also of dubious use in aviation. They are helpful in situations of high glare from reflected sources such as water, snow, and sand, but the trade-off is distortion of light rays passing through the windscreens.

ULTRAVIOLET RADIATION

A final consideration is filtering out, or absorbing, ultraviolet (UV) and infrared radiation (or light rays). UV radiation is in the 200–400 nanometer (nm) electromagnetic spectrum (a nanometer is 1 billionth of a meter). Wavelengths of 200–295 nm (abiotic UV) are all filtered out by the atmosphere but might become significant at high altitudes of flight. UV radiation from 300–400+ nm is abundant at Earth level. Sunlight falling on the Earth is about 58 percent infrared (760–1,000 nm), 40 percent visible light (400–760 nm), and only 2 percent UV; however, at higher altitudes, UV radiation can reach levels of 4–6 percent.

Long-term exposure to these rays can cause clouding of the lens (cataract). It is essential, therefore, that whether or not eyeglasses are needed, some protection is necessary, especially for long flights. This is easily accomplished by ensuring that the sunglasses, worn by just about every pilot, state that they filter out ultraviolet light, preferably the full range, as well as infrared radiation. Read the label to be sure you are protected. The coating is inexpensive. An interesting dilemma is the possibility of more UV entering the eye when wearing sunglasses because the iris opening is bigger.

8
Orientation

Mark had started his descent from 16,000 feet surrounded by blue sky and bright sunshine. That all changed a few minutes later. His destination was enshrouded in a solid cloud covering that went from 3,000 to approximately 14,000 feet MSL. As he continued the descent, ATC vectored him with a 20-degree right turn. About the time he entered the cloud deck, he reached behind his seat for the approach book. When he turned back around and saw that his visual reference had been lost, Mark was convinced that he was in a 20-to-30-degree bank turn to the left. He had been trained over and over again to get on instruments in this situation and trust what his eyes were telling him, but his first reaction was to respond to his other senses and try to "level" the aircraft. Mark's copilot at the time was busy on the radios coordinating the clearance and didn't notice the predicament or the incorrect way Mark was responding to it. After the fact, the copilot informed Mark that he assumed Mark was making a course correction at the time he banked the aircraft to the right. It wasn't until the controlling center queried the turn off course that Mark realized what was going on and returned to level flight as indicated by the cockpit instruments. It took every ounce of effort he had to believe what his eyes were telling him and ignore his other senses. Mark could easily have rolled the aircraft directly into the ground had he been flying solo, while at the same time believing he was in straight-and-level flight.

AT THE BEGINNING of this textbook, it was stated that the greatest physiological challenge to the pilot and his or her fitness to fly is the prevention of a medical incapacitation. If we could assure the pilot, the passengers, and the FAA that the flightcrew would not become incapacitated under any conditions of health or flight environment (providing the airplane works), the aviation community would have few concerns about a pilot's physiological safety. Except for one other flight situation: orientation.

Orientation is the key element in preventing aviation accidents and incidents. If the plane works, and the pilot is not impaired, then the only other significant ingredient in safe flight for the trained pilot is maintaining orientation of the plane and himself or herself. *Situation awareness* (or *situational awareness*), is similar to orientation, but includes awareness of what's going on with other aircraft, weather, ATC, cockpit communication, and the like.

Lack of orientation (or awareness of the situation) is *disorientation*, which is found in various forms. These include temporal, postural, vestibular, positional, rotational, auditory, and spatial, all of which, by themselves or in combination with others, can make an otherwise uneventful flight a memorable event.

Disorientation is a leading cause (not just a contributing factor) of more than 15 percent of reported accidents and incidents. Disorientation is a human factor, not helped much by the advanced technology of electronic guidance systems. All pilots experience disorientation at one time or another. Furthermore, most situations of disorientation go unreported; this is true for both military and civilian aviation. In most cases, the pilot and crew recover from being disoriented and no one beyond the crew is the wiser in terms of frequency of occurrence.

Because few such scenarios are officially reported, disorientation does not show up in human factors statistics as frequently as it actually occurs. But listen to pilots talk in the crew lounge, away from the ears of management or governing agencies. There are frequent descriptions of their periods of disorientation and how they recovered. Pilots know disorientation happens frequently.

It is the responsible pilot who respects this fact and continues to take measures to prevent disorientation. That means keeping current in flight proficiency and maintaining an awareness of what affects orientation in the flight environment. Experiencing the effects of disorientation through unusual-attitude flight or a Vertigon simulator reinforces the disrupting power of this human factor on maintaining control of the aircraft.

Therefore, the objective of this chapter is to raise that level of awareness, identify conditions that lead to the various kinds of disorientation, and discuss measures to minimize the occurrence of disorientation. It is essential to recognize that even the minimally physiologically impaired pilot greatly increases the risk of becoming disoriented because tolerance to confusing signals from the body is decreased. Put another way, the impaired pilot can easily become disoriented, which results in further incapacitation that eventually leads to a performance problem sufficient enough to cause an accident or incident.

DEFINITION OF DISORIENTATION

To be oriented is to be associated with or related to a particular point of reference in any situation. That's a basic definition. It follows that if one isn't oriented to a particular situation, then there is disorientation. A pilot usually thinks of one kind of disorientation, spatial disorientation. Furthermore, the term "spatial disorientation" is often and usually meant to represent disorientation relative to disruption of the vestibular system (the inner ear and semicircular canals). As will be recognized shortly, this is not an all-inclusive term and needs to be used in appropriate situations involving various kinds of disorientation.

The body uses several physiological senses to orient itself. Vision is the most important (and reliable) followed by vestibular, and proprioceptive (relative muscle position). These senses often work in harmony and will be discussed in terms of how they can give proper, accurate, and yet occasionally misleading cues. Often there is a conflict between these cues from the different sources, thus creating more confusion.

A pilot can become disoriented several ways, and these will be discussed individually. For purposes of clarification in this textbook, however, *spatial* disorientation refers to how the pilot perceives his or her *position* in space and relative to other objects, especially when in motion. It is related to and determined by vision, primarily peripheral vision. When the pilot is disoriented as a result of a disruption of the vestibular system (sometimes called *vertigo*), this is called *vestibular disorientation* and technically not spatial disorientation. These are two separate and distinct kinds of disorientation, as will be explained later.

Be advised that in discussions outside the classroom, the term *spatial disorientation* might be used to mean vestibular disorientation. Be sure to clarify that everyone is talking about the same kind of disorientation in discussions about orientation in flying.

There are a variety of ways to discuss disorientation. A traditional method is to use the common denominator of sensory illusions and their causes. Another is to explain the anatomy and physiology of the body's mechanisms of orientation and balance on Earth and then the problems associated with transition to flight. These all are appropriate descriptions, with the end result being a relatively comprehensive picture of how the pilot responds to various signals and cues from orientation organs.

The individual types of disorientation will be discussed first, explaining the causes. Then the cumulative and concurrent effects of different cues will be described as they pertain to illusions experienced by pilots. Furthermore, one can separate each kind of disorientation into recognized, unrecognized, and incapacitating, focusing more on how the pilot is affected and impaired in flight.

Situational awareness

Before discussing traditional disorientation and because it is gaining more respect in safety, some comments are appropriate about situational awareness (SA). As noted in chapter 4, SA has become the term to identify loss of where the pilot is and where others are, what is happening in the cockpit and the flight environment, during an approach and landing, or during cruise, among other situations.

For example, classic indications of loss of SA is no one looking out the windows, no one flying, being unsure of the aircraft's condition, expectations of an event not being met, two or more conflicting bits of information, confusion, attention of crew focusing on one condition, and the like. Pilots and crews become more susceptible to loss of SA when tired, stressed, or not being medically airworthy. Maintaining SA is crucial and includes prevention of other forms of disorientation, keeping alert to changing conditions, and using good CRM principles if loss of SA is even suspected. There is an increasing recognition that losing situation awareness can lead to some forms of disorientation even though a crewmember will be confident that he or she has SA, but really doesn't.

TYPES OF DISORIENTATION

Postural (proprioceptive) disorientation

A constant source of sensory input comes from the body's interpretation of the direction of gravitational force through *proprioceptive signals*, which are signals from various muscles in the body. Our minds are programmed to interpret these signals as being up or down, relative to the Earth and its gravitational pull. This in turn determines our posture (our position relative to the ground). Concurrently, there are proprioceptive sensors within our skin, muscles, tendons, and joints that detect changes in relative position, pressure, and changes from that of up or down posture.

Every time a muscle contracts or relaxes, tendons are pulled or released and joints move. Proprioceptive signals are those generated by these changes. All these inputs, which are continuously coming to the brain, tell the pilot what position he or she is in, usually relative to the foundation of gravity. It's similar to inertial navigation, where cockpit devices determine a plane's current position relative to where it started moving. We learned how to walk upright not because we were taught but by trial and error and our proprioceptive signals. These signals were processed by our brain over time until all muscles were working together to keep us from falling.

Another source of signals defining posture is part of the vestibular system, the otolith organs (utricle and saccule), located in the inner ear. They play a role, along with proprioception, when the body is subjected to changes relative to gravitational force and when subjected to linear accelerations. They will be discussed under vestibular orientation.

Because gravity is "down" to the mind, any input from proprioception or the otolith organs sensing this position is interpreted as either being "down" or a variation from that position. Our largest source of such information is when we are sitting down: "seat of the pants" flying. By being seated in flight, the positional forces acting on the pilot can tell the brain about the quite distinctive motions of the aircraft and body, both acting as one.

Because these senses are associated with the vestibular system's signals (including otolith organ signals), there is, under conditions other than flying, a comprehensive resource of a variety of information the mind uses to orient itself, much like a computer data bank. In flight, these signals can conflict and be confusing to the mind, giving false interpretations and leading to disorientation; however, they are somewhat predictable and one can train the mind and body to cope with these misleading cues, especially through experience and periodic practice in unusual attitudes of flight maneuvers.

Proper postural orientation in flight is noticeable in a well-coordinated turn, where all postural signals are equal to that noted when sitting upright on the ground and directly related to gravity. If the turn is coordinated and doesn't disrupt the vestibular system and the position is confirmed visually, there is no impairment to the pilot's performance. Postural orientation is particularly important in erratic and unfamiliar airplane movements such as aerobatics and high G-force maneuvers that conflict with the pull of natural gravity by duplicating that sense artificially.

However, in an improperly executed turn, especially during a climb, these signals become very confusing to the brain, giving false illusions as to one's true posture (plus other disorienting feelings to be discussed). Rather than being pushed directly into the seat, the body is being pushed sideways as in a skid or slip, giving the impression of tilting. In the absence of visual reference, the only sensation will be from the body being pressed into the seat, but not necessarily toward the ground. The vestibular system will also be sending signals, some being opposite to the proprioceptive signals. Recovering from such a situation in a climb gives the illusion of descending, causing the pilot to pull back on the stick.

To the physiologically impaired pilot (fatigued, dehydrated, hypoglycemic, etc.), these expected signals become illusions that lead to subtle incapacitation and potentially eroded performance in spite of being well-trained and current. Also, as we get older, these proprioceptive sensors become less sensitive and provide a less reliable source of postural information. Sitting for long periods of time also dulls these sensory signals, again providing inaccurate and unreliable data to the brain.

Positional disorientation

Positional disorientation basically means the pilot is lost, if even for a brief moment, and doesn't know her position. Realizing for a short period of time that she doesn't really know where she is, the pilot becomes disoriented until she resolves her position or lands at the wrong airport. In other words, the pilot doesn't know what to do: keep going, go back, call for directions, or what. Positional disorientation is also referred to as being geographic and directional. They all mean the same thing; the pilot is temporarily lost and unable to take effective corrective action until she locates her position.

All animals are creatures of habit, especially humans. Much of what we do during a typical day is a routine that we do the same way, day after day, and usually without too much thought. Interrupt or change that routine and one is disoriented, not knowing how to continue without having to think about it and then figure out the next step. Take for example how you take a shower or bath. Most people wash in the same way, each time starting at one part of the body and ending up at another. If you start with the front of your left shoulder, for example, that's how you start every time. The next time you shower, change the sequence. Start with your right foot. Now you have to think about what to wash next. This is no longer your routine and for a moment you are lost on your own body. You are briefly disoriented.

The same thing happens on familiar routes. You fly the same airways, talk to the same controllers, and make the same approaches on all the legs of the trip. Initially, when you fly the route for the first time, you concentrate on maintaining your awareness of where you are. Then the trip truly becomes routine because there is nothing to break the monotony of the hundredth trip. It now becomes easy to fall into the trap of flying past a checkpoint without knowing it. If ATC gives you a revised flight plan or changes your heading for traffic, you might miss the new instructions or clearance or continue on the same flight path you have always flown before.

The pilot becomes more susceptible to positional disorientation if he becomes distracted. He might be thinking about problems at home, concentrating on some activity in the cockpit unrelated to flying, such as reading a manual, listening to a commercial radio station, or playing with the computers in a "glass cockpit." Or he becomes involved with a political conversation with a crewmate. Now both pilots are out of the loop of maintaining positional awareness.

Another trap is the last flight of the day, or "get-homeitis." This becomes more risky than the initial leg because there is the added distraction of anticipating what needs to be done at home. In any case, checklists read many times before become superficially reviewed. Distances, checkpoints, and waypoints become committed to memory (or so you think). Even familiar frequencies are part of this virtually automatic flight.

Now throw in any deviation from the expected routine or a malfunction of the aircraft, and the pilot is easily distracted and becomes disoriented regarding his position. He has to spend time reasoning out what needs to be done to correct the problem and get back on course.

Pride often gets in the way of taking corrective action during these situations, with the pilot thinking that she can get out of this fix without having to refer to manuals or maps. Certainly she is reluctant to admit to ATC or even a crewmate that she is lost or forgot the controlling center's frequency and name. Consider how this situation can be amplified if it happens at the same time you enter actual instrument conditions. This combination of positional and potential vestibular disorientation is an incident or accident waiting to happen.

Listen to other pilots talk about their episodes and read accounts of honest and candid pilots who are willing to share experiences of being lost or temporarily out of touch with their surroundings or their situation, which are other examples of impaired situation awareness. Then consider the fact that this positional disorientation is another form of an incapacitating human factor. Even if only for a brief moment, the pilot is technically impaired, unable to function in a timely manner.

Temporal disorientation

Disorientation relative to time (also called *temporal disorientation*) is very subjective and is a direct function of how fast the brain is processing information. How long any given activity exists in the mind of a pilot depends on the level of activity and motivation occurring at that time.

During periods of high activity, time can be perceived as being expanded. That is, the pilot thinks he has more time than is actually available to accomplish some action because the mind is working in high gear, probably secondary to an increased flow of adrenaline. On the other hand, time might appear compressed, with the pilot feeling that less time has passed than actually has, while at the same time realizing that it seems to take forever to get to the next checkpoint. This is common in monotonous, slow, and uninteresting activities.

Little is known about why this happens in the mind. There is some evidence that the amount of adrenaline flowing through the body during periods of high and exciting activity will increase all metabolism, especially the brain.

The hormonal chemical adrenaline is instantaneously released into the blood stream in fight-or-flight situations, allowing the body to respond to unexpected and emergency situations. With extra "emergency" adrenaline, pulse rate and blood pressure go up; blood flow is diverted to crucial organs such as muscles, the heart, and the brain. More energy from blood sugar stored in the body is made available for metabolism. Sweating is increased to cool the body. Changes occur in the ear to intensify its ability to hear, and the threshold for pain and endurance increases. The brain therefore becomes more intensely alert and processes information at an accelerated rate.

Frequent stories from people in sudden emergency situations later revealed that events seemed to happen in slow motion. They vividly recall past events in their lives. Everyone has felt that "adrenaline rush" instantly after a near-miss at a street intersection, or upon realizing that the airplane engine has lost power.

It's the reason some pilots leave training simulators soaking wet from sweating due to the stress of the exercises. Professional athletes will know when they are having a good day because mentally the baseball or tennis ball appears extra large and very slow, allowing them to react quicker and more accurately with winning swings. Their adrenaline is helping their athletic performance.

Whatever the reason, the perception of expanded time, for example, is real in the mind of the pilot during the crucial event. It becomes a concern because the pilot, in an emergency situation, might take too long to react, thinking that there is plenty of time available to try some other course of action. In the military, it is a major concern with fighter pilots. In an emergency, activity level is high, and in the pilot's mind, there appears to be plenty of time to recover, and he or she does not eject in time. It's interesting to note that if the fighter pilot does eject and then is asked how long he felt that it took from the time he pulled the handle, the canopy released, he punched out, and the parachute opened, he will probably say at least a minute or two before he was hanging from the parachute. In fact, it takes but a few seconds. But he will swear that it takes longer. Furthermore, the pilot will be able to recall, in detail, every position of the instruments and controls, what was happening in the aircraft, and what he was doing during the ejection, all in those few seconds.

The opposite is true in minimal mental- and physical-activity situations. Here time drags on although the pilot feels it is moving faster. This is less crucial than time expansion because the actual prolonged time makes the trip even more boring but doesn't affect performance other than increasing the likelihood of positional disorientation and easy distraction.

Temporal disorientation, therefore, is not clearly understood from a scientific perspective, but we know it happens. And it can be an unbelievable experience when shared with others. There is no way to prevent the body's automatic response to emergency situations. How and when the pilot responds to that situation, despite his or her perception of misleading available time, is a result of the level of training and respect for known responses.

This might become a factor in the transition from traditional flying to that found in the glass cockpit. How will a pilot, experienced in stick-and-rudder controls and conventional instrumentation, react to an emergency in a high-tech airplane? Will the time distortion prevent him or her from following new techniques or will he or she revert to old methods?

Vestibular disorientation

Recall that vestibular disorientation is the term commonly used for *spatial disorientation* and is also compared to the feeling of vertigo. While they are similar and often occur with each other, they are separate entities. Spatial disorientation will be discussed after vestibular.

Vertigo, to the pilot, is the feeling recognized when actually experiencing vestibular disorientation. To the medical profession, vertigo has a different meaning. The symptoms might be similar: lightheadedness, dizziness, the sensation of either the room or the person spinning, and a feeling of instability; however, to the doctor, vertigo might be caused by a variety of abnormal situations. Certain abnormalities might be neurologically significant (such as tumors), others might be related to infections of the inner ear or central nervous system. Consequently, when telling a doctor about these symptoms, she or he will probably pursue additional testing to rule out any medical pathology. For aviation, it is better to keep with the term *vestibular disorientation* and leave vertigo to other medical problems. Vertigo from any source does cause vestibular disorientation.

Disorientation from dysfunction of the vestibular system (or apparatus) is probably the most severe and intense feeling of instability or unbalance. It is thought that this is the source of most motion-sickness problems. In fact, even experienced pilots can get airsick under conditions of flight that interrupt the vestibular system. Although vision is still the most important source of orientation signals, disturbances of the vestibular source of signals are the most difficult to cope with and the hardest from which to recover.

The vestibular system is located in the inner ear (Fig. 8-1). There are two distinct structures—the semicircular canals, which sense angular acceleration, and the otolith organs (Fig. 8-2), which sense linear acceleration and gravity (Fig. 8-3).

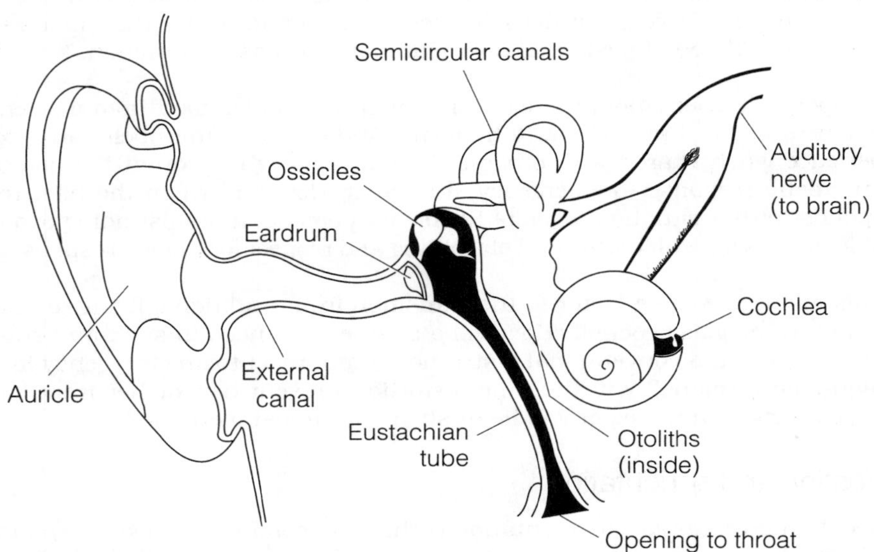

8-1 Anatomy of the ear.

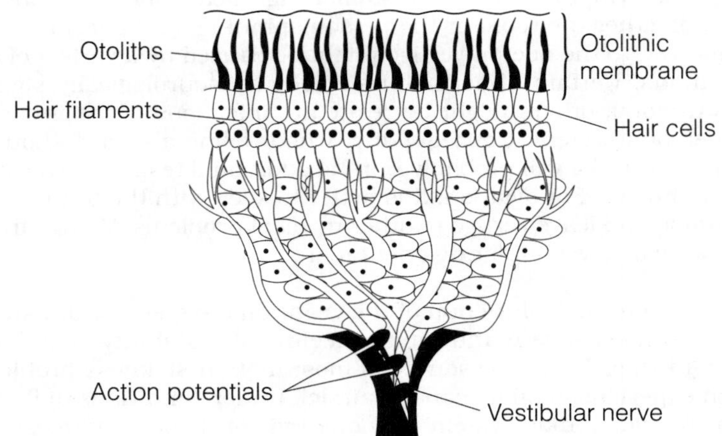

8-2 Anatomy of the otolith organs.

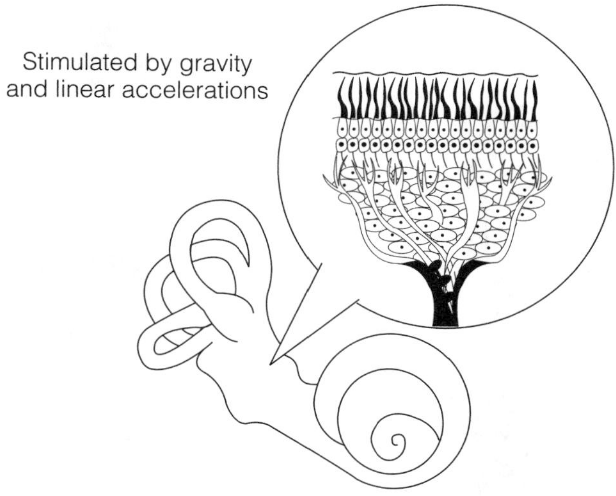

8-3 Otolith organ in upright position.

Semicircular canal

This organ is similar to three gyros in three geometric planes that are perpendicular to each other (Fig. 8-4). It is a series of three closed tubes filled with a fluid called *endolymph* (Fig. 8-5). This fluid is put into motion as a result of angular acceleration in the plane of the canal. Motion of the fluid exerts a force upon a gelatinous structure called the *cupula*. Bending of the cupula results in movement of the hair cells situated beneath the cupula. (These hair cells move in the same way that sea grass moves with the currents and wheat fields move in the wind. As the currents and winds change, so does the movement of the blades of grass or wheat as a unit.) This movement in turn stimulates the vestibular nerve, whose impulses are transmitted to the brain where they are interpreted as rotation of the head.

When no acceleration takes place, the hair cells do not move and remain upright, and a sense of no turn is felt (Fig. 8-6A). When a semicircular canal is put into motion, as during the acceleration of a turn, the fluid within the canals moves (Figs. 8-6B, 8-6C, and 8-6D), but there is a lag in the response along the canal walls. This bends the hair cells in the direction opposite that of the acceleration. The brain interprets the movement of the hairs as motion, or a turn.

Otolith organs

Two components of the otolith organs are the *utricle* and *saccule*. They are small "containers" located in the inner ear between the semicircular canal and the cochlea (refer back to Fig. 8-1). Lining both the bottom of the utricle and the wall of the saccule is a patch of small hairs called the *macula* (similar to what is found in the semicircular canal). What is different is that within the container's endolymph and on the container's wall there is an overlying gelatinous membrane containing chalklike crystals called *otoliths* (refer back to Fig. 8-2).

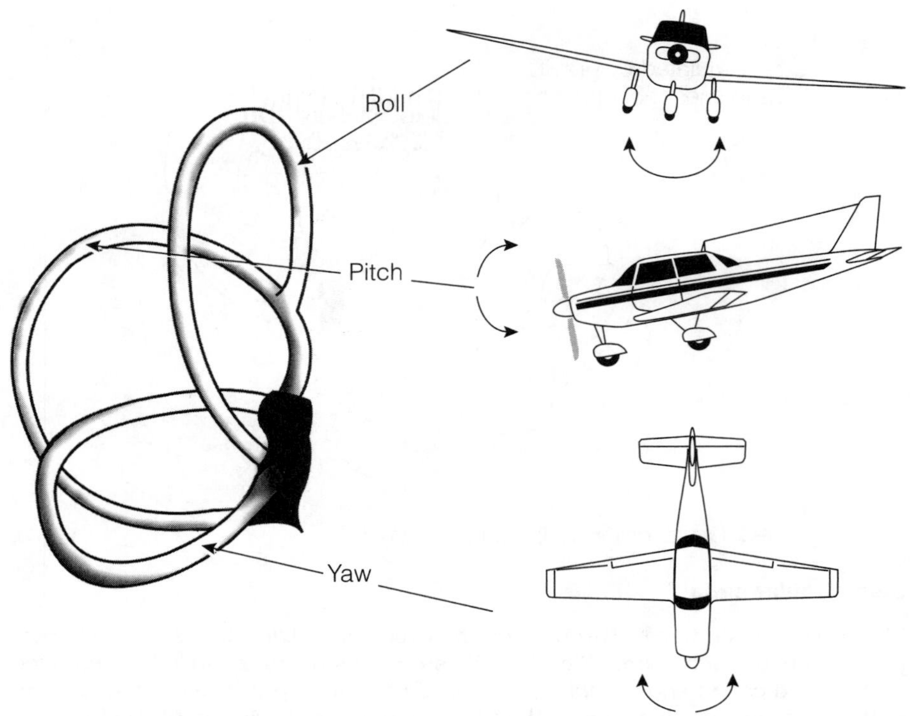

8-4 The semicircular canal responds to roll, pitch, and yaw.

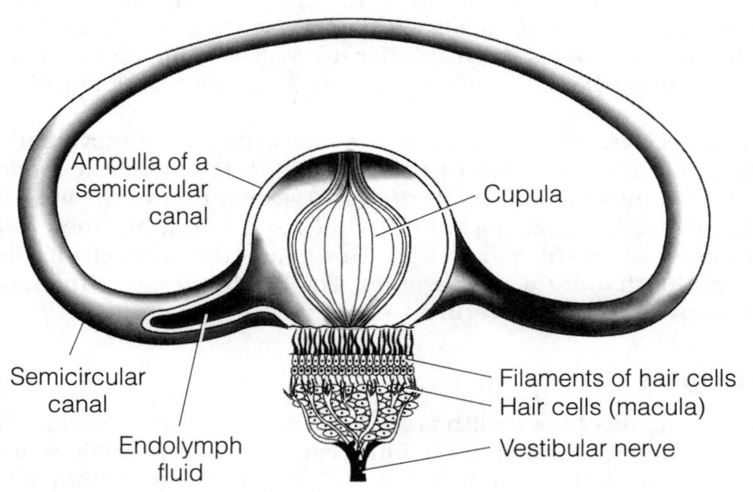

8-5 Semicircular canal.

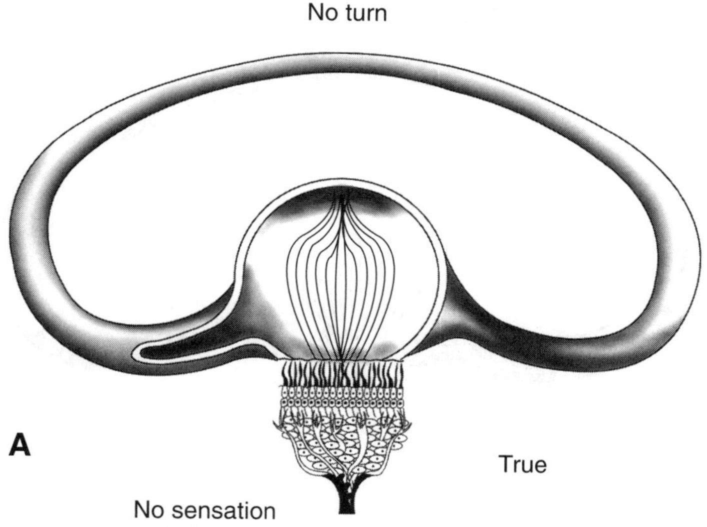

Hair cell position and resulting sensation under nonacceleration conditions.

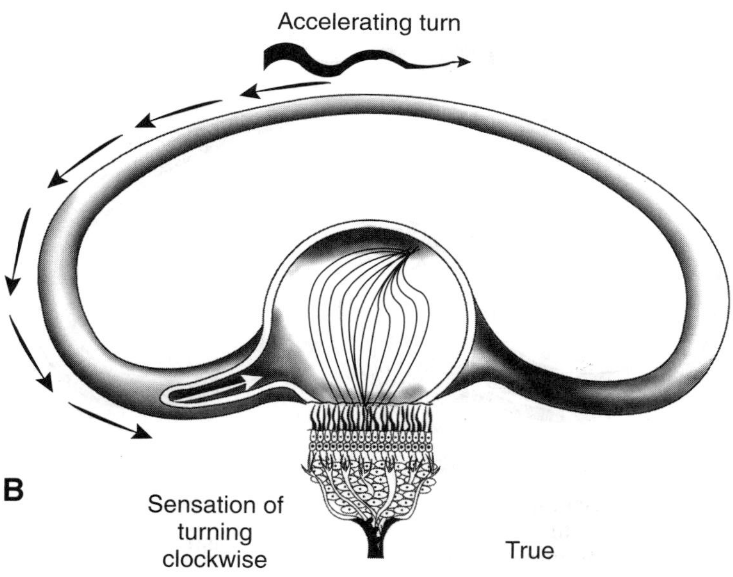

Hair cell position and resulting sensation under conditions of acceleration to the right.

8-6 Hair cells change position during turns.

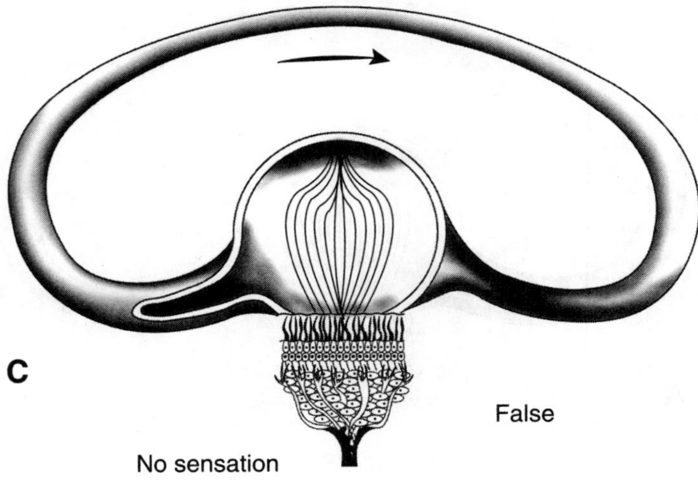

Hair cell position and resulting sensation under conditions of prolonged constant turn.

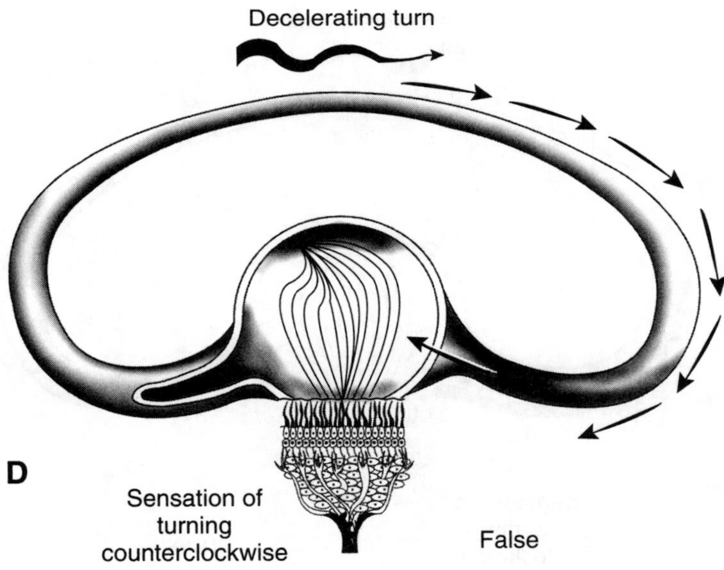

Hair cell position and resulting sensation under conditions of sudden stop or slowing.

8-6 Continued.

Changes in the position of the head relative to the gravitational force cause the otolithic membrane to shift position on the macula, thus bending the sensory hairs and signaling the change in head position. When the head is upright, a "resting" frequency of nerve impulses is generated by the hair cells. When the head is tilted, the resting frequency is altered to inform the brain of the new position of the head relative to the vertical.

Linear accelerations also stimulate the otolith organs because the resulting inertial forces cannot be distinguished physically from the force of gravity. A forward acceleration, for example, results in backward displacement of the otolithic membrane, which can create an illusion of backward tilt, if adequate visual reference is not available (Fig. 8-7).

A combination of semicircular canal, otolith, and proprioceptive signals enable an individual to determine body position and its movement in space, often with confirmation from visual references. This is a sensitive collaboration of sensory inputs to the brain under usual conditions on the ground. It's a different story in flight, and illusions called *somatogyral* occur.

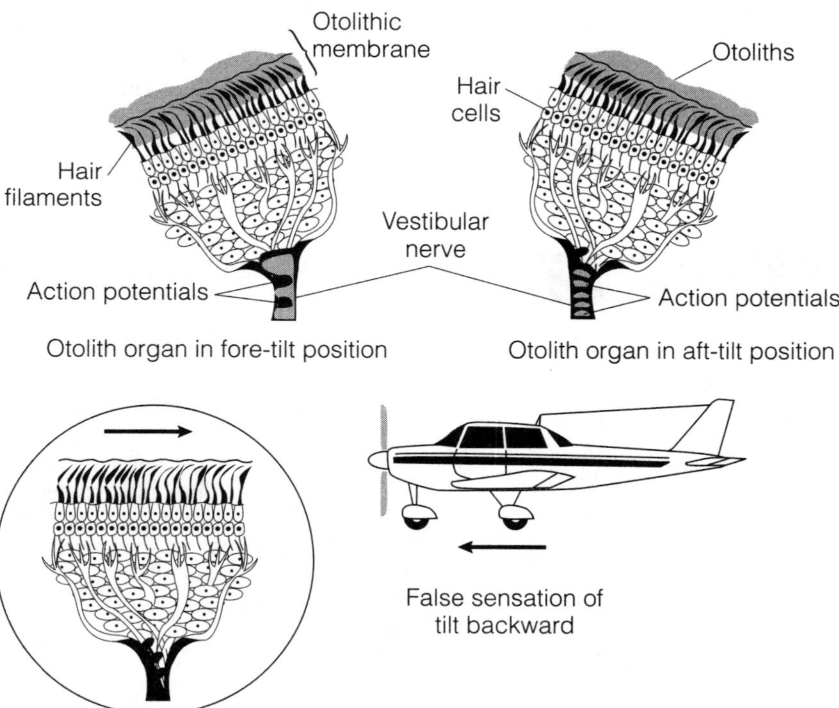

8-7 Otolith organ in fore and aft tilt position (top). Forward acceleration in an aircraft tricks the hair cells into perceiving a false sensation of tilting backward (bottom).

Vestibular illusions

The vestibular system, along with other sources of orienting sensations, works well in keeping the pilot "balanced" with the rest of the world; however, when subjected to the variables of flight's different forces on the body, these sources do not function in a manner that the body and mind are accustomed to and give the pilot misleading information. The strongest sense of disorientation comes from a vestibular system that is sending conflicting signals relative to the proprioceptive signals and what is actually happening, which results in erroneous and impairing perceptions for the pilot: illusions of orientation.

Coriolis In weather, Coriolis is the circular motion of wind as it passes over a rotating Earth. In physiology, it represents the "tumbling" of our internal gyros, or the semicircular canals (Fig. 8-8). This is the most dangerous illusion because of its often overwhelming sense of disorientation, and it can occur in any phase of flight where some type of climbing or descending turn is initiated. Some pilots described the sensation as being a giant hand taking over the instruments, fighting the pilot's urge to move the stick.

When the pilot is straight and level and without acceleration (likewise the aircraft), the fluid in the canals is in equilibrium, not moving. Now, when the aircraft turns and banks, there is an accelerative force on the fluid, which tells the body there is a change. If the pilot stays in that turn and at the same rate, the fluid once again reaches equilibrium of its motion and does not move. In other words, only when there is change does the fluid

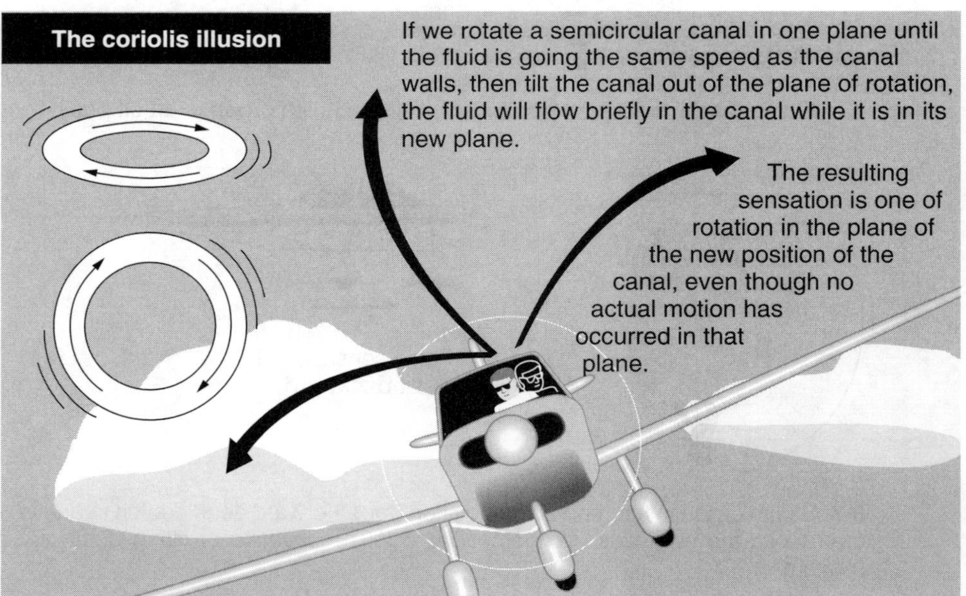

The coriolis illusion
If we rotate a semicircular canal in one plane until the fluid is going the same speed as the canal walls, then tilt the canal out of the plane of rotation, the fluid will flow briefly in the canal while it is in its new plane.
The resulting sensation is one of rotation in the plane of the new position of the canal, even though no actual motion has occurred in that plane.

8-8 If a pilot moves his or her head abruptly during a prolonged turn, the coriolis effect can cause an overwhelming illusion of change of aircraft attitude.

move and give out a signal. All three planes of the canals are affected to varying degrees, depending on the rate and degree of turn.

If the turn is coordinated and not abrupt, there is minimal misinformation going to the pilot. If during this maneuver of a turn, the pilot should turn her head in a different direction from the turn, such as downward, one of these geometric planes is disrupted and sends a misleading cue as if the airplane were in a roll or yaw. The Coriolis illusion makes the pilot perceive her airplane to be doing maneuvers that it actually is not doing. This illusion can also be induced in turbulence during a climb or a descending turn. The illusion is caused by the sudden change to the vestibular apparatus.

Without a visual reference to the horizon, this illusion can overpower the pilot's recognized need to recover, creating a conflict between the sensation expected in the maneuver and the requirement to control the aircraft against that sensation. It is easier to demonstrate this illusion than describe it, and this is why unusual attitudes are practiced in flight and why it is recommended that a ride in a Vertigon (simulator that demonstrates Coriolis) be experienced.

Leans This is the most common illusion and occurs when the pilot senses a bank angle when the aircraft is actually in level flight. If she maintains the level attitude of the aircraft (as she should), she will still feel compelled to align her body with the perceived vertical. In doing so, she actually leans in the opposite direction of the perceived turn (Fig. 8-9).

The leans can easily occur if the pilot's attention is removed from the instruments for even a short period of time. For example, if during this distraction the aircraft begins to turn slowly to the right, undetected by the pilot, the canals will respond accordingly. After a period of time in the turn, the brain "forgets" that it is in a turn. When the pilot's attention is again directed back to the aircraft and instruments and she returns to level flight, the pilot will have a false sense that the aircraft is banking to the left. If she reacts to her senses rather than to the instruments, she will roll the aircraft to the right.

Oculogravic illusion When an aircraft accelerates or decelerates in level flight, the otolith organs sense a nose-high attitude relative to gravity (Fig. 8-10). If that sensation is acted upon by the pilot without cross-checking instruments, he might pitch the aircraft down. Deceleration causes a similar sensation of a nose-low attitude.

Rotational illusion Also called *angular motion illusion*, rotational illusions result from misinformation from a constant-rate turn, as in a spin. If the turn is at a constant rate, the fluid in the canals returns to its original stable position of equilibrium. This results in the pilot's perception that the turn has stopped. When the turn slows (a change as in acceleration) or the aircraft levels off, the fluid is once again displaced but this time in the opposite direction.

The "graveyard spiral" is a condition most likely to affect fixed-wing aircraft pilots (Fig. 8-11). Upon recovering from the turn (or spin in this case)

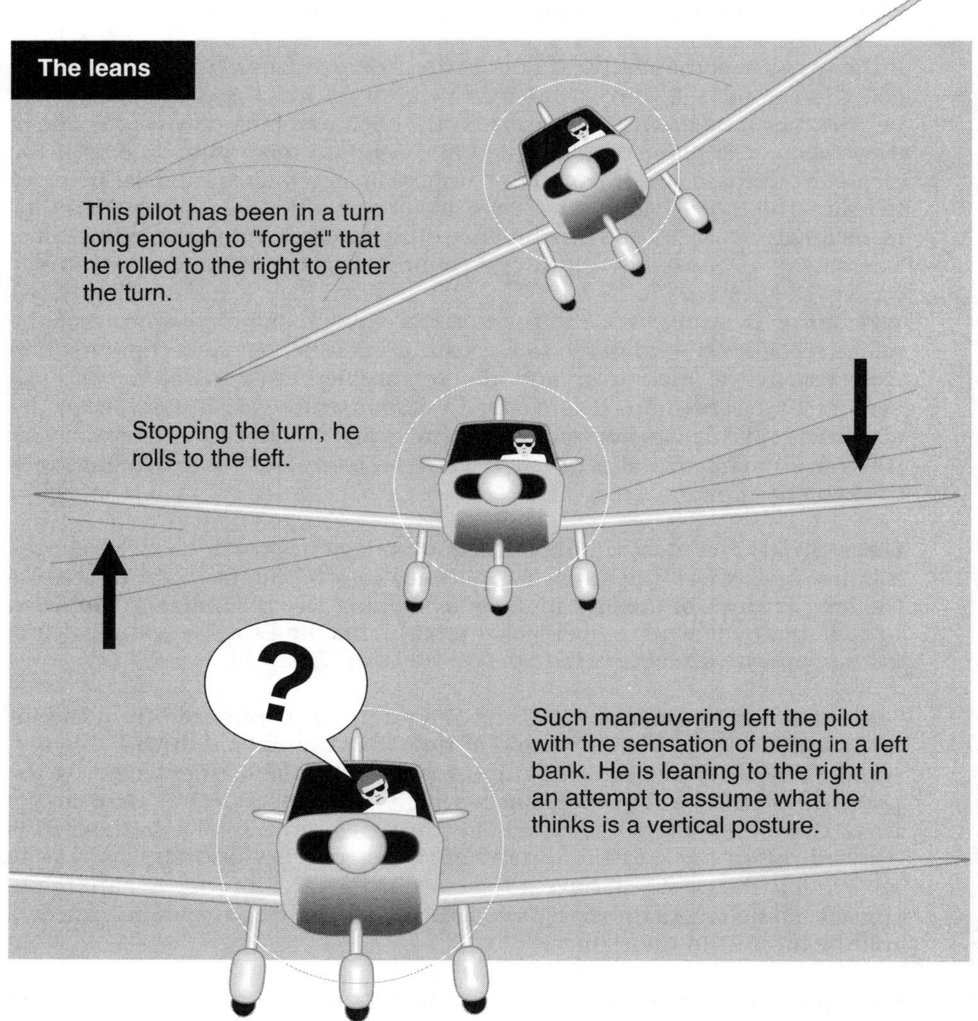

The leans
This pilot has been in a turn long enough to "forget" that he rolled to the right to enter the turn.
Stopping the turn, he rolls to the left.
?
Such maneuvering left the pilot with the sensation of being in a left bank. He is leaning to the right in an attempt to assume what he thinks is a vertical posture.

8-9 The leans.

as just described, the pilot will undergo deceleration, which will be sensed by the canals and interpreted as a spin or turn in the opposite direction. The pilot's instruments will tell him that he is not spinning, but he will have a strong sensation that he is.

If there is no outside reference to the horizon and the pilot disregards or doesn't believe the instruments, he will be tempted to make control corrections against the falsely perceived spin so that he reenters a spin in the original direction. Often the conflict is noticed, but the sensory input, even though erroneous, is overpowering, and the pilot is unable to control his actions.

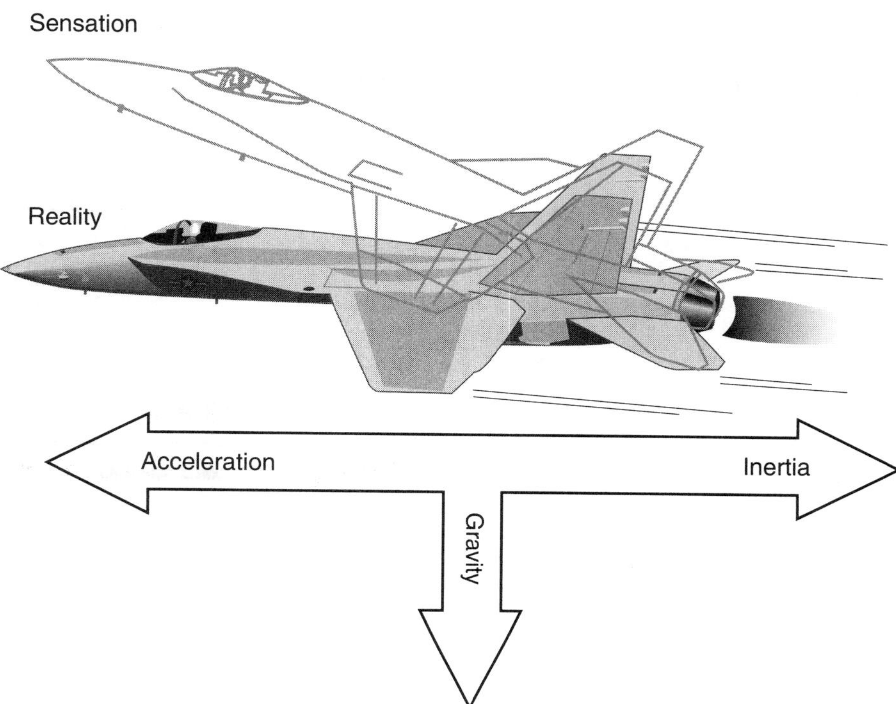

8-10 When an aircraft accelerates forward, inertia causes the otolithic membrane in a pilot's otolith organs to move. This results in the sensation of climbing and might cause the pilot to dive in an attempt to compensate for illusory change of attitude.

This action is compounded by the pilot's noting a loss of altitude as the spin develops and applying back pressure on the controls and adding power in an attempt to gain altitude. This tightens the spin to the point where recovery is nearly impossible. In a graveyard spiral, the angular velocity is in the form of a coordinated turn rather than a spin. Remaining in this turn long enough (perceived as a constant-rate coordinated turn), that sensation of turning will be lost, and the flight will continue as a spiral.

Spatial disorientation

For purposes of this book, *spatial disorientation* defines illusions or perceived position associated with relative motion, often called *vection illusion.* Furthermore, *spatial* infers a visual orientation in a given space: how the pilot is oriented to the horizon as either straight and level, in a turn, or climbing or descending. Visual illusions, separate from spatial illusions, will be discussed later.

Of all the senses available to provide input to the brain, vision provides the most valuable data. Some have determined that 90 percent of usable cues for orientation come from vision. The pilot's visual acuity comes primarily

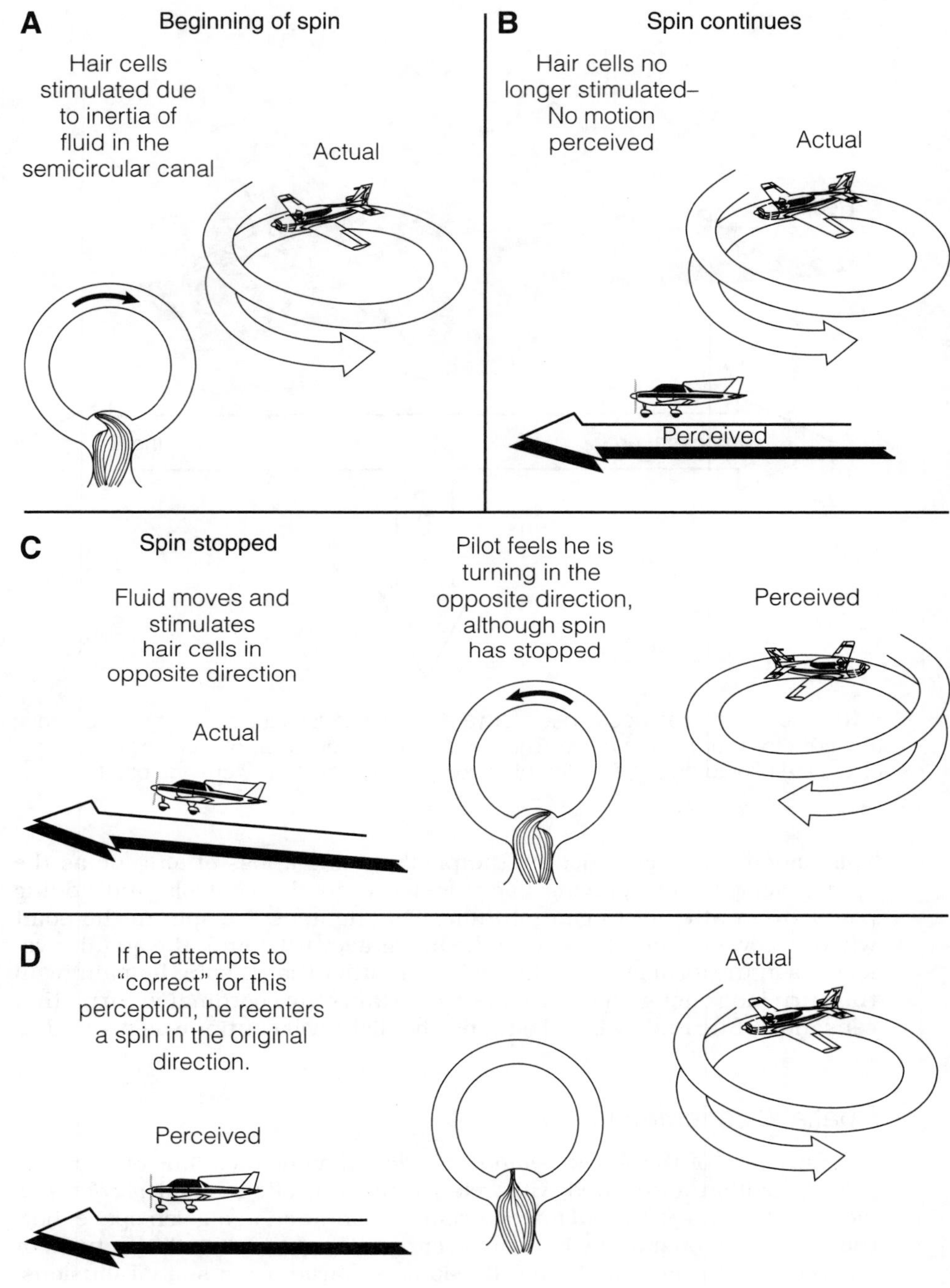
A
Beginning of spin
Hair cells stimulated due to inertia of fluid in the semicircular canal
Actual
B
Spin continues
Hair cells no longer stimulated–
No motion perceived
Actual
Perceived
C
Spin stopped
Fluid moves and stimulates hair cells in opposite direction
Actual
Pilot feels he is turning in the opposite direction, although spin has stopped
Perceived
D
If he attempts to "correct" for this perception, he reenters a spin in the original direction.
Perceived
Actual

8-11 A "graveyard spin" illusion is a type of rotational illusion.

from central vision (foveal), and peripheral vision (ambient) supports the perceived position in space, especially when in motion.

Peripheral vision is so strong in relaying body position to the brain that it is estimated that 90 percent of visual effects on orientation in a changing position are from the peripheral vision. One can see the potential conflict in instrument conditions, where central vision is used to read the instruments and virtually no peripheral vision senses exist to establish spatial orientation. This will be demonstrated later in this chapter.

Recognizing that peripheral vision is a powerful source of determining balance and orientation, it should be easy to recall the fact that everyone experiences spatial (vection) disorientation, whether in a car or an airplane. Looking straight ahead and concentrating on where you are going or scanning instruments maintains direction; however, if you put visible materials (fog, snow, or vehicles) in the area around you as you pass through that area, you will experience the effects. Remember how you feel in a car wash with your car stationary as the apparatus moves past, and you are staring straight ahead? You swear the car is moving.

For example, you're sitting at an intersection in your car, waiting for the light to turn green, and you suddenly perceive your car moving backward. You immediately exert more pressure on your brakes only to realize your car is stationary. It's the car next to you, creeping forward in anticipation of the traffic light. Your peripheral vision was telling you how to react with the relative motion of the two cars.

The same is true when flying through snow or rain. Your peripheral vision will mislead you into an illusion of abnormal interpretation of your own motion and spatial position. Indeed, in the cockpit, anything that moves in your peripheral-vision field is easily picked up, especially a flashing warning light, even though you are concentrating on data from your central vision.

For helicopter pilots, the same illusion is apparent when hovering over tall grass, for example. The wave action of the moving grass gives the pilot the sensation of moving, when in fact she is not. Flying in clouds with strobe and other anticollision lights generates the same confusing cues. These all lead, at best, to a sense of detachment from the cockpit, often creating some dizziness. There is a sense of wanting to correct the situation, but there is no situation that needs correcting. These confusing signals now become distracting and will often evolve into a more serious situation.

The power of peripheral vision over central vision can be easily demonstrated. Stand heel to toe, look forward, and concentrate on some small distant object. Close one eye and wait until your balance stabilizes. Reduce your central vision by holding your fist a few inches in front of your open eye. You still should feel relatively stable. Now take your fist away and hold a tube (like the inner tube of a roll of toilet paper) against the skin around your open eye, which reduces your peripheral vision. You will notice a balance instability because the brain no longer has peripheral vision to orient itself in space.

This instability varies, depending on many conditions. One of the intents of maintaining flying currency is to be able to suppress inputs from peripheral vision and increase dependency and reliance on central (or focal/foveal) vision. Unfamiliarity of the aircraft also increases vulnerability to peripheral vision cues.

Spatial disorientation, as described here, becomes a serious source of disorientation, especially in instrument conditions. The physiologically impaired pilot (tired, hypoxic, using OTC drugs, etc.) will lose his or her tolerance to suppressing inputs from peripheral vision. The pilot can quickly become incapacitated as a result of conflicting spatial senses.

As a final note, a flight simulator does not duplicate the aircraft movements in flight to give you the sensation of movement. You can't tell by looking from the outside what the pilot is experiencing inside. The simulator is fooling the vestibular and otolith system by cleverly creating motions that generate sensory perceptions of just about every activity of an airplane. Add vision, and the effect is very real.

Having defined the different situations leading to disorientation, it's now realistic to return to traditional terms of spatial disorientation and vertigo in pilot conversation. Just remember that these terms are now generic.

VISUAL ILLUSIONS

Visual illusions affect what the pilot perceives through vision, which in turn determines how he or she will respond. It is more of a misguidance in a flying activity (such as landing and judging distances), as opposed to the "illusions" of disorientation, where the body automatically responds to physiological cues from the vestibular system and peripheral vision. They are obviously related in that they both can lead to incapacitation especially when the events happen concurrently.

To appreciate the impact that the combination of the semicircular canals has on orientation, try this exercise. Hold your hand straight ahead of your eyes with two fingers spread apart. Now move your hand sideways, back and forth at an increasing rate. Note that it doesn't take much motion to see the fingers as a blur. Now keep the hand still and move your head sideways, back and forth at an increasing rate. Note how the fingers stay in focus no matter how fast you move your head.

The reason? Because semicircular canals act as a gyro, allowing the brain to lock the eyes on a moving object no matter how much or where the head and body moves. This ability of the eye to stay on an object is a powerful reflex to motion and difficult to overcome when subjected to some of the following illusions.

If the eye can focus and lock onto a fixed object outside, ideally the horizon, the pilot usually can overcome most of the disorienting situations. Vision is a reliable source of information; however, in the absence of outside visual cues, such as in instrument conditions, the chance for disorientation greatly increases.

Autokinetic Autokinesis is the perception of false movement when a static source of light is looked at by the pilot for a period of time (minutes) in the dark (Fig. 8-12). This moving reference point (an illusion) could lead the pilot to visually follow it. It is felt that the cause is the brain and eyes's attempts to find some other point of reference in an otherwise featureless visual field. Prevention is a combination of realizing the eye must focus on other objects at varying distances, not fixating on one target, and basic scanning.

Oculogyral This was previously described under disorientation. From the visual point of view, this illusion is more apparent at night, when vision is already compromised. The pilot perceives the apparent movement of a distant object in the visual field as a result of stimulation of the semicircular canals; therefore, in the early stages of a turn to the left, for example, the target appears to move rapidly to the left. After the turn is established and then stopped, the target object appears to move in the opposite direction, to the right. The pilot might not correct her aircraft accordingly unless she is scanning her instruments and confirming that she is tracking a moving illusion.

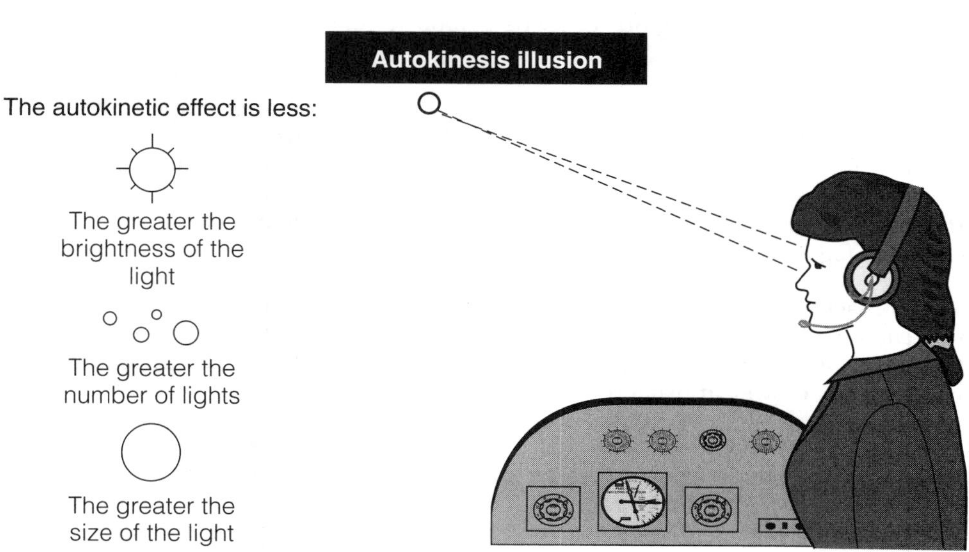

8-12 To a pilot suffering from autokinesis illusion, a stationary light that is stared at for several seconds in the dark will appear to move.

Oculogravic Also called *somatogravic illusion*, this is a perception of tilt induced by stimulation of the otolith organs. From the visual perspective, the illusion to the pilot is either a climb or a descent when he has no visual reference to the horizon. With a climb, the eyes try to compensate with a downward tracking, as is common when flying through an updraft. This illusion is also called an *elevator illusion*. The opposite illusion, familiar in helicopter flight in an autorotation, is the intuitive response of the pilot to change direction and/or altitude, which can decrease airspeed below a desired level.

Visual-cue illusions In addition to how the body responds (disorientation) to cues from a variety of sources, there are also illusions that are strictly visual that mislead the pilot into erroneous actions. This is not a physiological event as associated with the semicircular canals. This is the brain's misinterpreting an image as a result of the illusion of misleading visual cues as transmitted to and received by the eye. The greatest chance of these illusions affecting flight is in the landing phase and, to a lesser degree, in cruise where outside visual cues can be misleading.

Several illusions are explained throughout this text but here only those generated by what is seen, not affected by our vestibular system, for example, are discussed. The objective is to concentrate on just the illusions as they pertain to actual flight.

Fog and haze *Depth perception*, the ability of the brain to determine relative distance from visual cues, is compromised by any atmospheric conditions that interfere with light transmission. In addition to obscuring the ground and nearby outside obstacles for avoidance and orientation, fog and haze refract light rays differently than clear air, leading to the eyes perceiving the target object as being out-of-focus or poorly related to known characteristics of that target.

Contrast is reduced, which is important in defining size and shape. It becomes difficult to judge distances, especially height above ground and distance from the end of the runway. Light from ground lights, including REIL, VASI, and runway and taxiway lights, are somewhat diffused, losing the visual detail necessary for a precision approach and landing. Penetration into fog can create the illusion of pitching up.

Water refraction Rain on the windshield refracts incoming light rays, which misleads the pilot concerning the position of the horizon. The perception is that the horizon is below where it actually is, especially on an approach, resulting in a lower approach.

Landing visual illusions Runway width that is narrower than usual or expected tends to make the pilot think she is higher than she actually is and to therefore fly a lower approach (Fig. 8-13). A runway width that is wider than expected has the opposite effect, with the approach being too high.

An upsloping runway or surrounding terrain can result in the illusion that the aircraft is higher (the glide path appears high) than it actually is, resulting in a lower approach (Fig. 8-14). Downsloping runways have the opposite result; the glide path appears low and the approach might be high.

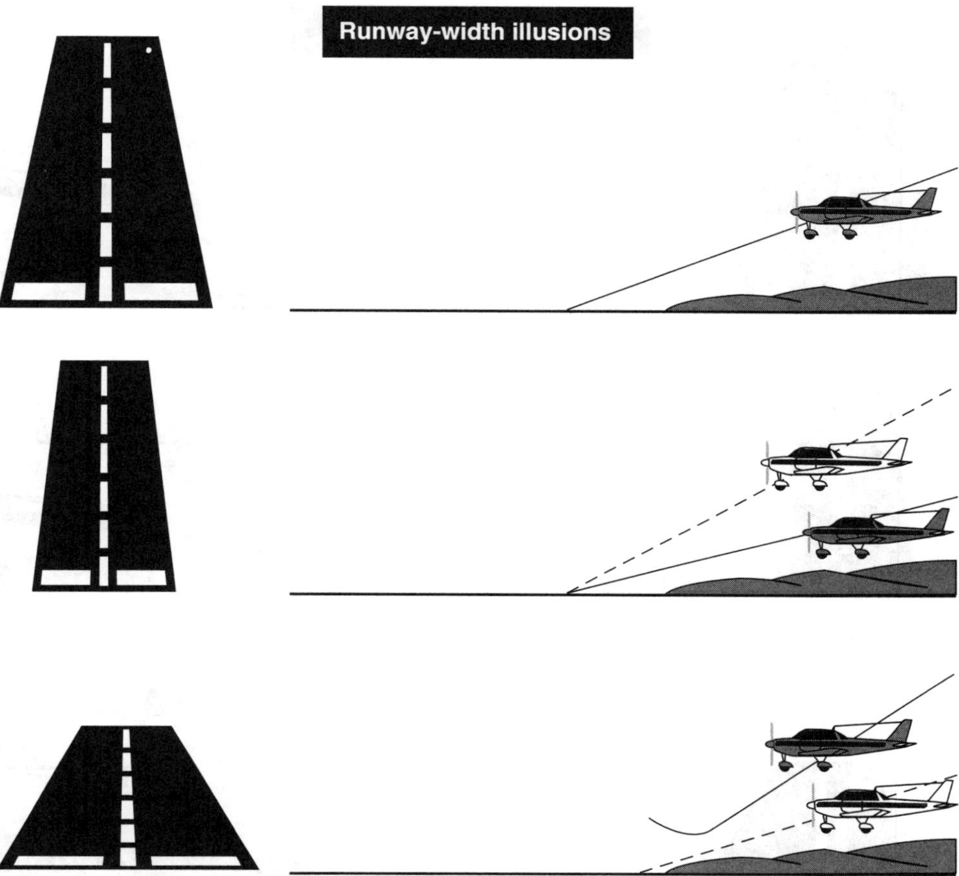

8-13 The effect of runway width on the pilot's image of runway (left) and potential effect on the approach flown (right). Accustomed width during a normal approach (top). A narrow runway makes the pilot feel higher, so he or she flies the approach too low and flares too late (center). A wide runway gives the illusion of being closer than it actually is, so the pilot tends to approach too high and flares too soon (bottom).

The absence of ground features such as buildings and trees with which the pilot can relate distance, speed, and position can cause the pilot to think he is higher than he is. Flying an approach over water, for example, without frequent referral to instruments will result in a lower approach height (Figs. 8-15 and 8-16).

Recognizing the importance and power of peripheral and central vision inputs to the brain, it should be respected that visual cues can be disorienting in addition to misleading the pilot in this position. It should be obvious that the scan and cross-check of instruments become a crucial part of any flight activity. These illusions are best understood by recognizing their presence in flight, not assuming that being informed of their possible presence will overcome the illusion. Only practice and currency will overcome these illusions.

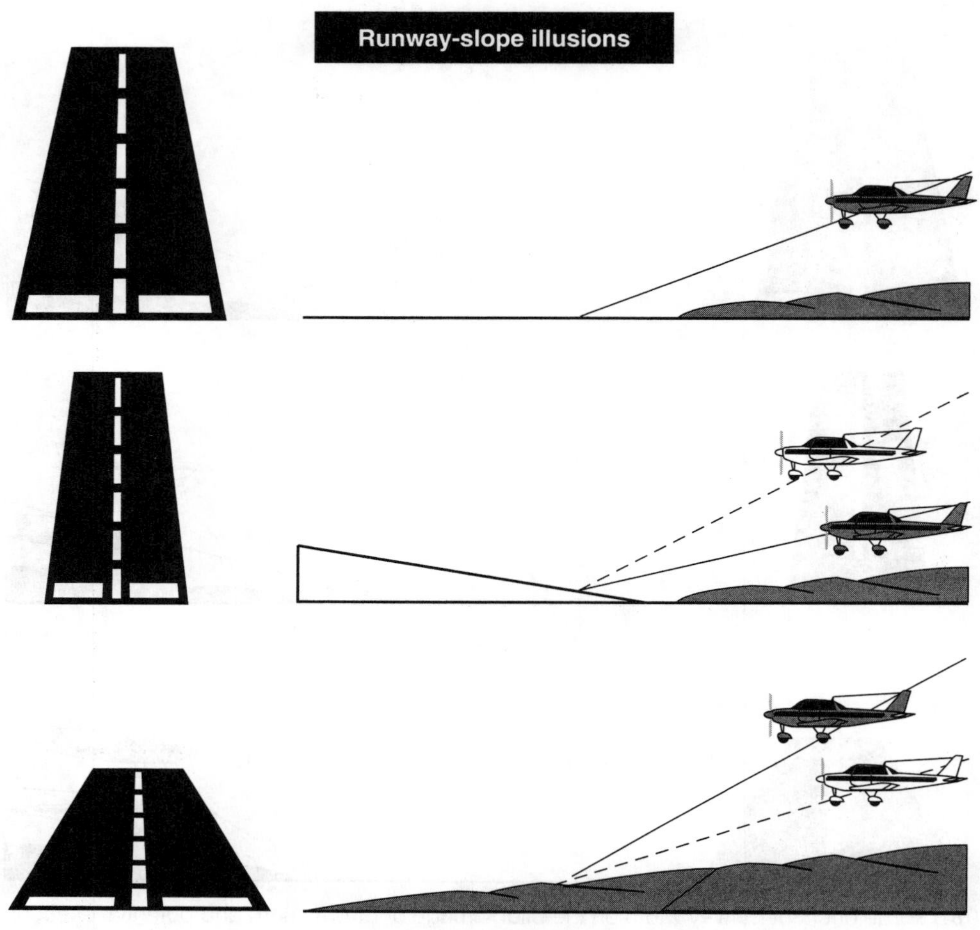

8-14 The effect of runway slope on the pilot's image during final approach (left) and potential effect on the approach-slope angle flown (right). Flat runway and normal approach (top). An upsloping runway creates the illusion of being high on approach, and the pilot flies the approach too low (center). A downsloping runway has the opposite effect (bottom).

TOLERANCE TO DISORIENTATION AND ILLUSIONS

If the healthy pilot can become disoriented and visual illusions affect everyone, then it must be recognized that the less-than-healthy pilot is at greater risk. Any one or combination of the following conditions will degrade a pilot's performance, especially during crucial phases of flight.

The following are all discussed in other chapters and are identified here to raise the level of awareness of their importance. Conditions that can reduce your tolerance to overcoming the effects described above include use of alcohol (even if it has been more than 24 hours since the last drink and

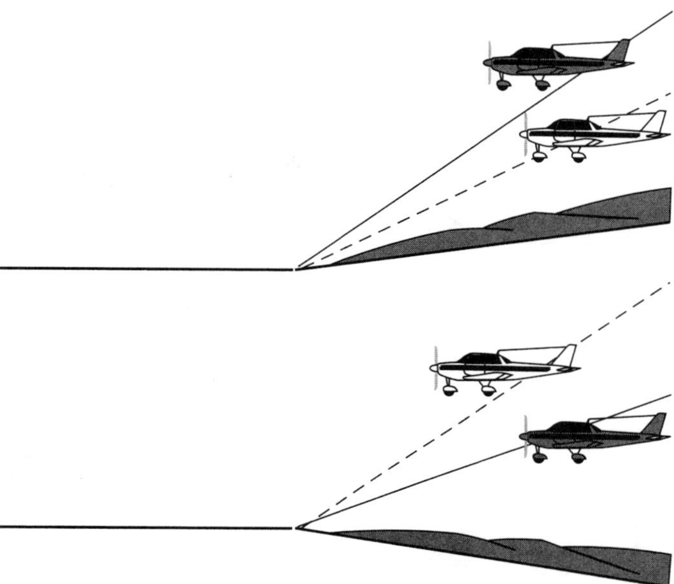

8-15 Potential effect of the slope of the terrain underneath the approach on the approach slope flown. The terrain slopes down the runway; the pilot thinks he or she is coming in too shallow and steepens the approach (top). Upsloping terrain makes the pilot think the aircraft is too high, so he or she corrects by making the approach too shallow (bottom).

the blood alcohol content has since diminished to zero), hypoglycemia, fatigue, use of medications, illness, and hypoxia.

If the pilot knows she is suffering from any of these conditions, she must be sensitive to being particularly careful about instrument scans, attentive to all communications, and maintain a high level of awareness and cross-checking to monitor performance. In a crew environment, it is advisable to request assistance in crucial phases of flight, possibly delegating some of your responsibilities to a crewmate. Informing the crew of your less than optimum performance is an essential part of CRM.

The pilot can cope with these situations of disorientation and illusions by maintaining currency in familiar aircraft. Flight proficiency is also absolutely essential. The lack of either or both reduces your tolerance to these potentially incapacitating events. Every pilot knows that he or she is "rusty" in his or her performance skills, often after only a few days of not flying. Flying instruments only once in awhile puts that pilot at high risk of unacceptable performance when eventful instrument conditions arise. When you have to consciously think about what to do in activities that

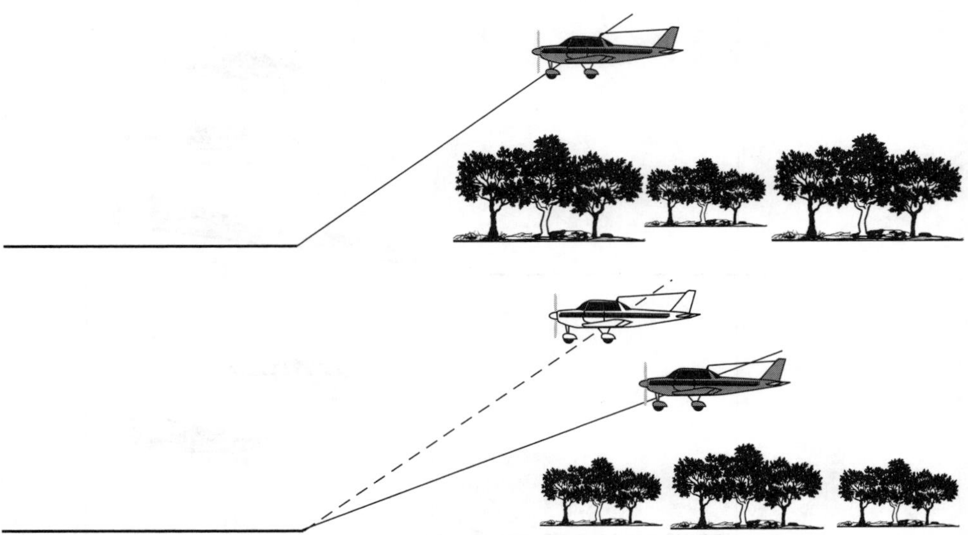

8-16 This shows the potential effect of unfamiliar terrain upon the approach slope that is flown. A normal approach over trees of familiar size (top). Unusually small trees underneath the approach path make the pilot think the aircraft is too high, so he or she makes the approach lower than usual (bottom).

have been done instinctively is a symptom of potential impairment and intolerance to illusions. There is no substitute for training and maintaining currency.

FLICKER VERTIGO

Flicker vertigo is a poorly understood sensation, probably not as serious or frequent as one would expect. Pilots, especially in helicopters, suffer a variety of symptoms from light that has a strobelike flashing or flickering to the eyes. This occurs as light from the Sun or other source is reflected off a propeller or rotor blades or interrupted in the same cycle by the props. The rate is about 4 to 20 cycles per second and can produce unpleasant and potentially dangerous reactions.

Nausea, vomiting, and fatigue can result. In extreme cases, in highly susceptible pilots, neurological reactions occur, such as convulsions and changes in levels of consciousness. The pilot who suffers from the previously mentioned disabling conditions is also at higher risk. There is a great deal of anecdotal information but little scientific data to confirm a frequent occurrence of "flicker vertigo."

MOTION SICKNESS

The vestibular system appears to be the source of the symptoms of motion sickness (airsickness). People who do not have functioning semicircular canals as result of a birth defect, surgical removal, or disease do not experience this problem. Any pilot, subjected to the right combination of unusual attitudes or flight conditions can become airsick. In other words, if the vestibular system can be "tumbled" in a way to which the pilot is not adapted, especially when there is a conflict with visual inputs, then he will become sick. There is no doubt that currency, self confidence, and familiarity with the aircraft and its typical maneuvers is the best deterrent to air sickness.

Passengers (or pilots not in control of the aircraft) are easy subjects. New pilots are also susceptible. Some of the additional common causes include heat discomfort, anxiety, observing or smelling someone else who is airsick, and eating foods that are nauseating.

Symptoms

Nausea and vomiting are the most common symptoms of motion sickness and are experienced in varying degrees. Often a slight queasiness is noticed, but the fear of becoming sick is all it takes to become sick. Even after vomiting, nausea often persists. One then becomes *apathetic*, not really caring about what's going on. A headache and perspiring are also common. At best, someone who is airsick is impaired.

Treatment

Most people know if they are prone to airsickness. Even so, there is no known prevention technique or adequate treatment to help. Much research is being carried out because motion sickness is a prime concern in space travel. Some suggest that there is more chance of becoming airsick if one flies on an empty stomach; therefore, one should eat simple foods that he or she knows can be well tolerated. This includes adequate fluids.

Various medications are available over the counter and prescribed; however, there is little reliability or predictability from one person to another. Side effects from the commonly used drugs do occur, especially over-the-counter medications containing scopolamine. The best tactic is to work with your flight surgeon, trying different combinations of diet and medications along with awareness of current studies on prevention and treatment of motion sickness. If you frequently have problems, learning biofeedback techniques and "desensitization" has proven to be effective. Training to help pilots break the cycle of airsickness is becoming more popular. Check with flying organizations for assistance.

9
Self-imposed medical stress

Joe knew he shouldn't have accepted the flight. It was the middle of the night, he had been in bed two hours, and the air cargo mission he had been assigned could wait a few hours anyway. But then, if he didn't take the job, someone else would. Oh well, a couple of cigarettes and a dose of coffee, and he'll be ready to kick the tires and get airborne. Once in the air, Joe decides to fly direct from Chicago to Dallas at 11,000 feet. Too bad he doesn't have any supplemental oxygen on board. Then he misses a frequency change and has to be reminded of a course correction. He finds it increasingly difficult to hold it straight and level because the autopilot isn't working. As the trip drones on, he becomes lethargic and keeps up with the cigarettes and coffee, trying to stay awake. Joe never considers his hypoxic condition at that altitude and doesn't care anyway. Only after he bounces the aircraft several times on his landing does he realize that he is impaired.

MANY THINGS ABOUT OUR HEALTH and safety are uncontrollable: temperature extremes, illness, noise, weather, and the like. We do have control in several areas, in actions we do to ourselves that we don't have to do but that could affect our performance. Our safety is compromised as a result of how we abuse our health or live the unhealthy lifestyle we choose to live. This includes playing doctor: making a self-diagnosis of a medical problem and then using self-medication. Also, subjecting ourselves to extremes of alcohol, nicotine, caffeine, and other easily obtained "abuses" can only impair our body's functions and our mind's capability to fly safely. They are self-imposed. That is, it is not required or even expected that we go beyond moderation in any of these areas. We choose to do these things to ourselves until our body rebels.

This chapter will cover each of these issues and describe in a general sense how they affect us. The other topics that are covered in other chapters are inherent to flight, but they, too, add stresses to our performance. These self-imposed stresses are not inherent, but they can complicate a tenuous flying situation to the point of being unsafe, even without problems typically associated with flight. For example, medications, as well as the conditions for which they are taken, can interfere with visual and auditory perception, decision making, judgment, and other mental processes, as well as our flying skills.

Although all drugs and medication, whether prescribed or over-the-counter (OTC), are evaluated for safety by the FDA before they are released for sale to the public, very few are evaluated as to their effect on performance in a flight environment; therefore, there is little scientifically proven insight about how any illness, medication, or habit will affect the pilot during flight. More and more data will be coming from the scientific community as a result of the recognition of human factors and flight physiology as an important part of safety. But studies are not necessary to prove what we already know from experience: These self abuses will affect performance at one time or another.

SELF DIAGNOSIS

The primary reason that we see a doctor is because we are in pain, uncomfortable, or unable to work or play. We are looking for relief so we can go back to our activities. The doctor, who is also eager to make you feel better, has the added responsibility of finding out why you feel bad, or more important, which conditions could be serious and which ones must be ruled out. Once this has been done, the doctor can treat the cause, if it's treatable, and then treat the symptoms.

Often, pilots feel they don't have to go to the trouble, time, and expense of seeing a doctor for simple symptoms such as a cough, a headache, congestion, and the like. The real world often doesn't permit a pilot to ground himself or herself with such basic and simple symptoms; however, it is recognized by aeromedical professionals that even an untreated cold or flu will erode a pilot's performance equivalent to a couple of beers and impair reaction time by up to 57 percent. It becomes a matter of how significant the symptoms are and whether or not they are related to some simple medical problem. These same symptoms could be something more serious, but you won't know that unless you can have a medical evaluation and appropriate tests. Avoiding this medical step is truly self diagnosing, comparing your symptoms with your insight of illness as read from some lay publication or hearing of someone else with the same problem.

The following are some of the common ailments for which one sometimes seeks medical attention, or more commonly, self diagnoses. One common strategy is to self diagnose and treat, and if that doesn't do the job, then see the doctor. Time spent avoiding the doctor can vary from a day to several weeks, depending on how you feel. Another strategy would be to check with the doctor and get an opinion before you take any action. This is more scientific but somewhat unrealistic. Somewhere in the middle is good judgment with safe and prudent choices.

Colds and flus

Very universal conditions, colds and flus are commonly caused by viruses, which means there is no cure. Antibiotics do not cure viral infections as they do bacterial infections. Time and allowing the body to fight off the infection is the most realistic course of events.

The symptoms of a cold or flu range from a simple sore throat and cough to generalized aches and pains, fatigue, headache, and congestion. Such symptoms, which are often limited to the respiratory system, are sometimes caused by bacteria; therefore, if symptoms don't begin to clear on their own in a few days, seeing a doctor is the next step. Antibiotics could help in this situation.

Flus often have the same symptoms as a cold, if the flu virus affects the respiratory system; however, many flus are associated with the gastrointestinal (GI) tract with resulting nausea, vomiting, diarrhea, and stomach cramps. Sometimes these symptoms include a generalized *malaise* (achy, tired, "feeling rotten"). A common observation is that even after the infection is gone you will still tire easily, have decreased endurance to be active during the day, and have little resistance to picking up another virus, which can start the process over again.

These same symptoms could be the result of other problems, such as a bacterial infection of the sinuses or throat, pneumonia, early appendicitis, and so on. In most cases you will recognize whether or not your symptoms represent something more than a simple cold, for example, and you will base that conclusion on your previous experiences; however, if the illness is different from what you would expect, then seeking medical attention is advised.

For whatever reason, these are the variety of symptoms associated with those conditions for which some pilots will seek self help. The marketing statements on the packages of OTC medications/drugs imply that they will clear up those symptoms and cure the problem. If you feel that you aren't too sick, then it is very tempting to assume you're okay as long as you can take care of the symptoms with these OTC medications. But now you have two strikes against you—the effect of the cold or flu and the unpredictable side effects of the medications.

Hay fever

Hay fever affects just about everybody to varying degrees, depending on geographic location and type of activities. Even though someone might not consider that he or she has hay fever (allergies), mild allergic symptoms are often considered to be caused by a mild cold. The significant difference between the two is that a cold is an actual infection and has additional symptoms such as fatigue and malaise that are potentially unsafe. Those who know (or think they know) that they have allergies or hay fever will often self medicate, using OTC antihistamines to relieve symptoms, as suggested on the package. If you truly have hay fever as diagnosed by a physician, then you should consider other treatments, such as desensitization, to control the symptoms. Using OTC allergy medication and then flying is not acceptable or safe.

Headaches

Headaches are common in some people, although some never have any. Headaches can be caused by many problems. If you know from previous experience what is causing your headache, then take the measures you took before; however, if it is different or if you usually don't have headaches, then a doctor's evaluation is necessary. Because of the variety of causes, some being potentially serious, a headache should not be minimized in terms of significance. If simple aspirin takes away the headache, it might be insignificant. But if the headache doesn't go away, see a doctor.

Injuries

Injuries that are not serious (minor broken bones, cuts, sprains, etc.) are often minimized by pilots because of their eagerness to fly. Arthritis and tendonitis, although not injuries, are experienced by many people as they get older. All are painful. In such cases, the diagnosis of arthritis is relatively apparent for the lay person. Persistent symptoms, however, are often an indication that there is more going on.

A variety of OTC medications are available to treat the symptoms of pain of these and other conditions. Another problem with discomfort is that pain pills are often used with the thought that in time the cause of the problem will go away anyway and that relieving the pain will allow you to return to flying. This could be a setup for an unsafe situation because the injury or other source of pain might by itself compromise performance, such as if you are unable to put adequate force to rudder pedals, too weak a hold on controls, or have a restricted range of motion in reaching instruments. Furthermore, taking away the symptoms might mask the only signal the body has to alert you to a continuing problem.

Pilot's attitude

There is a conflict between a pilot's desire to fly and her willingness to accept that she might not be safe. A simple statement from a doctor saying that your medical condition shouldn't interfere with flying might be misleading because many doctors are not familiar with the unique physical and mental requirements for flying. This is especially true in larger, more complex aircraft with numerous controls, switches, overhead panels, and black boxes. That statement from a well-meaning doctor might give you a false sense of safety when, in fact, you are not safe. For example, a sprained ankle after several weeks is still weak, but you will not realize that until you try pushing a rudder pedal in a crosswind landing. Or your endurance after a two-week bout with the flu will be markedly decreased and won't be recognized until halfway through a trip.

There is also the philosophy that because others have been known to put up with the illness, use self medication, and then go fly without apparent difficulty, it's okay for you to fly under the same conditions. This reinforces the rationalization that one is safe to fly. Looking at the labels on OTC medications complicates the situation because of their claims of being able to make your symptoms go away: the quick fix. The eagerness to fly the trip now overrides the potential of becoming impaired.

Other factors contribute to the willingness to justify flying when medically unsafe: pressure from the company, taking that checkride before a deadline, building up flying time, making use of opportunities for advancement, not wanting to use sick time, and many other self-imposed traps.

A rule of thumb to keep in mind is if you have any medical situation for which you feel you need any treatment, including OTC medications, then you probably aren't fit to fly. Mild symptoms from known causes might be the exception, but only if you are sure of your diagnosis and the associated safe treatment.

OVER-THE-COUNTER MEDICATIONS

Most people seek medical help only when they don't feel well. The reason a visit to the doctor is made is because of symptoms that interfere with that person's activities or the fear that something serious is occurring. Apathy and even denial regarding the potential of "finding something" often interferes with a prudent decision for that person if medical intervention is anticipated. The threat of the doctor finding anything serious often prevents the very act that could rule it out or find a cure. When FAA medical certification is on the line (and hence a career), less than responsible choices are often made.

Therefore, since people initially are looking for symptom relief, there is a huge market for medications and drugs that the public can purchase without a prescription or visiting a doctor. Over-the-counter (OTC) medications are a multibillion dollar business in the United States alone. Marketing departments in the drug companies have a great time creating new ways to package and sell the same medication and for reasons that sometimes are not even safe or healthy, especially to aircrew.

Although the Food and Drug Administration (FDA) is charged with protecting the public from medications and chemicals whose labels mislead or imply false results, there is a continuous struggle to keep ahead of the misleading claims with factual information. Current enforcement by the FDA to ensure legitimate labeling is helping the consumer. The choice still remains with the pilot.

Pilots especially are vulnerable to the hype of the many ads for quick-fix medications. Although not deliberately seeking unsafe or unhealthy therapy, the pilot is often swayed into judging that he knows how ill he is and thus self diagnosing. This becomes a setup for the pilot who is anxious to return to flight if he could just get rid of the symptoms; it's a case of denial. The catch is that although the symptoms might be relieved, the cause goes untreated. Plus, self medication might mask the very symptoms the doctor needs to know to evaluate your problem. Drug stores and supermarkets have bountiful shelves of a wide variety of chemicals, drugs, and medications to meet the need of the pilot looking for the quick fix.

Flight surgeons are often asked if it is okay to take a certain medication and still fly. The initial response has to be "No." There is no list of approved drugs or medications that the pilot can use, prescription or self medication, without approval by a flight surgeon or a clear understanding of its effects. Technically, pilots are forbidden from any self medicating. Obviously, there is a place for all medication and treatments, even OTC, but it must be with the full knowledge of both pilot and doctor as to why the medication is needed and the effect of this medication on safety and performance.

The military by regulation forbids any self medication. All treatment must come from the flight surgeon. Because civilian aviation does not have the close relationship to the aviation doctor, the pilot must assume responsibility for determining what course of action to follow, as stated in FAR Part 61.53. This regulation states that the pilot must not have any disqualifying medical problems to be legal (see chapter 17). The pilot must assume a medical problem is unsafe until proven otherwise. If in doubt, check with the AME.

Marketing claims

Reading labels on the containers of these OTC medications is often very revealing, and sometimes the claims are quite comical. It's uncommon if there is any comment about cure or dealing with the cause of the symptoms. And, of course, there is the ultimate disclaimer to see your doctor if symptoms persist. The assumption is made that the purchaser has made the correct diagnosis or has ruled out more serious causes. It's like soap ads; there is just so much you can get out of soap, yet there are bigger and better ads coming out every day touting how much better one soap is over another in removing stains or grime.

Consider some of these claims for OTC drugs:

- Symptomatic relief! Some symptoms are relieved, but sometimes at the expense of causing significant side effects. "Symptom relief" is an effective statement because that is precisely what you are looking for in the medication.
- Temporary relief! This could mean a few minutes to a few hours. It does not mean permanent relief as is expected if the medication were curative.
- Extra added ingredients! This implies greater potency or more reliability in meeting its claims, even though that extra ingredient could simply be caffeine. Here is where the FDA is cracking down on claims that these added ingredients really do what they are claimed to do.
- New and improved! Compared to what? The ingredients probably are the same, but the amounts might be slightly different. Or perhaps the only difference is the shape of the pill or the color of the liquid. It doesn't mean the drug is any more effective than the previous drug.

It would be a good exercise to read labels just to see how creative marketing people can be to get you to buy their product. Yes, the side effects are listed, in small print, and so are the ingredients. But let's face it, you're more interested in what the large print says that it can do for you. So, read labels, know what the ingredients actually do, and then make your own decision as to what you should do. A good reference is the annual edition of *Physician's Desk Reference* or "PDR." One edition of the PDR is dedicated to OTC medications. As an aside, reading labels on food products is also revealing, a good example being whole milk, skim milk and low fat. Check the actual weight of fat present in each.

Basic ingredients

Over 450 different drugs are made from a combination of one or more of the chemicals listed in Fig. 9-1. These eight ingredients make up practically all of the usable chemicals that are allowed in OTC drugs. There are others, of course, but these are certainly the most common. In addition to discussing the effects and side effects, look at the half-life for each chemical. *Half-life* in this case refers to the amount of time that half of the chemical needs to get out of the body. For example, if you ingest 1.0 gram of a drug and its half-life is 10 hours, then in 10 hours, 0.5 gram will still be in the body.

Chemical name	**Half life***	**Actions and side effects**
Caffeine	Varies	Stimulant (adrenalinlike) Diuretic (urine loss) Fatiguing Withdrawal "anxiety"
Aspirin	20 minutes	Analgesic (pain killer) Antipyretic (fever) Anti-inflammatory Tinnitus (ringing ears) Gastritis
Acetaminophen	3 hours	Analgesic Antipyretic
Destromethorphan	Varies	Antitussive (cough)
Ephedrine and Phenylphrine	3 hours	Decongestant Stimulant (like caffeine)
Chlorpheniramine	6 hours	Antihistamine (allergy) Sedative Drying agent (secretions) Vertigo Blurred vision Increased heart rate Decreased coordination
Phenylpropanolamein	6 hours	Decongestant Stimulant Anorectic (appetite suppressant) "Amphetamine"-like Increased blood sugar Increased blood pressure

*=It takes about five times the "half-life" to clear the body

9-1 Ingredients contained in a majority of over-the-counter medications.

Antihistamines

Antihistamines are the most popular ingredients in allergy, cold, and flu OTC medications. Several kinds of antihistamines are available over the counter, and other antihistamines are available only by prescription. This is the drug for which most people buy cold and allergy medications. Some of the more common antihistamines used are brompheniramine, chlorpheniramine, doxylamine, and diphenhydramine. Note that each generic name ends in "*ine.*"

Effects Antihistamines primarily help dry up secretions from upper respiratory mucous membranes (nose and sinuses) that cause sniffles and post-nasal drip, often leading to a mild cough. In addition, some of the symptoms associated with an itchy nose and eyes and sneezing are controlled; however, some of these symptoms are a result of the body's attempt to cure the problem, whether a cold, flu, or allergy. That's one of the trade-offs in self medication: Are you interfering with the natural course of healing?

Side effects Drowsiness/sedation is by far the most serious problem with any OTC medication that contains antihistamines. Countless stories relate pilots who were incapacitated in some way as a direct result of taking such medication. Some prescribed antihistamines do not cause drowsiness, according to certain studies. If such treatment is necessary to keep you flying comfortably and safely, then your doctor must be a part of that treatment and monitoring. Furthermore, the FAA will certify most prescribed antihistamines.

Many other, often subtle, side effects of OTC antihistamines are unacceptable in flight. These include decreased coordination, reduced visual accommodation (see chapter 7 regarding vision), dizziness, depression, and some cardiovascular effects (rapid heart rate, reduced ability to divert blood where needed, etc.).

Sleeping medications

It is important to note that virtually all OTC sleep medications are antihistamines. Drug companies have simply changed the name of the product and put a different label on the bottle and box, thereby transforming the antihistamine into a sleeping pill. Would you take a sleeping medication to treat a cold so you could fly? That's what you're doing if you take OTC antihistamines to treat a cold. *Check the labels.*

Because getting quality sleep is such a major problem and there is no good resolution without some trade-off to safety, there is continued interest in any cure that shows up in the press. L-Tryptophan, which is found in dairy products and appears to assist in sleep, was sought after in health food stores in pill form. Because of concerns of its effects on metabolism and its purity, there is continuing controversy as to its use. The same problem and dilemma exists with melatonin, a natural hormone within the body known to be a part of the physiology of sleep. The assumption is that it can be singled out as a solution to assisting sleep, adjusting circadian rhythms, and other concerns. Again, the complexity of its role in the body and its purity are concerns, yet anecdotal stories of success continue to add to its credibility. The best advice is to wait until melatonin is proven safe and effective.

Decongestants

Generally, decongestants are well tolerated by the body, have few undesirable side effects, and offer more flexibility for use in flight. These are found in OTC medications (separately and in combination with antihistamines and other chemicals) and are the ingredients found in many nose sprays. Common decongestant chemicals are phenylphrine, phenylpropanlamine, ephedrine, and pseudo ephedrine.

Effects Mucous membranes usually swell during colds, flus, and allergies. This swelling causes congestion in the nose, eustachian tubes, and sinuses. The congestion creates the problems in flight (ear and sinus blocks), and decongestants are effective at reducing this swelling; they do little to reduce secretions.

Side effects The major problem with these drugs is their stimulant effects, similar to what caffeine can do. Included are such effects as making you nervous, jittery, and overactive plus keeping you awake when you are trying to get needed sleep. Most people are quite tolerant to these effects but it is difficult to predict what happens to an individual in flight or on layovers. An already heavy load of coffee plus the symptoms for which you need a decongestant might increase impairment. Pilots who have borderline high blood pressure should be aware that decongestants can aggravate blood pressure elevation; however, OTC decongestants can be used in some situations. It is wise (and expected of the pilot) to confirm the use with your flight surgeon or AME.

Nose sprays

The active ingredient in most nose sprays is a decongestant (i.e., phenylphrine), which can also be a *vasoconstrictor*. That is, blood vessels get smaller in the area of the spray. This helps reduce the congestion of the mucous membranes of the nose. Frequent use of short but fast-acting (less than three hours) nose spray decongestants can lead to *rebound*. The spray can be an irritant to the mucous membranes of the nose; therefore, as the expected favorable action begins to diminish in a few hours, one takes another spray. Over a period of a few days, the irritation of frequent use causes its own swelling and congestion for which more spray is used. Longer but slower acting sprays (over eight hours, as with oxymetazoline) might be better tolerated, especially if used for no more than a few days. Rebound can also occur from overuse. A prescription nasal inhaler is available with a mild cortisone that can be very effective if long-term use is expected.

Another issue is the recommendation to keep a long-acting nose spray in your flight bag as a "get me down" treatment, when you get caught in flight with an ear or sinus block (see chapter 5 regarding altitude physiology). This is not to recommend using the nose spray prior to flight to overcome current congestion because, as stated elsewhere, if you think you need treatment before flying, then maybe you should not be flying. Keeping such medication in your flight bag might be challenged by some flight surgeons, and for good reason because such a practice could lead to some pilots abusing an implied "okay" and flying with the assumption that the nose spray can be used if they have a problem even before flight. All things considered, keeping a nose spray available during flight probably outweighs the risk of imprudent use by the pilot. If there is doubt, don't fly and ask your AME.

Analgesics (pain relievers)

Analgesics are sold alone as a reliever of pain or in combination with other cold and flu medications. They are marketed in a variety of ways, all leading you to believe that a company's OTC medication solves many problems. Discomfort, whether pain, achiness, or general malaise, is generally why you seek help from an analgesic, but your symptoms could also be nature's way of telling you that you are impaired; therefore, it is more important with analgesics that you consider why you need the medication. The possible reasons include headaches, toothaches, sore throats, pains from injury, and other similar symptoms that could represent potential problems made worse in flight. The pain might affect specific pilot functions unique to flight, such as pushing a rudder after a blown tire or trying to reduce the discomfort of a sore back while seated. Pain or discomfort is also distracting, a common cause of incidents.

Aspirin As one of the most common pain relievers, aspirin is a medication that has several other effects, both good and bad. It is also an *antipyretic* (*fever reducer*), an anti-inflammatory medication for arthritis and other inflammation problems, and is currently used to thin blood in the prevention of heart attacks, among other medical situations. It is relatively safe if used as directed, which is taken with food or an antacid, not taken more often than recommended, and avoided if there is any history of previous stomach distress. Aspirin has been associated with ulcers in some people and ringing in the ears with very high doses.

Acetaminophen This ingredient often competes with aspirin in marketing OTC medications. It is an effective pain reliever for some people and is often found in combination with other chemicals in cold and flu medications. It is not as effective in anti-inflammatory treatment. It appears to be tolerated better than aspirin in those who have problems with gastritis. It is also relatively safe in usual doses.

Ibuprofen This was once a prescription medication, but now OTC pills containing comparatively small doses are sold. The prescribed medication is usually used for inflammation, and the OTC is meant for pain (the analgesic properties were found as a side effect when it was used initially as a prescription). This is also relatively well tolerated in most people, but others do notice side effects such as stomach distress. Other such similar products that were initially by prescription only are becoming available as OTC medications, such as naproxen, which is a pair killer like ibuprophen.

Cough suppressants (antitussive)

In most cases, cough suppressants contain *dextromethorphan*, a fairly innocuous chemical that can assist in relieving coughs in some people. It has few if any significant side effects. One should keep in mind, however, that derivatives of codeine are often found in some OTC cough medication, especially when purchased in a foreign country. Codeine, usually only available by prescription in the United States, is a very effective antitussive agent, but it is considered a narcotic, a mood altering drug that can cause drowsiness and will show up in drug-testing programs.

Extra added ingredients

Extra added ingredients include a variety of other chemicals that might enhance or not even affect the other ingredients and their effectiveness. The FDA is watching closely to see if these additions do any good. Caffeine is one of the most popular added ingredients and is commonly found in combination with aspirin and diet pills. Another is an antacid, which basically helps overcome some expected side effects such as stomach distress. Some of these ingredients are considered inert or inactive, but they are still present, sometimes in significant quantities. Alcohol is a common example and is used as a solvent for the other ingredients. Up to 25 percent of some cold medication is alcohol. Consider also the synergistic effects of alcohol and medication on the brain. Even this small amount of alcohol can show up in a breathalyzer test. The rules deal with this problem by allowing the pilot to repeat the test 20 minutes later, which should allow for a negative test.

Phenylpropanolamine (PPA)

Phenylpropanolamine (PPA) deserves a separate section. It's a common ingredient in a variety of OTC medications: diet pills, cold and flu relief, and allergy control. It acts as a decongestant and an *anorectic* (appetite suppressant), depending on how it is packaged; however, PPA is also a very strong stimulant to some people, giving some a real "high" similar to amphetamines. It can have a severe effect on blood pressure, often elevating normal blood pressure to the point where the diagnosis of hypertension is made. PPA can interfere with sleep and cause headaches, dizziness, rapid heart rate, mood changes, and nausea. The FDA is taking special interest in this chemical as a result of its many adverse side effects.

There is an increasing availability of prescribed medications as over-the-counter medicines. And they won't follow the "-ine" rule. The most common at this printing are medications prescribed for ulcers and now packaged as treatment for "heartburn," to compete with antacids. The problem is that the symptoms might be that of an ulcer, which a pilot cannot fly with until fully treated. It's best to check with your AME if you have symptoms suggestive of ulcers before you assume it's just indigestion. It's very likely that the future will see more medications once under the control of a physician now put into the hands of the layman.

Herbal medications are also being promoted more. The danger is that they are not regulated and the user is not sure of what the herbs really do, nor can the consumer be sure of its purity. Just keep in mind that there is no "quick fix" to medical problems that are perceived as difficult or impossible to treat in traditional ways.

Review

Some concerns bear repeating when discussing OTC medication or self diagnosis. The most important is, *read the label.* You will find the preceding chemicals and ingredients in varying doses in just about all of the common drugs. In other words, by using the same ingredients in different combinations and different advertisements, one can see there is little to remember when determining what you are really ingesting. Most of the chemicals listed that could be a problem end in *-ine*; use that suffix to develop a better idea of what you can expect in terms of helping or hindering.

The bottom line is this: If the symptoms that you choose to diagnose as being a result of a cold, or flu, or whatever are sufficient to treat with OTC medication, you probably are potentially or currently impaired and not safe. In flight, the side effects of the medication might be worse than the symptoms and the causes that you are treating. Read the label. If in doubt, either call your AME or ground yourself for a few days. Halfway through a trip is not the time to realize you should have stayed home.

CAFFEINE

When considering safe flight, caffeine must be considered a drug, no matter where you find it. It also ends in *-ine*, which indicates it could affect your performance as a pilot. Its use is a mild addiction for some people, or, at the very least, a strong habit. It is one of the most common drugs that can be purchased over the counter as part of a beverage. It is a part of our family, business, and social life style. For many, coffee is a staple of life, without which they are incapable of functioning properly, or so they think. For some pilots, a cup of coffee is a part of their after-takeoff checklist. But caffeine, like alcohol, when used to excess in some people can cause problems resulting in a variety of physical and mental dysfunctions. Caffeine from any source is one of the most underrated, readily available, mood-altering drugs that causes potentially unsafe problems, especially in and after flight. It is also considered to be addictive, that is, the body needs the extra stimulation to function.

Caffeine increases heart rate and can disturb pulse rhythm, constricts blood-vessel diameters and coronary circulation, increases urine production and excretion (diuretic), and disrupts normal mental processes. Tolerances vary among different people, some noticing few if any side effects, others noticing significant unwanted symptoms, even with only one cup of coffee. In other words, it is a habit-forming stimulant with unpredictable side effects to an individual as well as problems of withdrawal when stopped.

Sources

Perhaps the most common sources of caffeine are coffee and tea (Fig. 9-2). Regular brewed coffee has an average of 100 mg of caffeine per cup. Decaffeinated coffee has only about 4 mg, but it is still present and will affect those that are very sensitive to caffeine. Tea has about 70 mg per cup (or glass of iced tea). Others include a cup of cocoa (50 mg), cola drinks (40–70 mg per 12 ounces), and chocolate (15 mg). And recall that caffeine is present in many OTC medications.

Substance	Amount (mg)
Coffee, 5 to 8-oz. cup	
Decaffeinated instant	2
Decaffeinated brewed	2–5
Instant	40–110
Percolated	65–125
Drip	110–150
Tea, 5 to 8-oz. cup	
Bag, brewed for 5 minutes	20–75
Bag, brewed for 1 minute	15–50
Loose, black, 5-minute brew	20–65
Loose, green, 5-minute brew	15–60
Iced, canned	25–35
Cola drinks, 12-oz. glass	
Pepsi-Cola	38
Coca-Cola	33
RC Cola	30
Diet RC	25
Diet-Rite	17
Other soft drinks	
Mountain Dew	54
Dr Pepper	38
Coca, 8-oz. cup	50
Milk chocolate, 1 oz.	3–6
Bittersweet chocolate, 1 oz.	25–35

9-2 Caffeine content.

It is easy to compute how caffeine intake in a 24-hour period could reach significant levels when considering all these sources. Intake in excess of 250 mg is symptomatic in most people who have not built up a tolerance over the years, with symptoms becoming more noticeable in very sensitive people. Some pilots are not aware that some of their fatigue, headaches, insomnia, and nervousness could be secondary to their caffeine intake. Others notice a change in their heart rhythm (skipped beats) in addition to heart rate after just one cup, and often this will show up on their FAA EKG (if the EKG is required).

Coffee is a strong stimulant, which is the reason most people use caffeine. The morning ritual of waking up often includes coffee. It's often the first thing offered to you in the morning at a restaurant. It is estimated that Americans drink more than 370 million cups of coffee per year, which means that nearly one in five Americans drink more than five cups of coffee a day (that's about 500 mg); however, one to two cups is usually well tolerated and for some people does diminish fatigue, improve mood and alertness, and temporarily increase mental and physical work capacity. It is one of the drugs of choice in trying to control circadian change and fatigue on long flight days. The withdrawal effects after the trip however, can be counterproductive.

Side effects

Because caffeine is a stimulant to the central nervous system, some people are unable to sleep if they ingest caffeine as long as several hours before retiring. This can be a true dilemma if the pilot uses coffee to stay awake on a long trip and then subsequently tries to get to sleep on the layover. For some people, just one cup of coffee or tea in the afternoon or evening is enough to keep them awake.

Caffeine is also a strong cardiac stimulant, acting like adrenaline to the heart muscle and often increasing pulse rate. Some evidence states that blood pressure is raised, but that is less of a fact than its effects on pulse and rhythm. *Arrhythmias*, such as "skipped beats" or *premature ventricular contractions* (PVCs), and other irregularities, might not be clinically significant, but if they show up on an EKG, such as with a company or FAA exam, they have to be explained. This means other possible causes must be ruled out. Pilots have been grounded for this situation until the evaluation is completed satisfactorily and no other problems are discovered.

Caffeine is also a diuretic (increases urine production and excretion). In other words, one cup of coffee could result in urinating two or more cups of water, creating a negative fluid balance that makes dehydration even a bigger problem than what is usual in the cockpit (due to reduced humidity at altitude and in a pressurized cabin or cockpit). Because of the additional chemical *theophyline*, tea can be a stronger diuretic than the same amount of coffee. One can see that iced tea, especially during the summer, can be refreshing, but as a diuretic, it can also be more dehydrating in an already dehydrating environment. Cola beverages are very popular and are consumed all day, even for breakfast.

One of the most serious yet subtle side effects is on mental functions, often called *caffeinism*. It can cause anxiety and panic disorders, worsen premenstrual syndrome, and creates headaches and nervousness. This becomes particularly true during withdrawal from caffeine. Swings of heavy intake, as on a trip or studying for exams or checkrides, to minimal amounts during routine days often generate problems not usually considered to be associated with caffeine. "Cold turkey" withdrawal for someone who knows she is a heavy coffee, tea, and cola drinker and wants to quit can result in even more significant degrees of the same symptoms. Withdrawal effects, primarily headaches and irritability, can result from sudden cessation of only one to two cups of long-term coffee intake.

This could be a diagnostic problem for your doctor or flight surgeon when you show up in the office with symptoms of headache, disruptive sleep patterns, nervousness, and anxiety. The doctor will probably not consider caffeine as the sole cause and must evaluate you more carefully to determine what's going on and rule out a psychological problem. Pilots have been hospitalized with these symptoms only to be discharged with the final diagnosis of caffeinism.

Tolerance

Low amounts of caffeine are generally well tolerated by most; however, over a period of time—years—the body builds up a tolerance to greater levels. That is, you might not be aware of symptoms of too much caffeine, but the body is still reacting to those high levels and expects this added boost every day. It's similar to an alcohol tolerance; you don't even know you are drinking a lot. If you are drinking more than 3 cups of coffee (or iced tea or cola products) per day, then you can assume you are at undesirable levels, especially in the environment of flight, even without noticeable impairment. Swings of high to low intake during and after long flights are common causes of posttrip discomforts.

A good policy to follow is to have your one cup in the morning but then either dilute subsequent cups or substitute decaffeinated coffee or tea. Actively increase your water intake, especially during flight. Substitute juices, herbal teas known to be safe, or just plain hot water (or hot lemonade). Recognize that you probably don't need that third cup of regular coffee (it's probably more of a habit than a need), and find something else to take its place. Being creatures of habit, humans often need something to sip on.

Also consider the other sources of caffeine. It makes no difference to the body where the caffeine comes from (OTC medications, for example), and it is easy to accumulate large doses from a variety of sources over a short period of time.

MISCELLANEOUS ABUSES

Hypoglycemia

Hypoglycemia is also mentioned elsewhere and is also explained here since it is a self-imposed stress (as a result of poor or inadequate nutrition). Often the situation is the inability to find or take the time to ingest adequate nutrition, limited many times to calories from a vending machine. Restaurants, coffee shops, and fast-food places aren't always open when you need them, you're in a hurry and don't have time in the morning, or the expected meal is not available on board the airplane.

Missed meals is a problem, and you might not be able to control the unexpected lack of food; however, there are creative ways of "bagging" food that you can keep in your flight bag or suitcase. This can range from a roll of candies to bags of peanuts or "trail mix" (peanuts, raisins, and candy). This is not as nutritious as a regular meal but does resolve the symptoms of hypoglycemia for the short term, a necessary trade-off to prevent impairment. You might consider keeping packaged or canned microwavable soups, stews, and the like, in your suitcase that can be at least partially heated in hot water if you end up at your hotel without a meal.

By recognizing symptoms of hypoglycemia, such as headache, stomachache, nervousness, or shakiness, you can take appropriate corrective action. This includes some form of instant energy: calories, such as candy. Prevention is still the best therapy, respecting the need the body has for adequate calories that will last several hours: fats and proteins. Even if you miss a meal, you can prevent hypoglycemia by planning ahead and keeping an emergency source of packaged calories always available.

At the crew base with just vending machines, the combination of peanuts, whole milk, a candy bar, and an apple can suffice for an "emergency" meal. Avoid "empty-calorie" food (sweet rolls and colas) to take the place of any meal. The trade-off of not being nutritious is minor compared to the chance of becoming hypoglycemic. The cola and candy bar, especially for breakfast, can lead to sugar rebound effect (a sugar-high in the blood then followed in a few hours by a sugar-low). Try to find foods high in protein for snacking. These provide energy over a longer period of time.

Smoking

Aside from the risks to your health, smoking also puts carbon monoxide into your blood system and cells. Recall that carbon monoxide interferes with the absorption of oxygen, leading to even higher levels of hypoxia. Smoking one cigarette prior to flight increases your body's hypoxic symptoms as if you were 2,000–3,000 feet higher. This becomes a significant factor with night vision, which is dramatically affected by hypoxia. Visual acuity impairment is more pronounced.

In addition, smoke can be very irritating, especially to your nonsmoking crew. Symptoms of congestion, eye irritation, and general discomfort are common. Fortunately, there has been a decrease of or even a ban on smoking on the flight deck with most companies. Practically all airliners are smoke-free now. Secondhand smoke is another issue. Few will disagree that there is a problem, but there are still mixed feelings about cabin air quality, and not just from smoking. This is a changing situation as more information from studies becomes available. Smoking doesn't fit with aviation whether the flying is commercial or recreational (Fig. 9-3).

"Smokeless tobacco" (chewing tobacco, snuff, etc.) is another source of nicotine. It, too, is highly risky for medical problems. It puts more nicotine into your body than smoking. A wad of tobacco in the mouth generates a lot of saliva, which must either be swallowed or spit out.

Drugs
Exhaustion
Alcohol
Tobacco
Hypoglycemia

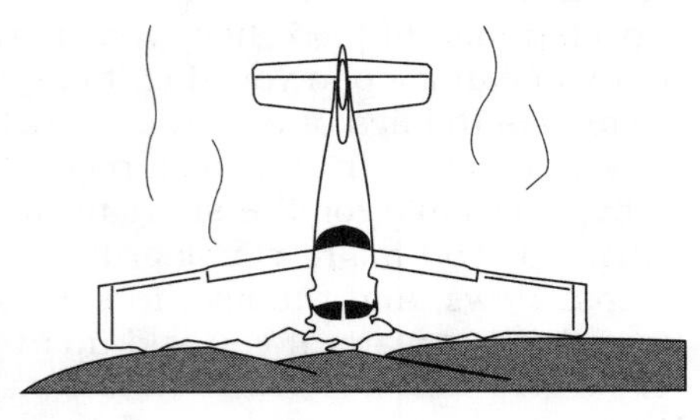

9-3 Self-imposed stress can spell DEATH for pilots.

Prescribed medications

Although not self-imposed, because your doctor feels you need to be on this therapy, it's important to know that there are situations when one can fly and be on prescribed medications. Many medications are certifiable if the FAA is given a complete medical report. (This is covered briefly in chapter 17 regarding FAA medical certification.) Essentially, when determining if you are safe, it is a combination of why you need the medication along with any side effects that would be present.

Remember that at altitude your symptoms might be different than at ground level. It is wise, therefore, to ground yourself for several days, determine if the medication is doing its job, and determine whether or not there are side effects. Some effects will disappear after several days on the medication. If the illness is under control and there are no side effects, then returning to flight status might be okay. In any case, this must be coordinated with your treating physician and your AME or flight surgeon before you can go back flying.

Another problem is your schedule interfering with your timeliness in taking the medication or just plain forgetting to bring your medications along on a trip. If you are supposed to take a pill four times a day, that doesn't mean one pill every hour for three hours in the morning before a trip that will last for the rest of the day. It is essential that you follow the required dosage and frequency and fully understand the potential side effects so that you know what to expect. Pilots on antibiotics will frequently quit taking them when the symptoms are gone. This is unhealthy because the infection still might be present; the lack of symptoms is not diagnostic of a cure. Always finish the medication if so directed. Seek necessary information and advice from your personal physician, and let him or her know that you are a pilot.

PHYSICAL CONDITION

A pilot's flying schedule and social life are not conducive for a nutritious diet and regular exercise. It takes a commitment to participate in a health maintenance program. The result is being in poor shape, not only in appearance but in performance and endurance. The body is relatively tolerant to these self abuses, lack of adequate diet, and no exercise, but in time, the body will be impaired. The immune system will be impaired, increasing the risk of infections. Tolerance to long days and jet lag will be compromised. The risk of occurrence of hypoglycemia increases when a meal is missed.

Improper health maintenance will cause future problems (heart disease, diabetes, obesity, etc.) that could reduce a pilot's competitiveness during pilot selection for a job. Being in poor physical condition is a self-imposed medical stress, and there is no viable excuse for not following the principles of health maintenance. "Cramming" for a medical exam by eating and exercising properly just before the evaluation, whether for the FAA or for employment, is not realistic. It's easy to be complacent and rationalize that you will get into a good program later, but many medical problems are initiated by poor physical condition. Plus, you will not be fully fit to fly.

ALCOHOL AND DRUGS

Before continuing with the topic of alcohol, it is essential that there is a clear understanding of FAR 91. Drugs could include over-the-counter medications in addition to the traditional definition of illicit drugs. Because "under the influence of" refers to alcohol and drugs, this FAR takes on more significance for the pilot even though no illegal "drugs" are being used. It reads:

- No person may act or attempt to act as a crewmember of a civil aircraft—
  ~Within 8 hours after the consumption of any alcoholic beverage (new alcohol testing rules state that no "alcohol" be consumed, not just "alcoholic beverages." See the drug testing subsection at the end of this chapter.)
  ~While under the influence of alcohol
  ~While using any drug that affects the person's faculties in any way contrary to safety or
  ~While having 0.04 percent by weight or more alcohol in the blood.

The remainder of this regulation refers to passengers and submitting to testing for alcohol and drugs. Keep these regulations in mind as you consider whether or not you and your crewmates are truly fit to fly and legal. An important issue to keep in mind is the impact on the pilot's life and career if caught with an alcohol-related offense, whether it be from driving or flying—they are both extremely deleterious. Consider that most states have removed legal loopholes in their driving laws regarding alcohol. There is simply no tolerance for driving and alcohol. Practically all states have heavy fines, often imprisonment, and loss of driving privileges all resulting in very large increases in auto insurance. The FAA is actually more lenient after the first alcohol-related offense and will usually certify if it can be proven that there is no problem with alcohol and the pilot was unlucky and got caught: more a matter of poor judgment.

Alcohol is found in a variety of places: food additives, medications, and beverages. Alcohol is alcohol, no matter where it is found, and it has the same effect on you whether it is in highballs, beer, wine, punch, or other sources. This might seem simplistic, but there are many who feel that beer is only a beverage to quench thirst and do not equate their beer intake to the fact that several beers is the same as several highballs. The saying "if you drink a lot of beer, you drink a lot" is very true, although few summer beer drinkers would admit that they drink a lot of alcohol. The same holds true for wine. The same amount of alcohol is in 12 ounces of beer, 1.5 ounces of vodka, or 5 ounces of wine. Alcohol is also found in many over-the-counter drugs, some up to 50 proof (25 percent of content). Alcohol in any product is a legal drug, yet it can be as harmful as illicit drugs.

The real problem in our society is the accessibility of alcohol at affordable cost to an aviation community that not only condones but often expects its population to participate in social gatherings such as attending "happy hours" or meeting for a drink after a trip or mission. Couple this with the fact that the use of alcohol varies from abstinence in some (which is becoming more frequent and acceptable) to dependence in some others. Furthermore, usage and the effects of use vary from individual to individual, depends on the social setting, and will change over the years (for better or worse). Misusing, abusing, or depending on alcohol are sure signs of an unsafe and irresponsible pilot. The bottom line for a personal policy is responsible use of alcohol that in no way impairs your flying and respects or exceeds current regulations.

The United States leads the world in production of beer and is 13th in consumption. It ranks eighth in consumption of liquor. Furthermore, peer pressure, advertising, visibility in the press, TV programs, and movies increase the awareness and acceptance of alcohol consumption, much to the joy of the beer, wine, and liquor industries. Unlike many jobs, one can't legally go to work as a pilot when under the influence of alcohol or with a hangover. The pilot might be legal eight hours after drinking, but he could still be unsafe. New alcohol regulations that went into effect in January 1995 add that any alcohol, not just that found in beverages, is in violation if consumed within 8 hours priors to flight. Yet it happens and is a real challenge to the aviation industry.

A clear understanding of the variations of the impact of alcohol on the body and mind is essential to being safe and productive. Listen to the facts, not what's observed socially and discussed openly in crew lounges. Respect the fact that most of the effects about to be discussed are occurring at the same time.

Physiology

Ethanol (or *ethyl alcohol*) is the chemical name associated with alcoholic beverages. It is readily absorbed in the stomach and small intestine. It immediately gets to the brain, where most effects are noticed; certain effects are expected and desired, some are unexpected and undesired. Other tissue cells become "intoxicated" with alcohol, sometimes replacing water in the cell structure. Ninety-five percent of the ethanol then is ultimately metabolized by the liver into carbon dioxide and water. Prior to being broken down, it is changed to glucose in the body, as a source of energy. Alcohol, therefore, is the ultimate empty calorie because it has no other elements of nutrition.

A full stomach often delays absorption, although the alcohol might last longer in the body. Liquors are absorbed more quickly than beer. It is also recognized that carbonation, as well as sugar in the mix, will increase absorption. These variables make it difficult to determine how an individual responds to any given amount of alcohol. To say the body can tolerate greater amounts because some of these variations are controllable is naïve. A person is still under the influence, and crucial organs are affected.

Because alcohol is fat soluble, it can enter into fat tissue in addition to organ cells; thus, the obese pilot can probably take in more alcohol before it shows up in his or her bloodstream. By the same token, it will take longer to metabolize out of the body. Under normal conditions, alcohol is metabolized at a rate of about 1.0 ounce for every three hours. If the intake exceeds this breakdown, the body becomes more intoxicated and impaired for a longer period of time.

Some alcohol is excreted unchanged in the urine and some from the lungs. The smell of alcohol on one's breath is only a small indication of the total amount being taken up by cells and therefore indicates that there is still a significant amount in the blood stream. (This is the basis for checking exhaled air (breath) for alcohol with a breathalyzer.) Even after your blood level reaches zero, there is still impairment from the effects of hangover and other physiological changes.

Basic physiological effects

Ethanol depresses the central nervous system and is probably the most dangerous kind of impairment to a pilot. Fatigue and sedation are closely related. How one feels and reacts with alcohol is a result of a loss of inhibitions, euphoria, and a general sense of well-being. One talks more and is more sociable. Performance is adversely affected: even at low doses (less than 0.02 percent), loss of discrimination and fine movement of muscles and touch, and a diminished memory and concentration.

Ethanol is also a diuretic, especially in early stages of drinking. As one approaches intoxication, the kidneys tend to make less urine. The color of your urine is a good indication as to how much water is being lost through *diuresis*, which means the lighter the color of the urine, the more fluids you are losing. This variation in the amount of fluids remaining in the body after the party is over might have some effect on the severity of the immediate hangover. Some studies indicate that fluid retention as well as blood vessel dilation causes the headaches and other symptoms.

Vasodilatation, or widening of the arteries, is another effect. This makes you feel warm and gives some drinkers the ruddy, or flushed, complexion;

however, alcohol is counterproductive in coping with temperature extremes because it interferes with the body's attempts at controlling internal body temperatures by diverting the blood flow as needed. You might feel warm, but there is no conservation of heat.

Effects on the body

Alcohol can cause hypoglycemia long after ingestion since the liver quickly metabolizes the alcohol first and then available food, which isn't usually available the morning after. Ethanol is a toxin to the body and to the liver, altering and killing tissue cells. For example, heavy drinkers lose about 100,000 brain cells per day, which is about 10 times the normal attrition. We have billions of brain cells, so there is only minimal damage being done in most cases, but some pilots need all the brain cells they can keep. The same is true with the liver, in that the cells are altered, eventually turning into scar tissue. This is called *cirrhosis*, and in severe cases it is incompatible with a productive life, let alone normal metabolism. These changes to the liver and kidney can be detected by changes in blood tests that include liver and kidney function measurements.

The cardiovascular system is affected by an increased heart size and elevated blood pressure and heart rate. Blood lipid levels (cholesterol, LDL, etc.) are also increased. A doctor seeing a patient with unexplained hypertension, rapid pulse, and elevated cholesterol will suspect alcohol abuse. Irregular rhythms are common with intoxication and become chronic with heavy drinkers, even after the last drink. These include heart blocks, extra beats, conduction defects, and tachycardia (rapid heart rate). Some of this impairment can last up to 24 hours or more.

Alcohol is an irritant to the esophagus and stomach, sometimes resulting in gastritis and possible bleeding ulcers. A medical emergency exists when one begins vomiting blood because that blood might also come from a ruptured esophagus as a result of vomiting from drinking too much; it's a vicious circle. Gastritis is common with drinkers, often requiring large doses of antacids to overcome.

The lungs are also at risk for problems, with irritation to lung tissue and bronchioles. Because alcohol is a cough suppressant, the body will not cough up material that needs to get out of the lungs, and the lungs can't be cleared. This becomes a more serious problem with smokers who already have bronchitis. (The combination of smoking and drinking, a common social combination, results in a synergistic increase in risks for medical problems.)

Ethanol also induces hypoglycemia after the last drink and for several hours later because it is of no nutritional value, only empty calories. This effect, in combination with other nutritional shortcuts usually taken the morning after consumption of alcohol, makes the pilot even more susceptible to the symptoms of hypoglycemia. Consider the scenario, even with just a few beers the night before, of an early departure, no source for a decent breakfast except the candy bar vending machine, and a busy flight during the morning. This is a perfect setup for problems.

Histotoxic hypoxia, where the cells are unable to use oxygen, even if available at the cellular level, is often caused by alcohol. Within the cell, alcohol is a toxin, interfering with metabolism, especially with oxygen. This is the explanation for the fact that the body acts as if it were at a higher altitude with alcohol in the system. Some have said that two drinks add between 4,000 and 8,000 feet to the altitude in which the body is trying to perform, the physiological altitude.

The vestibular system is affected by alcohol, even for many hours after the blood level is zero. The sense of balance and orientation is impaired. Dizziness is noted, especially during intoxication, and makes one feel as if the room is going around. When trying to lie down, the dizziness gets worse because the semicircular canals are truly "tumbled." These later effects include a marked reduction in the tolerance to G forces (even mild) and linear forces as experienced in a turn.

Effects on performance

The most significant effect is impairment of the fine skills and mental processes unique to flying. Pilots are taught to fly straight and level, but the greatest emphasis in training is how to react in emergency situations. This requires the pilot to react both instinctively and at the same time respond to rapidly changing situations and an increase of multiple tasks. Rapid decisions and actions are necessary in a potentially deteriorating flight scenario. Alcohol and past-alcohol impairment (hangover) interfere with every mental and physical activity associated with the skills necessary for a safe recovery.

Some of the impairments include increased reaction time, such as taking longer to make an evasive maneuver or correct an unsafe trim setting. Tracking is impaired, which equates into an inefficient and unproductive scanning of instruments. Judgment of distances and heights is compromised, leading to inadequate situation awareness on approaches.

Visual acuity is diminished, making charts difficult to interpret and increasing the likelihood of using improper radio settings; closing traffic will be difficult to see. Coordination skills are seriously eroded, preventing the pilot from doing several activities at once, as in making a simple turn while adjusting the throttle setting.

The most subjective problem is an inability to think clearly and to make sound decisions. Like hypoxia, the pilot is unaware that his performance is impaired. Because pilots are always monitoring themselves and seeking performance perfection, they are prime targets for unsafe situations when that self-monitoring process is diminished. Alcohol interferes with that process and allows the pilot to do dumb things without even recognizing the seriousness of his actions or inactions.

When under the influence of alcohol, the usually conscientious and careful pilot loses her normal attitude of caution when she sees a change in her performance, especially in evasive maneuvers. The pilot under the influence will take chances not ordinarily even considered when sober. Poor judgment is a by-product of this progressively dangerous situation and the pilot becomes a risk taker.

All these impaired performance-and-judgment skills can be occurring at the same time and at blood alcohol levels less than 0.04 percent, which is the legal FAA allowance. The pilot might think she is doing better when in fact she isn't. Even experienced pilots are impaired when faced with multiple tasks. And again, like being hypoxic, the pilot is unable and unwilling to determine or recognize how impaired she is and thus cannot and will not take corrective actions or let someone else fly.

Many more unsafe situations occur even when blood alcohol content returns to zero, and for many hours afterwards. The FAA is providing studies regarding how there is impairment with a blood alcohol content of 0.02 and lower. One review of studies concludes by saying: "The idea that there is a safe level of alcohol consumption below which there are no adverse effects remains simplistic when based on the evidence we have accumulated to date. What is safe for one individual might not be for another—safety continues to be a relative matter in any discussion of alcohol use." Couple this with the additional effects of a hangover and you are looking at an accident waiting to happen. In other words, being legal might not ensure that you are fit to fly.

Hangover

Scientists call this *postalcohol impairment*. The results, however, are the same. Those who have been drunk, and some who have just "sampled" alcohol, do not need to be reminded of the symptoms of a hangover. These symptoms often are so bad that many have sworn off alcohol during their fight with a hangover, only to return to that same situation at a later date and with the same results.

Hangover effects are distinguished as separate from the direct effects of alcohol in the body as previously described. These symptoms include tremors, thirst, nausea and vomiting, heartburn, sweating, dizziness, and of course, headache. More subjective effects are increased irritability, anxiety, and mental depression.

With heavier drinking, these symptoms are more severe, and mental function impairment is less subtle. Performance is affected in crucial phases of takeoffs, landings, and emergency situations. The pilot's ability to perform nonroutine tasks is markedly reduced. These impairments, including hypoglycemia, are noted even 14 to 24 hours after the blood level is zero.

Fatigue is a major problem even after blood-alcohol content (BAC) is zero. Alcohol is known to reduce rapid eye movement (REM) sleep, which means that the drinker is not getting a restful sleep. He might be passed out or appear to be sleeping heavily, complete with snoring, but it is from depression of brain activity, not sleep. It is believed that alcohol increases adrenaline flow, making one more active. The use of this increased energy is also fatiguing as an aftereffect.

Some of the more classic symptoms of hangover are a result of *congeners* in the alcohol. These are basically impurities and by-products of distillation or fermentation and give each beverage their unique flavor and aroma. This is the difference between vodkas, rums, and whiskeys. Some wines and beers have different kinds of alcohol other than ethanol. Alcohol other than ethanol generates greater problems. Other ingredients are also used as preservatives that can cause hangover symptoms. Red wines especially have a reputation for causing hangovers because of impure alcohols. Less-expensive beverages are often more liable to cause problems because of their less-than-pure ingredients.

Some foods make a hangover more noticeable to some people, especially the headache. These include any food with monosodium glutamate (MSG) (as in many oriental foods), food with preservatives, such as sodium nitrite (processed meats, such as hot dogs, salami, ham, and bacon), fermented cheeses, and chocolates.

Prevention of a hangover is related to decreasing the amounts and rate of ingestion of alcohol consumption and reducing the absorption. High-protein foods such as dairy products and meats stay in the stomach longer and slow down the absorption of alcohol. Gulping drinks increases not only the total amount consumed in a given period of time, but also forces the alcohol's entry to the bloodstream and hence toward the brain more quickly. Carbonated beverages, including beer, tend to make the stomach absorb alcohol quicker and more efficiently. This means that using juices and water instead of soda water are less likely to lead to hangover.

It should be obvious that 8 or 12 hours "between bottle and throttle" is not nearly adequate to prevent impairment from alcohol, especially from the effects of hangover. This is dependent upon the total amounts, the time over which alcohol is ingested, and the tolerance built up by the drinker. The pilot can still be under the influence even if his or her blood alcohol is zero.

The only cure for a hangover, or postalcohol impairment, is the passage of time, sometimes up to 36 hours. Coffee and oxygen have no effect on reducing the symptoms or improving performance. Caffeine might make you feel better, but it still disrupts skills. Oxygen is not part of the metabolism of alcohol; however, the act of "sucking on oxygen" might psychologically make you feel you are doing something to overcome the effects. There is no end to the anecdotes claiming to overcome postalcohol impairment. None of the traditional cures make the pilot safer. A good rule of thumb: If you're going to sip, avoid the trip.

Tolerance to alcohol

The body can develop, over time, a tolerance to very high levels of alcohol. Someone who brags that he or she can "drink you under the table" probably can but only because that person has built up this tolerance. It is a significant indicator of someone who has a serious problem with alcohol.

Studies also indicate that one cannot tell when impairment is present or when blood alcohol reaches zero. In fact, as noted above, some problems are unique to the status where alcohol is no longer present in the blood. Many pilots actually think they are safe to fly just four hours after their last drink, despite facts to the contrary. Often this is their perception based upon their increased tolerance.

There is also a conditioned phenomenon in experienced drinkers. A *placebo* (a drink without alcohol) ingested by this person will result in many of the same indicators of intoxication for that person: talkativeness, aggressiveness, and the like. The nonexperienced drinker will not notice any changes when drinking beverages without alcohol.

Withdrawal

Withdrawal from excessive prolonged use also generates unique and significant symptoms. Lesser symptoms are noted even with smaller amounts. Within 6 to 8 hours, there is a noticeable anxiety not usually present, increased pulse, and fine muscular tremor. Twelve hours after stopping, the drinker becomes agitated, with a markedly increased heart rate, blood pressure, sweating, and tremors. He or she cannot lie or sit still and will change positions frequently. There are lapses of attention and memory. More severe symptoms occur 48–72 hours later. In addition to all the previously mentioned problems, now there are irrational behavior and fears, a high susceptibility to noise, and often hallucinations and confusion. Severe withdrawal symptoms include *delirium tremens* (DTs)—the drinker is out of metabolic control and only through ingesting more alcohol can the DTs or even lesser symptoms be controlled, unless with medical help.

A good example of unexpected withdrawal symptoms occurred during the initial phases of Desert Shield, the military build up in the Persian Gulf in late 1990 that was a precursor to Desert Storm. A majority of medical problems seen by medics that often required an airlift to a hospital were related to withdrawal from alcohol. The reason was that alcohol is not allowed in the Arab states. This was withdrawal in its simplest form and in large numbers at the same time.

Recognizing a problem drinker

It really isn't that difficult to recognize a problem drinker; however, few want to be a snitch, nor do they feel competent to pass judgment on someone that they feel has a problem. After all, what is the definition of a problem drinker or one who is truly dependent on alcohol? Is it someone who drinks more than you, or are there distinct criteria? Anyone with a blood-alcohol level of 0.10 percent or more is considered to be legally intoxicated in most states, but how often must a person be intoxicated before being considered dependent?

All of these indications apply to a certain extent. You must keep a high index of suspicion when observing someone you know is developing an increasing drinking habit. And don't count on the flight surgeon or personal physician to intervene. It is estimated that less than 10 percent of referrals of alcoholics to treatment come from the doctor. Usually they come from concerned family, friends, business and flying colleagues, or managers.

Specific guidelines used by professionals who deal with alcohol dependency can be used in determining if there is a potential problem. Here are a few danger signals. In the right combination, these would also be diagnostic of alcoholism and cause for FAA medical disqualification. These criteria are used by the FAA to determine if disqualification should be considered when alcohol is known to be a part of these criteria:

- Any one of the following factors:
  ~A clinical diagnosis of "alcoholism."
  ~Evidence of withdrawal from alcohol.
- Any two of these major criteria which indicates that alcoholism is definite and a diagnosis of a history obligatory:
  ~A medical history of liver disease.
  ~A medical history of central or peripheral nervous system involvement.
  ~Psychological dependence on alcohol demonstrated by either admission or by history.
  ~More than one conviction for an offense involving alcohol, including non highway arrests.
  ~Known to various health and social agencies in the community through contacts related to alcohol problems.
  ~Treatment at an inpatient facility devoted to alcohol rehabilitation.
- Minor criteria from which any three or four fulfilled or one major and two minor criteria:
  ~Accidents involving injury or property damage that may be presumed to be related to alcohol.
  ~Known to have a history of troubled relationships in employment.
  ~Known to have a history of troubled relationships within family.
  ~Known to have troubled relationships with banks or creditors.
  ~Intoxicated more than four times a year.
  ~At any time a blood alcohol level of greater than 0.2 percent.
  ~A history of blackouts.
  ~Odor of alcohol on the breath at the time of medical certification.
  ~Evidence of social and behavioral disruption.

These criteria will be used as a guide to assist in determining the degree of use of and dependence upon alcohol and whether or not there is a problem that could interfere with meeting FAA standards. Many of these criteria are a sign of poor judgment in the use of alcohol.

The two biggest indications for suspicion of alcohol dependency are increased tolerance and denial. The denial is the hardest to define because the alcoholic is a master of denying that there is a problem, even when confronted by other people.

A variety of definitions and terms are used to explain the degree of an individual's alcohol tolerance. These terms are partially subjective because of the possible indication of dependency, the real problem. Often, to an observer, a heavy drinker is someone who drinks more than the observer. "Social drinking" is the occasional use of alcohol during specific situations. Consuming wine for a special meal, or a cocktail after a strenuous day or while at a party is not considered heavy unless the frequency of use increases. The abuse of alcohol is related to frequent binges, where intoxication is the intent. Some deliberately drink for the purpose of getting drunk with no other definable reason.

Drinkers will often rationalize their reasons for using alcohol. Situational conditions arise when life isn't going well: divorce, business going bad, tax time, school problems, and the like. Some people find solace, relaxation, and escape from the realities of their stress through alcohol; however, any misuse or abuse is unacceptable in aviation.

Driving records become an indication of a potential problem. The number of alcohol-related arrests is evidence of increasing poor judgment and a sign of growing dependence. Alcohol becomes so important to that person's lifestyle that the risk of getting caught or causing problems at home or work is minimized. It is for this reason that the FAA will check the driving records of pilots, especially those suspected of having problems with alcohol.

Dependency

Addiction, dependence, and alcoholism mean somewhat the same thing for our purposes. The drinker cannot control his intake or is unable to say "enough." He is unable to stop after the first drink. Frequency of use is often daily, with careful avoidance prior to work or flying. The amounts consumed are often excessive; a quart of vodka or a case of beer per day is not uncommon. The deception to the drinker and those around him is the perception that he doesn't appear drunk or impaired, but he is. He has developed a tolerance to large amounts of alcohol and the sources become varied. Addiction relates to the physiological need, the body rebels if it doesn't get the daily dose, so more alcohol is used to keep things under control.

Dependency is the main concern when discussing alcohol use, especially in aviation. The combination of being impaired and not recognizing that fact and the denial that there is a problem leads to the pilot often flying under the influence, even after the BAC is zero; therefore, identifying the problem drinker by keeping a high index of suspicion that dependency is present becomes the responsibility of everyone. You can't count on the pilot with the problem to come forth and admit his or her condition; neither can you wait for the flight surgeon to intervene. Others around the problem drinker are reluctant to be the one to turn in the pilot and, in effect, destroy a career. This fear need not exist if proper action is taken on everyone's part. A suspicion of abuse or dependency can be confirmed by watching for telltale signs. Here are some clues to recognizing someone in trouble; the clues are in addition to the clear observations of increased drinking and noticeable changes in performance:

- Increased tolerance to alcohol (probably the most important) resulting in increased drinking and drinking more rapidly than others.
- Gulps the first drink or two.
- A personality change over a period of time, usually several months.
- Increased absenteeism and use of sick time.
- Increase in complaints about family life or management, often trivial and of a suspicious nature.
- Frequently drinks alone and avoids nondrinking friends.
- Increased irritability and more argumentative.
- Denial of increased alcohol intake and of the problems just noted.

Treatment of dependency

The first step is recognition, by both pilot and those around him or her. This was just discussed, but the challenge remains; few will take the step of recognizing that there is a problem. This combination of denial is often called *enabling*, which means allowing the problem to continue by "enabling" the drinker to continue through reasoning that someone else will intervene and denying the need to upset the status quo.

Once a problem is recognized and action is decided upon, the next step is intervention. Although it is hoped that the problem drinker will seek help, this is usually not the real-world scenario. An intervention is where concerned individuals participate in a structured confrontation, an action that essentially corners the drinker into seeing and hearing what is happening to him plus what needs to be done to deal with the situation. There is only one way out, only one solution for the drinker, entering treatment (which must be prearranged). Intervention is often done by professionals along with family and colleagues. More information regarding this step is best obtained from treatment centers, AA representatives, or company-employee assistance programs.

Treatment for a dependent pilot must be at an FAA-accepted treatment center and usually as an inpatient admitted into the facility for the full term of treatment and not allowed to "commute" home. The techniques of treatment vary from center to center but generally follow the 12-step program developed by Alcoholics Anonymous (AA). Comprehensive medical and psychological testing is available that may be coupled with professional counseling and group encounter sessions with a facilitator. The details can be obtained from the same resources just mentioned. It is recommended that you become familiar with the scenario to better appreciate how you can help someone who needs help.

Returning the pilot to flying is usually successful if the treatment is completed satisfactorily and there are no remaining underlying psychological problems. The FAA is eager to get the pilot flying again (especially commercial pilots) because the desire to fly and return to work is so strong in pilots. There is less than a 10-percent relapse rate after treatment. This is probably the best success rate for any professional group; however, the

FAA expects continued total abstinence, indefinite involvement in AA, a structured monitoring program, and annual reviews by treatment professionals. This is all spelled out in the FAA's expectations when returning the pilot to flight.

Cooperation from company, union (if present), colleagues/peers, support groups, management, and family is essential to a rewarding and successful program. If the program is followed and documentation attesting to this fact is compiled, the FAA can return a pilot to flying as soon as three months after treatment.

Because this program is built around cooperative monitoring that is already available within a larger flight department, recertification becomes more difficult for noncommercial pilots. It can be accomplished, but it takes more time of proven abstinence, sometimes up to 2 years.

Alcohol, especially when associated with dependency, is a major problem within all aviation. Often the only way the pilot with a problem can be identified, treated, and returned to flight is through the combined efforts of those around him or her. The FAA will participate in this program if it sees the proper efforts and results; therefore, one is not ruining a pilot's career or flying status if the above mentioned procedures are followed. There is no reason, in reality, to avoid intervening. If you suspect he or she has a problem, seek advice from your company's employee-assistance department or union, or call the FAA to obtain advice. (No names need be mentioned, you're seeking recommendations on whom to talk to for directions toward professional help.)

Illicit drugs

Illicit means *illegal.* Pure and simple. A pilot using these drugs is in violation of the law and federal regulations. It's not simply a matter of when drugs are used, how much is used, or the level of pilot impairment. If the pilot is caught, he or she has no excuse. He or she broke the law!

That aside, there are some facts that the pilot needs to know, especially to recognize why these drugs are dangerous before and during flight. A prime problem with these drugs is their highly addictive properties. Furthermore, a person can become more dependent on these chemicals than with alcohol, which means increased difficulty in achieving successful treatment.

Kinds of drugs

The *mood-altering* drugs described in this section are often associated with being a stimulant, depressant, or hallucinogenic; hence, drug tests typically target amphetamines, marijuana, cocaine, PCP, and opiates.

Marijuana is probably the most common, although cocaine is equally common in some cultures and parts of the United States. The active ingredient is THC (Delta 9 tetra hydrocannibinal), which is tested for in urine. THC causes temporary euphoria and relaxation; however, more than alcohol, it distorts perception, weakens critical judgment, and interferes with the ability to concentrate. THC also increases heart rate and the incidence of cardiac arrhythmias. Short-term memory is impaired, IQ is thought to be diminished, and there is a decrease in reaction time and tracking (similar to alcohol). In severe cases, marked depression and some hallucinations occur.

Physiologically, THC is stored in fat tissue, which means it can be slowly released at unpredictable rates long after THC is first taken into the body. The same effects can then develop, with the greatest threat to the brain, and can occur up to 30 days after taking in THC. Because most THC is taken in by smoking, it should be noted that this smoke is far worse in producing cancer than regular smoking. THC might also cause genetic changes and damage and interfere with the body's immune system.

Cocaine is used because of its stimulant effects, somewhat similar to amphetamines. The stimulating effects take place 15–30 minutes after use. Also, there is euphoria, a profound sense of well being, alertness, pleasure, and increased self-confidence; however, these effects are often short lived, requiring another dose to keep the effect going. Additionally, there is an increase in heart rate and blood pressure. The adrenaline seems to flow freely as the user becomes addicted to cocaine. And cocaine is highly addictive once started. Intoxication with cocaine produces virtually all of the psychiatric symptoms recognized by therapists treating any psychiatric condition.

If this "high" cannot be sustained with more cocaine, there are three different stages of withdrawal. The crash occurs about one to four hours after the last hit of a binge. The symptoms include a craving for more cocaine, depression, agitation, and increased anxiety. There is a strong desire to sleep and eat. Actual withdrawal occurs about one to five days after the crash with increasing inactivity, listlessness, boredom, and lack of the pleasurable feeling most enjoyed while being used. This whole scenario further increases the craving for more cocaine. The final stage occurs months later; keeping in mind that no cocaine has been used, there continues to be a craving but to a lesser degree, and especially if there are no opportunities to reuse cocaine. Cocaine derivatives are often used by doctors for legitimate reasons.

The other drugs are fairly well recognized by professionals and are not used to any great extent. PCP is rarely used, especially in any professional community. Its side effect of hallucinations is the greatest threat to safety. Amphetamines are still used in medical situations as a stimulant for depression, and sometimes in diet-control programs. Opiates include morphine and codeine, the most common. In clinical settings, they are used for pain relief and sometimes for cough suppressants. Their side effects include euphoria and drowsiness. Codeine is less addictive than morphine.

Drug testing

Using the current testing programs, there are no false positives. The common procedure is two-phased tests for amphetamines, marijuana, cocaine, PCP, and opiates in the urine. The first part is a screening test, and false positives can occur in the first part; some other chemical might show up as being positive for one of the five drugs. This positive is not reported, but it automatically leads to other tests; confirmatory gas chromatography and mass spectrography analysis generate a unique curve for every known chemical.

Furthermore, it is necessary that the amount of the drug (or its metabolite) found must exceed a predetermined level. In other words, the test is not only qualitative but quantitative. For example, if 90 units of THC were suspected in the screening test and the level to be considered positive is 100, then that part of the test is reported as negative. If above 100, then the confirmatory test is done. Again, there is a level above which that drug must be present before being reported as positive. That level is determined by those mandating the testing, usually Department of Transportation (DOT).

If the final result is positive for that drug, the report must be reviewed by a medical review officer (MRO), who is a doctor knowledgeable in drug testing, to determine if there is a credible reason for that drug to be present in those quantities. This could include something prescribed by your doctor or some other situation that would be a legitimate reason. There are few such reasons and it has been found in some cases that once the person is confronted with the positive result, he often resigns rather than be further questioned.

The rules are in a continuous state of change because it is recognized that illicit drug use in aviation is very rare; however, there are still exceptions. If there is to be a drug-free aviation community, then everyone must be suspect. Be cautious of the reasoning behind the variety of groups that oppose testing or the amount of testing. Percentages are usually used and they are very low; however, there are hundreds of thousands of tests being done and even a minute percentage means a fairly high actual number of positives.

Because of the high level of addiction with these drugs, it is much more difficult and probably not realistic to be favorably considered for recertification by the FAA, even after treatment.

As of January 1995, testing is done for alcohol use and the rule is different from FAR 91.17. It pertains to Part 121 and Part 135 and includes all flight crewmembers. The "Alcohol Misuse Prevention Program" basically states that alcohol in any form, not just that in beverages, cannot be used while working in a safety-related job nor within 8 hours prior to working. That could include "no-alcohol beer" or some OTC medications that used alcohol as an inactive ingredient. Violation of this rule means immediate and permanent grounding for the first offense of consuming alcohol *on the job*. The second offense of drinking 8 hours *prior* to work will result in the same action. No matter how the rule is ultimately intended, it is unwise to consume any product with alcohol in it sooner than 8 hours before flight. Because this whole issue of alcohol and drug testing is evolving, it is imperative that you keep in touch with changes in interpretation and proposed new rules.

10
Environmental stresses

It was a balmy 72° February day when the crew left Kansas for a three-day trip to Alaska. The reality of where they had landed hit them square in the face when they opened the hatch upon landing and experienced –50°F—which was actual temperature—no wind-chill factor added. After unloading his gear, the captain had to sit in the cockpit while the aircraft was towed (power off). It was the worst 15 minutes he had ever spent in a cockpit. His body began to shiver almost uncontrollably. He couldn't move switches because he couldn't grab them. All of his mental and physical actions slowed down as he tried to move as little as possible. After returning to the hotel the pilot remembered how exhausting those 15 minutes were; he could hardly get his winter clothing off. Then, on the return trip, the crew was in such a scramble to get through the preflight and get airborne that after takeoff the captain noticed the fuel switches on the panel had been improperly positioned, too—all because they were rushed.

FEW PLACES ON EARTH have a constant weather system. Temperatures, winds, dew points (or humidity), and rain or snow are continually changing. Some parts of the world have weather extremes that last for months, allowing inhabitants to acclimate to the season. People living in the far reaches of the Northern Hemisphere adapt to cold winters, whereas those living in the hot and humid areas during winter would find the northern conditions not only unbearable but dangerous as well.

Our bodies and minds have the metabolic resources to respond to these variations and extremes and to allow the person to continue to perform. On the other hand, the body also reacts in adverse ways, which interferes with expected performance, again, both physically and mentally. This interference equates into impairment of the pilot's skills, making him or her less than safe and, in effect, incapacitated.

Aircraft take us great distances, whether across a continent or across oceans around the world, all in a matter of a few hours or one or two days; therefore, the pilot is exposed to variations of weather condition extremes without the advantages of adapting or becoming acclimated. Furthermore, the pilot is often unprepared to cope with these extremes, flying without adequate protective clothing and being in less than ideal physical condition. At any time of the year, it is not uncommon for a flightcrew to depart from a cold, dry, and windy airport and arrive several hours later at a hot, humid, and calm destination. And they probably will stay there only a few days before heading back to the originating point or going to yet another location of variable extremes.

This chapter is meant to raise the level of awareness of how the body copes with weather and environmental conditions and to alert the pilot about ways to prepare for these extremes by anticipating the needs of the body. Because the pilot's flying schedule will often change dramatically during the years, she must recognize and respect that these conditions do affect performance and are detrimental to safety.

BASIC PHYSICS OF HEAT

The first thing to realize is that the definition of cold is the absence of heat, which is an important law of thermodynamics. Absolute zero temperature means there is no heat and no energy. As the energy of heat is added, the temperature rises; therefore, in any discussion of temperature-related situations, it's important to understand that we are talking mainly about the transfer of heat (Fig. 10-1).

What happens in our bodies is essentially the same physics as occurs in our homes. Warming up the building means adding heat to the circulating air. Cooling means removing heat. How this heat is exchanged depends on other physical properties of heat transfer, with special differences for our body.

Conduction

Heat that is transferred by direct contact with another object is called *conduction.* This can happen within the same object, like an airplane, by going from one part to another (heat from the engine warming the airframe). The more common situation is when a part of the body, such as hands or feet, comes in contact with another object that is colder or warmer, like sitting in a cold seat or leaning against the frame of an airplane sitting on the ramp in the middle of summer. Burns and frostbite can occur to a localized part of the body when touching the hot exhaust pipe or handling luggage outside without gloves.

Convection

Convection is the most common heat transfer method. Heat is moved (carried) within gases or liquids as it flows from one object to another. This happens in a building or an airplane cabin. Air is heated by a flame or hot object (engine) and the airflow carries this heat to another part of the building or aircraft. We feel warm as the heat is transferred from the air to our skin as it flows past.

Depending on the ambient temperature (the temperature of the air around us), the flow of air as it moves across the body also becomes a major factor. This flow can be from a fan or from the body itself as it moves. If there

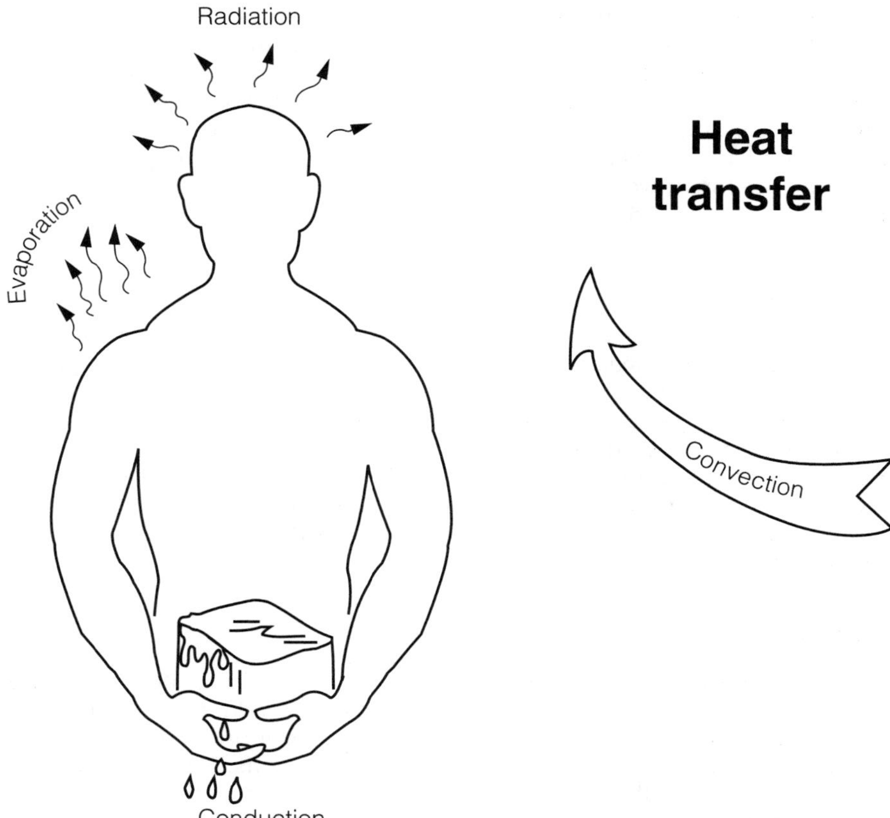

10-1 The first law of thermodynamics states that heat is transferred from hotter to colder objects.

is no difference between the temperature of the moving air and the body, the body will still feel colder because of wind chill, which is subsequently described. This can occur in a room or airplane cabin with a temperature of 72°F. If there is a slight movement of air at the same temperature, the body will still feel cooler.

The usual definition of wind chill refers to extremes of below freezing. How one perceives any temperature comfort or discomfort is dependent on the same process but to a lesser degree; it might not be significant but it can be distracting. Within our body, warm blood carrying heat is then transferred by convection from one part of the body to another.

Radiation

When heat is transferred from one object to another of different temperature, and without being carried by a gas or liquid (and a person is not touching anything), this is *radiation*. The transfer is in the form of electromagnetic waves. The sun is the most obvious example. We often feel the warmth from the sun on our skin even though the ambient temperature is low. The amount of heat transferred depends on the size of the radiating object, the relative difference of temperature, and, like light, (exponentially) the distance from that object. Radiation also initiates the "greenhouse effect" in the cockpit of an airplane sitting on the ramp in the sun. Here, radiated heat warms the cockpit air and the interior structures, but the air is trapped, and the temperature rises.

Evaporation

As a change of state, evaporation requires energy or heat to change a liquid to a gas. Mechanical refrigeration with a compressor uses this physical property when heat is taken away from the container (refrigerator) as the refrigerant expands. The same holds true with the body. Sweat, for example, evaporates off the skin and in doing so must take heat away from the skin. It requires about 580 calories of heat energy to change 1.0 gram of water into vapor at body temperature; therefore, the same amount of heat is removed from the body when the vaporized sweat is taken away by the air surrounding the body.

Comfort factors

How we feel (how comfortable we are) also determines how effective we will be when performing our job. If we feel too hot or too cold, our performance will be substandard. This comfort factor is dependent on all these forms of heat transfer and to varying degrees and combinations. For example, in a small cockpit or small airplane, radiant solar heat warms the air and body without heat from any other source. The "greenhouse effect" can raise the temperature 10–15°C higher without any other source. Convected heat (or lack of heat) from air vents and heaters supplements the radiant energy. Sitting in a hot or cold seat will conduct heat to or from the body. And if one is sweating from the preflight activities, evaporation will add to the discomfort (or comfort while cooling).

Personal comfort levels are subjective, and each pilot will respond differently and unpredictably. In temperature extremes, it is necessary for the pilot to identify how he functions and how his crewmates are functioning under the same conditions. When talking about temperatures, everyone has a different response with one tolerating heat and humidity and another suffering great discomfort.

THE BODY'S TEMPERATURE CONTROL SYSTEM

In review, cold is the absence of heat, and heat is the only thing that the body can generate and distribute in controlling the temperature of the body. But there needs to be a means of distributing heat: our circulatory system. As described in chapter 2 on anatomy of the body, the circulatory system is similar to a typical airplane hydraulic system with a pump, pipes, and peripheral "organs" (flaps, gear, ailerons. etc.). The big difference is the unique ability of the "pipes," or blood vessels (primarily the arteries), to dilate or constrict, thus restricting or increasing blood flow, in effect, shunting (diverting) blood where it is most needed from where it is least needed.

When discussing temperature control of the human body, it is easier to think of a core temperature. The "core" defines critical organs such as the brain, heart, liver, lungs, and some of the senses of vision and hearing. This core of organs must be kept at a constant temperature (average of 98.6°F), which is often measured orally. The requirement for maintaining constant internal temperature distinguishes warm-blooded animals from cold-blooded animals whose core temperatures stay close to ambient temperatures. The failure of warm-blooded animals to keep the core temperature within this range (96.6°F to 99.5°F) results in depressed performance and ultimately total incapacitation. These effects will be described later.

Therefore, the goal of the body's temperature-control system is to maintain a constant body-core temperature, often at the expense of other less essential parts and organs of the body. This also requires a properly functioning human body in good physical condition. Temperature control incorporates all the physics of heat transfer and flow, heat generation, "air (blood) conditioning," and heat/cold regulation.

Generation of heat

Heat is generated by virtually all activities of the body and is the ultimate form of body energy. Other energy (about 10–15 percent) goes into the actual functions of the body, such as metabolism, muscle use, and organ activity. For our purposes, the majority of heat generated by the body and used as heat comes from muscle activity. In fact, it is estimated that so much heat is created that if it weren't dispersed, the core temperature would rise to 190°F! The body, therefore, has a vast resource of available heat.

During strenuous exercise, the heat generated can increase by a factor of 20. It is in these situations that the body's need to control temperature is crucial. When heat is needed and not externally available, the body begins to shiver, which is the body's way of involuntarily increasing muscle activity and generating more heat. All muscle activity requires internal calories from the energy of ingested food. It follows, therefore, that during periods when additional heat is needed, more food must be eaten to provide adequate fuel.

Vascular system's role in temperature control

The skin (the body's largest organ) is considered the most important heat-exchange component because of its large surface area. The more blood that reaches the skin, the more heat that can be dissipated, provided the heat is taken from the skin either by convection or evaporation. The amount of blood reaching the skin is controlled by the ability of the arteries to shunt blood away from or toward the skin. Here is where vasodilatation and vasoconstriction of arteries play an important role.

For example, in a hot environment, the body must get rid of excessive heat in the core to prevent the core temperature from rising. It does so by opening the vessels to the skin and within the skin's tissues, getting more heated blood to the surface area of the body and allowing the heat to be dissipated. The skin becomes red and flushed as a result of the superficial tiny blood vessels expanding. The reverse is true in a cold environment. To preserve the core temperature, blood is shunted away from the cool skin and concentrated in the core, resulting in pale and blanched skin color. This leads to poor warming of extremities and other outer-body parts with reduced blood supply (such as the nose and ears).

It should be apparent that efficient blood flow is dependent on an efficient circulatory system and a healthy heart; therefore, those in poor physical condition are less effective in tolerating any temperature extremes. Also, self-induced stresses (alcohol, self medication, dehydration, poor nutrition, etc.) will seriously compromise effective coping with temperature extremes and reduce effective acclimatization.

Evaporation's role in temperature control

The body, therefore, is capable of generating heat and getting excess heat to the surface of the body to be dispelled away; however, heat removed by radiation and by convection of moving air is not efficient enough. Heat also has to be taken away from the skin by evaporation of sweat on the skin. Sweat is produced by sweat glands located within the skin. Because evaporation requires heat, this process is effective in using the heat in the skin. The more sweat that can be made and the faster the evaporation, the more heat that can be blown away in the form of the vaporized water from sweat.

Just as additional fuel is needed for generation of heat through increased muscle activity, the body obviously needs more water under conditions of increased sweat production. This is an internal need, which means fluids must be available within the body to be effective. Because sweat includes some electrolytes (mainly sodium and potassium chloride), one must be aware of their depletion during high losses of sweat in unacclimated individuals.

Environmental conditions interfere with adequate evaporation of sweat that the body manages to produce. It serves no purpose if sweat doesn't evaporate or the evaporated sweat isn't effectively blown away to make room for more sweat to be evaporated. These situations occur in environments of high humidity where evaporation cannot take place. The partial pressure of water vapor in the air and on the skin are equal, and no transformation to vapor can occur. Also, enough skin must be exposed for the ambient air to blow away the evaporating sweat. Yet there must be protection of the skin to prevent overheating from radiation, conduction, and convection present in the working environment.

The body's thermostat

The body has a thermoregulatory center (like a house's thermostat) in the hypothalamus of the brain that coordinates the activities of the body for generating heat, increasing sweat production, and managing the shunting of blood flow to or from the core. This regulatory center reacts directly to local ambient temperatures detected by receptors on the skin and to the effects of warmer blood reaching the thermostat.

The thermoregulatory center takes input from the body's receptors and then generates heat by either increasing muscle activity, which warms the blood, or increasing sweat production, which cools the blood. The center then regulates the flow of blood between the core and the skin to either conserve heat or to dissipate heat from the core. Most of this is done *insensibly*—that is, without conscious effort on the part of the pilot. Some conscious effort must be taken, however, by the pilot in extreme conditions to assist the body's own actions. These conditions will be discussed separately. It must be kept in mind that there are varying intensities of heat and cold, plus the pilot might be subjected to both extremes during the same trip; therefore, one must consider all variations of temperature control to keep ahead of any incapacitating results.

COPING WITH HEAT EXTREMES

Body's physiological response to heat

Recall that when the body's thermoregulatory center senses a rising core temperature, certain functions begin to take place. Although these responses vary from individual to individual and other factors can also interfere, the overall response is the same. Blood supply is shunted to the core where heat is accumulating. The heated blood is quickly taken to the skin where the dilated blood vessels warm the skin relative to the amount of heat being dissipated. More heat means a higher skin temperature. Concurrently, sweat volume is increased from the sweat glands for evaporation.

Sweat production increases abruptly at an ambient temperature of about 83°F, although other forms of heat loss are in effect (radiation dissipates about 60 percent and convection and conduction about 15 percent). As the temperature reaches about 95°F, virtually all cooling comes from sweat evaporation. As much as 4–16 quarts of water per day can be lost through sweating and respiration in an active person in a hot environment. This fluid must be replaced, ideally *before* the body requires and uses it.

For example, from a comfort point of view, a person given enough water can function all day at a temperature of 115°F with a relative humidity of 10 percent. If, however, the relative humidity is raised to 80 percent, that same person might be incapacitated within 30 minutes. Dew point is sometimes a better indicator of anticipated discomfort. If the dew point is above 60°F with outside air temperatures at any level above that, the person will be uncomfortable, which might cause physiological impairment that will be discussed in another portion of this chapter.

Some of the conditions that make a person less tolerant to heat extremes include increasing age, excessive alcohol intake (alcohol dilates blood vessels and prevents efficient shunting), lack of sleep, obesity, and previous history of heat stroke (the more serious kind of heat stress). Lack of being adequately acclimated makes a person much more susceptible.

General symptoms of heat stress

As body temperatures increase (hyperthermia), and the body struggles to maintain a constant core temperature, there are some symptoms that have a direct effect on a pilot's performance. Generally, effective human performance is impaired above 100°F. These symptoms include less reliable short-term memory, increase in error rate, erosion of motor skills, and a general decrease in performance skill. Heat stress increases the pilot's irritability and diminishes insight into judgment skills, creating a tendency to overreact and make more mistakes. Heat also increases the susceptibility to motion sickness, hypoxia, and any degree of G forces. In addition, fatigue becomes apparent after several hours of coping with heat.

Specific heat stresses

In addition to the previously mentioned generalized symptoms, the following are physiological stages of heat stress that need to be recognized. As each stage evolves without recognition and intervening treatment, the pilot becomes increasingly impaired to the point of becoming completely incapacitated.

Heat cramps

The heat cramps condition is also called *minor heat stress* due to a body temperature from 99.5°F–100.5°F. Painful muscle cramps of the extremities, abdomen, and back are common in heat cramps. Some feel this is a lack of adequate electrolytes, but it is related more to fluid depletion in early stages. Identification of this stage is important because it is the first indication of the body's struggle to maintain temperature control. The skin feels cool and moist. Treatment consists of resting, getting out of the heat environment and into shade, and forcing as many fluids as can be tolerated. "Athletic" beverages with electrolytes might be helpful, but *water replacement* is essential. If the heat conditions are not too severe, returning to work and flight is okay, providing adequate fluid replacement has been accomplished beforehand.

Heat exhaustion

The body temperature of heat exhaustion is 101°F–105°F. This becomes the first serious stage where the body begins to be unable to keep up with controlling core temperature. The thermoregulatory center is still working, and the body is responding with proper shunting of blood and enough sweat to evaporate. Outside conditions play a crucial role in determining the severity of this stage. If the conditions are severe with high heat, high dew points, no air movement, little or no shade, and high levels of activity, the body will be unable to keep up without rest and more fluids.

Symptoms include headache, confusion, uncoordination, loss of appetite, nausea, and cramps. Because the heat control system is still functioning, there is sweating, which means the skin is moist and still relatively cool. But the cardiovascular system is working harder to keep up with the increased heart rate necessary to overcome the dilated vessels during the transport of heat from blood flow.

The treatment is to help the body's heat-exchange functions and get out of the heat. Resting in the shade or in a cooler setting is mandatory. Fluids are especially important. Water is probably all that is needed at this stage. Do not be overly concerned about salt replacement. A person usually cannot take in too much water; however, too much water consumed too quickly can cause nausea and vomiting. Continual sipping of cool water is needed. If unable to take in fluids orally, medical attention is necessary since intravenous (IV) fluid replacement might be necessary.

This person has reached the limits of the body's attempt to control core temperatures. He or she is "behind the power curve"; therefore, returning to work or to flying is not safe for several hours. Even after the individual has apparently recovered, he or she will be weak and tired. More important, he or she is far more susceptible to a recurrence of heat exhaustion if stressed again too soon after recovery. This serves as a warning stage to the body.

Heat stroke

The last phase of inadequate temperature control is the most serious when body temperature rises above 105°F. This is *a medical emergency* when identified. The body is no longer capable of defending itself. The body's thermoregulatory center has broken down and is unable to manage the body-temperature control functions and keep core temperatures acceptable. This person is already significantly impaired and might become unconscious in a short time.

The symptoms should be apparent; he or she is sick. There is headache, confusion, dizziness, weakness, and often nearly a coma. The key to this diagnosis is the lack of any body-cooling activity. Sweat is no longer being made; therefore, no evaporation is taking place. The skin is dry and hot. The temperature might reach as high as 106°F, which is life threatening. The body is literally burning up.

The first step is to get this person to a cool place and off of hot surfaces. Treatment includes calling for medical assistance along with taking immediate measures to cool the body down, such as immersion in cool water or spraying or pouring any liquid over the body to help cooling by evaporation. This also means providing some sort of air movement. It probably will be difficult to get him or her to drink because there is confusion and early unconsciousness. If fluids can be taken orally, keep the fluids cold. In most cases, IV fluids are the only effective treatment, and delays initiating this fluid replacement further risks permanent problems.

In general, when in hot conditions, whether on the ground or in the air, be suspicious of extreme conditions, and look for potential symptoms in others as well as yourself. If there is any doubt, and the environmental conditions are suspect for causing heat stress, treat the situation as if it were heat exhaustion. Recovery must be noticeable within an hour; if not, seek medical help.

Prevention of heat stress

As with any other medical problem, prevention is the best cure. Heat stress could be considered self-induced because you can prevent it. The most important part of prevention is drinking more than enough fluids. That means drinking *water* before subjecting yourself to heat extremes. In other words, maintaining hydration is the single most important goal. In hot weather, an average person requires 2–3 quarts of fluid per day in addition to that supplied in a normal meal.

In more serious extremes of hot, windy conditions while being active, 3 or 4 gallons of water are necessary per day. Kidneys effectively regulate fluid balance if enough water is present in the body. By the way, fluids do more good in the stomach than being poured over the body to cool off. It might make you feel cooler, but the body's heat control system needs the water internally to replace evaporated sweat. Thirst is a poor indicator of need because a person is already dehydrated when the feeling of thirst becomes apparent.

Avoid those fluids known to contain diuretics—caffeine and alcohol—especially prior to being subjected to heat extremes. Tea, as in iced tea, is particularly bad because there are two diuretics present: caffeine *and* theophyline. Although subject to some debate, sugary fluids and carbonated beverages could decrease water absorption from the stomach, which further justifies the need only for water. If one is not acclimated to the environment, salt can be added to food but salt tablets are not generally required. (Most Americans consume more salt than is healthy anyway.)

Protective clothing is also important, but air circulation over the skin must be maintained. Darker clothing absorbs radiant heat and should be avoided. White, loose, lightweight clothing is preferable. Protection of the head is also important to protect the skull and brain from a heat load. Prevention of sunburn is also a consideration when choosing what to wear.

It might seem overly stated to consider heat stress a problem to pilots who are in flight. In a way, this is true if not considering the problem before taking off; however, consider the fact that the pilot also spends a lot of time out of the airplane, such as preflight, loading, and waiting, often on a hot ramp in humid weather. She might even be dehydrated before showing up for the flight. The pilot who now jumps into the hot cockpit is behind in fluids, and the body will continue to be heat stressed with the same symptoms as if she were still outside. After sitting in a hot cockpit (as a result of the greenhouse effect), it could take more than 20 minutes just to recover after the cabin finally cools.

Acclimatization

Acclimatization is essential in the prevention of temperature problems. Before being acclimated, the capacity of an individual to work in heat or cold is impaired. The body senses the change in climate and adjusts its thermoregulatory system, which alerts the body to prepare for more efficient use of body fluids and heat generation. For heat, the sweat becomes less concentrated with electrolytes, and the diluted sweat is produced in greater quantities. In addition, the cardiovascular system becomes more responsive to blood flow and necessary shunting of blood.

Subjectively, as acclimatization continues, the person copes better in hot environments and can be more active both physically and mentally without undue impairment. Comfort level is increased and a man or woman is less aware of the temperature extremes. A variety of other physiological functions become adapted to the expected needs of the body, and the body can control core temperature very effectively if the body is not prevented from functioning properly.

All this takes two to three weeks to completely adjust to the environment. This means that the pilot who lives in a climate that turns cool in the fall and travels to a hot and humid location for a few days does not become acclimated and is therefore at high risk of developing heat stress. Conversely, living in warm climates and trying to cope in below-freezing conditions is equally impairing. Some people feel that it's a good idea to participate in a good aerobic exercise program for several days prior to going into a hot environment, but don't overdo it to the point of exhaustion prior to departure.

COPING WITH COLD EXTREMES

Coping with cold is basically generating and conserving heat. Enough heat is generated by the body if the body is active and there is enough caloric fuel. The primary goal, then, is to protect the body from losing too much of the heat that is already present. Generally, there are two forms of cold impairment: hypothermia, where the body core temperature gets lower than 98.6°F, and frostbite, which is localized cold injury to the tissues (freezing).

Hypothermia symptoms

In the early stages of hypothermia, there is obvious discomfort leading to distraction, a form of impairment. As one gets colder, there might be muscle stiffness and weakness, fatigue, and sometimes drowsiness. Thought processes are dulled and stuporous. As the cold increases, shivering begins uncontrollably. These early symptoms are the first signs of clinical hypothermia, and corrective action must be taken. As heat is taken away from the core (by the futile attempts of the body's heat-conservation system plus the loss of heat from lack of body-surface protection), the body begins to become generally impaired and unable to keep up. It's just the opposite of trying to keep the body core cool.

The same heat-transfer process is taking place. In order to conserve heat, blood flow is shunted away from the surface (the skin) at the expense of the extremities and other nonessential body parts. Muscle activity is increased involuntarily. As the body loses this battle, the core temperature continues to go down.

The next stage is when the core temperature goes below 96°F. Drowsiness increases with a general loss of awareness of what's happening. Shivering becomes intense and by itself is incapacitating. Further cooling (below 90°F) can lead to loss of memory and unconsciousness. Shivering actually decreases with the muscles becoming rigid: total incapacitation.

Hypothermia symptoms are less subtle than those of hyperthermia, and in most situations, more easily controlled if the person is prepared. The degree of heat loss is dependent on the physical condition of the person, the environmental conditions (temperature and wind), and the protection of the person's body.

Loss of body heat

All the physical properties of heat transfer are in effect in cold extremes, but with variations from that associated with heat. Radiation through any uncovered skin now becomes the leading cause of heat loss. The scalp is especially vulnerable because of its highly vascular anatomy (that's why there is so much bleeding when the scalp is lacerated). With ambient temperatures about 40°F, nearly 50 percent of heat loss is from the head and increases to 75 percent when the temperature reaches 5°F. Hats, therefore, are an essential piece of cold weather clothing even though the cold is tolerable by some.

Conduction is also noteworthy because a pilot is in contact with cold seats, instruments, and other unheated objects. Cold and wet clothing removes heat 25 times more rapidly from the body than warm and dry clothing. Even evaporation of sweat, which can occur when very active even in cold conditions, causes some heat loss. But it's much more important to recognize that sweating soaks the clothing next to your skin. This moisture is cooling as it evaporates, resulting in heat loss from the body first through conduction then convection. Loose clothing is therefore important.

Convection plays a big role in heat loss especially when surrounded by cold air or cold water. When cold air passes over exposed skin, heat is lost rapidly: the *wind-chill* factor (Fig. 10-2). Not only is heat lost in a short period of time, but local areas of exposed skin on the extremities and other locations where blood has been shunted away (ears and nose) can freeze.

The situation is worse in cold water because water conducts heat about 25 times faster than air. Convection then enters in by carrying that heat away as the water moves over the body. *Exposure* is another term used to describe being in extreme conditions of cold wind or water. Any liquid can cause the same problem, including aviation fuel.

Wind speed		Temperature (°F)																				
Calm	Calm	40	35	30	25	20	15	10	5	0	-5	-10	-15	-20	-25	-30	-35	-40	-45	-50	-55	-60
Knots	MPH	Equivalent chill temperature																				
3-6	5	35	30	25	20	15	10	5	0	-5	-10	-15	-20	-25	-30	-35	-40	-45	-50	-55	-65	-70
7-10	10	30	20	15	10	5	0	-10	-15	-20	-25	-30	-40	-45	-50	-60	-65	-70	-75	-80	-90	-95
11-15	15	25	15	10	0	-5	-10	-20	-25	-30	-40	-45	-50	-60	-65	-70	-80	-85	-90	-100	-105	-110
16-19	20	20	10	5	0	-10	-15	-25	-30	-35	-45	-50	-60	-65	-75	-80	-85	-95	-100	-110	-115	-120
20-23	25	15	10	0	-5	-15	-20	-30	-35	-45	-50	-60	-65	-75	-80	-90	-95	-105	-110	-120	-125	-135
24-28	30	10	5	0	-10	-20	-25	-30	-40	-50	-55	-65	-70	-80	-85	-95	-100	-110	-115	-125	-130	-140
29-32	35	10	5	-5	-10	-20	-30	-35	-40	-50	-60	-65	-75	-80	-90	-100	-105	-115	-120	-130	-135	-145
33-36	40	10	0	-5	-15	-20	-30	-35	-45	-55	-60	-70	-75	-85	-95	-100	-110	-115	-125	-130	-140	-150
Winds above 40 have little additional effect.		Little danger					Increasing danger (Flesh may freeze within 1 minute)						Great danger (Flesh may freeze within 30 seconds)									
		Danger of freezing exposed fleshing exposed flesh for properly clothed persons																				

10-2 Windchill is a product of temperature and wind velocity.

Treatment should be obvious. Get out of the cold and immediately remove wet and chilled clothing. Gently warm the body in a bath or shower. Drinking warm water would also help. Put on warm clothes and, when safe, gently work the muscles to generate more internal heat. As with extremes of heat, serious hypothermia must be treated by doctors.

Frostbite

Frostbite is a more extreme condition of heat loss from a local area. Tissues are actually frozen, and the degree of injury is related to how deep the freezing goes. Tissues that are frozen die as a result of the crystallization of tissue and cellular fluids.

The symptoms vary. The first sensation is numbness but not pain. Grayish or white spots might show up on the skin before freezing actually occurs. After awhile, there is no feeling, and the tissues have hardened to the touch. All this occurs because there has been no internal warming by the blood that has been shunted to the core.

Frostbite to fingers and toes is common to pilots who are on the ramp in slush and spilled fuel. Doing a preflight without gloves, hat, or other additional protection leads to other areas at risk for frostbite or hypothermia.

Treatment includes immediate warming. Do *not* rub with snow! Ideally, the affected parts should be immersed in water that is warm to the touch. Be advised that as the tissues thaw, the pain becomes severe. This is not an indication to stop warming. If the injury is this severe, medical attention should be obtained.

Prevention of cold stress

Prevention remains the most important action the pilot can take. A prudent pilot will practice not only prevention by keeping warm and preventing heat loss, but also preparedness (wearing adequate cold-weather clothing when the destination is colder than the departure point). Most "pilot" uniforms are not designed for warmth and prevention of heat loss.

A small bag containing at least gloves, a knit hat, extra socks, a scarf, and a sweater or jacket would be ideal. Add to the bag as your needs dictate. Wearing this protection before being subjected to the cold is the most important aspect of preparedness. Racing through a preflight on a cold and windy ramp not only subjects the body to early loss of body heat, but the pilot is probably taking shortcuts and even missing some items on the checklist because he or she wants to get out of the cold. A cold cockpit adds to the distress and distraction.

The specific and obvious measures are as follows:

- Be aware of the temperatures, winds, and wind chills at your present location and your destinations. Is rain or frozen precip forecasted?
- Limit the amount of time of exposure.
- Keep the body and clothes dry. If you're sweating and spill water or fuel on your clothes, change into dry clothes.
- Use the "layered look," wearing several layers of loose clothing that can be removed or added as comfort levels change.
- Be sure to watch your crewmates for signs of hypothermia. Better still, encourage them to wear proper protection if you notice they are out in the cold unprotected. (Set an example.)
- Pace your activities to prevent unnecessary perspiration. Don't work up a sweat.

A word on alcohol: Remember that alcohol is a vasodilator and is apt to make you feel warm (and flushed) because it keeps surface blood vessels wide open. This is just the opposite of what the body is trying to do, which is shunt blood away from the skin. Nutrition is equally important. The body needs calories to generate heat in cold conditions. This means eating high-calorie meals. Persons participating in wintertime activities know that 4,000–6,000 calories a day are necessary to maintain an adequate source of energy.

DEHYDRATION

Dehydration is very common in flying and not just when related to temperature extremes. The life style of the traveling pilot and the cockpit environment lend themselves to increased risks of becoming dehydrated. Hours of increased dehydration is one reason kidney stones are relatively common in pilots who fly great distances in a dehydrated condition.

Why? Because most pilots are not drinking water en route. More often than not, coffee is the "fluid of choice," although the trend is toward decaffeinated coffee. Pressurized cockpit humidity approaches less than 5 percent within 20–30 minutes after takeoff. Even if the pilot is physically inactive during flight, there is an insensible water loss from respiration (breathing out moist air), urination, and skin (minimal sweating) of 3–4 pints of water per day in ideal conditions, more so in dry conditions. This is a loss over which we have no control and must be replaced in addition to that lost with increased activity and sweating.

Often the pilot will show up for a flight already dehydrated for a variety of reasons associated with her personal activities. These reasons might include a strenuous workout before a trip or the effects of alcohol consumption even 12 hours or more before departure. Consuming a lot of cola, coffee, hot tea, or iced tea (diuretics) before the flight and urinating more fluid than has been consumed is another problem. This, by the way, can be monitored by observing the color of urine. Darker urine indicates worsening dehydration; nearly colorless urine indicates that adequate amounts of water have been ingested.

Dehydration (for any reason) has similar symptoms to those experienced with heat stress. Headache, dizziness, and especially fatigue are the most noteworthy. Imagine the combination of jogging early in the morning, drinking several cups of coffee, rushing to the airport for a preflight on a hot, dry, and windy ramp, and then asking for iced tea shortly after takeoff. Combined with other causes, this pilot will be significantly fatigued after that trip.

Unfortunately, pilots tend to wait for symptoms instead of practicing the commonsense approach of prevention. Thirstiness is an indicator, but thirst occurs only after dehydration is present. Drinking lots of water is a simple activity and all that is needed. This means at least a glass with every meal; drink more water between other beverages in flight. A good investment would be a large water container from which you can continuously sip. Add some lemon or lime juice to add some flavor. Don't hesitate to drink up (water) because the kidneys will control the desirable body-water content.

RADIATION

The term "radiation" discussed in terms of a hazard, implies radioactive properties; however, *radiation* actually means *electromagnetic radiation* (EMR), which covers the entire spectrum of waves, or rays, all traveling at the speed of light. Each type of EMR is further defined by its wavelength, ranging from cosmic rays as short as 4×10^{-12} cm to radio and power transmission rays that extend several miles in length. Frequency in cycles per second (or hertz) also describes each kind of radiation (Fig. 10-3).

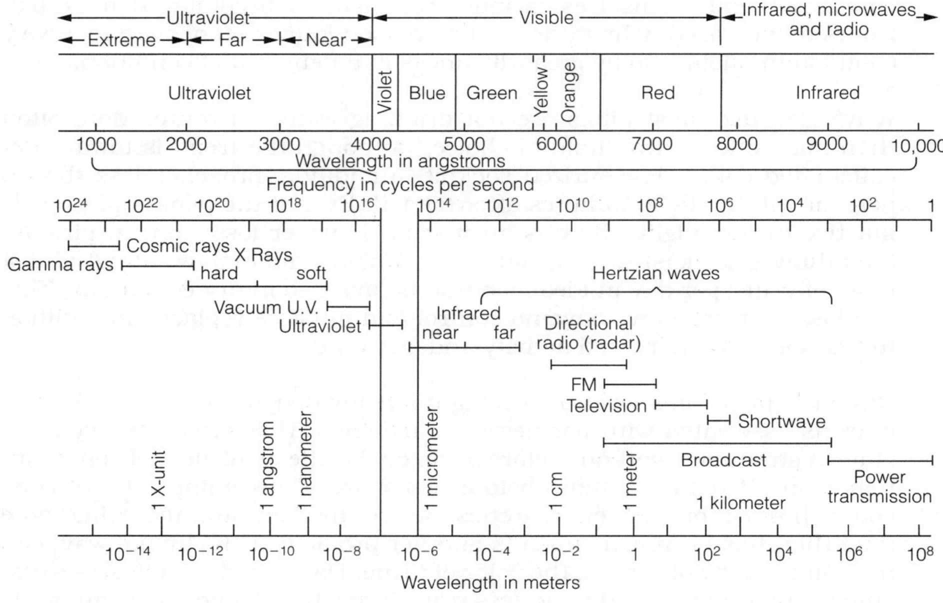

10-3 Electromagnetic spectrum and radiation.
R.L. DeHart. *Fundamentals of Aerospace Medicine.* Philadelphia, Lee & Febiger, 1985. Reproduced with permission. Mosby-Year Book, Inc., St. Louis.

When discussing hazards to flight and general health, radiation falls into two categories: ionizing and nonionizing radiation. By definition, *ionizing radiation* includes alpha rays/particles, beta rays/particles, gamma rays, radon, and X-rays and is associated with the radiation of ions.

Nonionizing radiation

Nonionizing radiation includes the major portion of the EMR spectrum, from ultraviolet through visible, infrared, and microwave. Laser light is included in this group. These EMRs, which are the most important to pilots from the perspective of health, include ultraviolet and to a lesser degree microwave.

Various portions of ultraviolet light, or radiation wavelengths, can affect the eye, specifically the cornea (UVB 200–315 nanometers), lens (UVA 315–400 nm), and to an uncommon degree, the retina (UVC 400–700 nm). Most ultraviolet radiation from the sun is filtered out by the ozone layer in the atmosphere, especially UVC waves. There appears to be increasing evidence that the ozone layer is being depleted, which could increase the risk, especially to pilots flying at high altitudes for long trips, of ultraviolet radiation damage. This damage is limited to eye changes, such as cataracts. This is easily prevented by ensuring that all glasses, sunglasses, and contact lenses have an ultraviolet coating that filters out the ultraviolet radiation. It is important that the full ultraviolet range (200–400 nm) is filtered out. Check the label on the glasses.

Exposure to ultraviolet radiation also affects the skin, resulting in the all too common sunburn. UVB causes most of the damage, but studies now show that UVA (found in tanning booths) adds to the effect of UVB but is not as severe as UVB. Few pilots have avoided the symptoms of sunburn, and most can remember that a bad burn is at best distracting. Despite increasing evidence about the relationship between ultraviolet radiation and sunburn with skin cancer, tanning still is a summer ritual for some. Skin can be acclimated by gradually increasing exposure.

The best protection from sunburn, especially in the unacclimated body, is to use a sun screen. Many on the market come with various combinations of lotion ingredients and levels of protection. The key ingredient is *para amino benzoic acid* (PABA). The more PABA, the greater the protection. Skin that is very sensitive to the Sun's rays requires greater protection. Obviously, this is not a problem if proper clothing is worn to protect the skin, but exposed areas such as the hands, face, and ears might require screening.

Microwave radiation is of minimal danger, the worst being the generation of heat within the cells. Radar is in the area of microwaves and results in the same symptoms. Because the eye is vulnerable to any radiation, direct microwave pulses to the eye should be avoided. If the pilot feels she has been exposed (by having walked in front of an aircraft radome with the radar operating), she should be checked by an ophthalmologist.

Ionizing radiation

Ionizing radiation is potentially serious, and many studies are being done to determine its effects. To commercial pilots on long flights, the main concern is the development of cancer in the pilot and other aircrew as they spend more time at altitude. A lot of variables are involved, including length of exposure, altitude, latitude, and time of year. Most of this kind of radiation is filtered out by the atmosphere; however, there is some scientific indication that the risk might increase over time.

Although cosmic radiation does not reach the levels considered to be toxic, any level of radiation that causes problems is considered unacceptable. The scientific community does not know yet what those levels are and whether or not there is a true hazard. Considering that there is background radiation at ground level from the Earth itself, any suspected problems must be relative to the background level.

There is no proven increased risk from ionizing radiation at altitude. This could change as more results from studies are compared. For the moment, concentrate on known risk factors such as smoking, ultraviolet radiation, poor nutrition, and other factors described throughout this text.

CABIN AIR QUALITY

For years, a controversial discussion topic has been the quality of the air both in the cockpit and in the passenger cabin. Cigarette smoke was a major issue but has been virtually banned from commercial aircraft. Ozone, carbon dioxide, low oxygen, toxic fumes, carbon monoxide, perfumes and colognes, and other more exotic contaminants have been implicated in causing a wide variety of symptoms and illnesses.

Additionally, selected claims state that infectious organisms such as bacteria and viruses are easily spread to others; however, many studies for and by the FAA have ruled out any significant degradation of cabin air despite claims made by many individuals and some organizations. Cockpit air is usually exchanged independently from the cabin air, and the rates of either or both will depend on the aircraft and how the crew maintains the level of air conditioning. In most cases, the air is exchanged more often than in other closed environments, such as a car, home, or office building.

While exceptions are possible, air circulation, especially in larger aircraft, adequately exchanges the air enough to keep it safe. Newer aircraft, however, can reduce this circulation rate to optimize aircraft performance; monitoring must continue to ensure that occupants are not subjected to unhealthy conditions.

TOXIC CHEMICALS AND FUMES

The most important problems with toxic fumes occur in flight when there is an acute exposure to any of a number of fumes that can occur. Like hypoxia, it is important to maintain a high index of suspicion if there is a strange smell, or a change in performance in yourself or in your crewmate.

Likewise, there is a chronic consideration of long-term exposure. This occurs mainly with ground-support personnel, but many pilots also perform ground functions. In any case, the degree of toxic involvement is defined by the time/dose relationship. Some fumes are toxic in low doses over a period of time, and others are serious in higher doses over shorter periods of time. The impairment of any body organ, or the lack of being in good physical condition, increases the risk of developing a problem.

Probably the most noteworthy toxic fume is carbon monoxide. As described in chapter 5 regarding altitude physiology, carbon monoxide competes with oxygen for hemoglobin in blood cells. This leads to anemic (hypemic) hypoxia because the blood cell is unable to transport oxygen to the cells. Hemoglobin has 250 times more affinity for carbon monoxide than it does for oxygen. Carbon monoxide is found in cigarette smoke and frequently in aircraft engines (Fig. 10-4).

Symptoms are similar to hypoxia plus a more severe headache, weakness, nervousness, and generalized impairment. Another potential problem is the pilot who comes from an environment (home or car or other work) where she has already been subjected to carbon monoxide. Exhaust gases, in addition to containing carbon monoxide, are also irritating to the eyes and lungs. Like hypoxia, there is no indication of its presence because there is no smell.

Percentage of circulating Hb saturated by CO	Symptoms (resting state, at ground level)
0–10	None noticeable.
10–20	Tightness across forehead and slight headache.
20–30	Headache and throbbing of the temples; breathlessness on exertion and perhaps nausea.
30–40	Severe headache, weakness, dizziness, dimness of vision, nausea and vomiting, collapse.
Over 40	Increasing likelihood of collapse, increasing pulse rate, irregular breathing, coma, convulsions, respiratory failure.

10-4 Symptoms of carbon monoxide poisoning.

Cargo containing toxic and hazardous material is a growing concern. Leaks can occur at altitude and go unnoticed until symptoms are incapacitating.

Aviation fuel is another toxic fume, but its smell is obvious. Aviation gasoline is at least twice as toxic as ordinary gasoline. Its vapors can cause a loss of sensation and reduction in consciousness. It is also an irritant to the nose and pulmonary system. Typical symptoms include headache, nausea, blurring of vision, and mental confusion.

Carbon dioxide is not commonly present but is a factor in hyperventilation.

11
Sleep, jet lag, and fatigue

This was the first time the flight engineer (FE) had flown the long-haul trip from Chicago to Tokyo with a layover in Honolulu. He was eager to see the sights of Hawaii and had already made plans with the first officer to be shown the nightlife. They had hoped the captain would join them but he declined, saying "I've got to plan ahead and be alert for the approach into Tokyo." Not much of a socializer, thought the FE. By midnight, the sightseers had gotten back to their rooms, tired but excited about all that they had done that night. The wake-up call came at 6:30 a.m. with their departure for Japan set at 9. Halfway through the trip, the FE began to nod off. The captain had to ask him several times to work out the fuel aboard computation. Even the first officer missed reporting their position to the company—twice. It was a real struggle for them to stay awake and be ahead of the airplane. The captain, who was alert, made an uneventful landing although the FE couldn't remember the ATIS and the first officer missed calling out speed and altitude. On the ground, the two fatigued "Hawaiian tourists" recognized what their captain had meant by planning ahead. He knew how he was going to feel even under the best of conditions and wasn't going to compound his anticipated fatigue by not getting plenty of rest on the layover.

PROBABLY THE MOST FREQUENTLY TALKED ABOUT human factor and physiological situation is fatigue. In fact, most physiological events (noise, hypoxia, dehydration, and temperature extremes to name a few) include fatigue as a symptom. Lack of sleep is commonly thought of as the most frequent cause of fatigue, and there are many reasons a restful sleep is not achieved. Other causes of fatigue will be discussed, but the lack of *restful* sleep is a cornerstone and can create a cumulative problem.

Aviation is often the source of virtually all causes of fatigue. This mandates that all pilots must be familiar with the various causes of fatigue and the related symptoms. Some causes cannot be controlled (length of flight, circadian change, illness, climate, etc.), but the knowledgeable pilot can cope with these situations by being better prepared, avoiding controllable causes of fatigue, and maintaining a high index of suspicion of his or her performance and that of the other crewmembers.

Because lack of sleep is the situation most common to all pilots, it will be discussed first. Circadian changes, which often lead to poor sleep, will be discussed next. Fatigue, the ultimate symptom of these and other causes, will be defined at the conclusion of this chapter.

SLEEP

Even though we spend over a third of our lives in some stage of sleep, there is still little known about why we need sleep and how we can better control this essential function of the body. It is widely believed that sleep revitalizes and prepares the body for the next day. There is plenty of research, and more is always in the works, but the practical aspect of how a person deals with his or her sleep habits and problems remains elusive. This is particularly true with pilots, who are subjected to sleep schedules that are comparable to interns in hospital training and worse than shift workers. A basic understanding of the sleep process is important, but even more important is how individuals deal with the variations in their own work and rest requirements.

No magic diets, medications, or techniques ensure a restful sleep. Only through awareness of the physiology of sleep—how sleep is influenced by our activities and what we can control to achieve sleep—can the pilot remain safe and productive.

Physiology of sleep

Everyone goes through the same basic stages of sleep; however, there are individual variations as to the length and depth of each stage, the number of hours of sleep needed in a 24-hour cycle, and how a person uses the benefits of sleep. Only through reviewing her sleep habits and patterns can the pilot determine where she fits. And these patterns will change with age. In addition, there are situations, such as a drunken sleep, where a person appears to be in deep sleep, but this sleep is not restful, and upon awakening, the person will feel unrested. Closer to the world of aviation is the fact that many pilots manage their rest periods poorly and begin their next trip in sleep debt.

During a normal night of sleep, humans alternate repeatedly between two distinct types of sleep: rapid eye movement (REM) and nonREM (NREM). NREM sleep has four stages. Both types of sleep have different and independent effects on the overall quality of sleep, determined by levels of subsequent daytime sleepiness and performance.

Stages of sleep

Essentially, the four stages of sleep define the depth of consciousness (or sleep) (Fig. 11-1). The deeper the sleep and length of time in deep sleep, the more the body and mind are getting rest. These four stages occur within a series of cycles over the period of sleep, occurring about every 90–100 minutes. REM sleep is another event in each cycle; sometimes REM sleep is considered to be another stage. REM sleep often occurs after sleep has be-

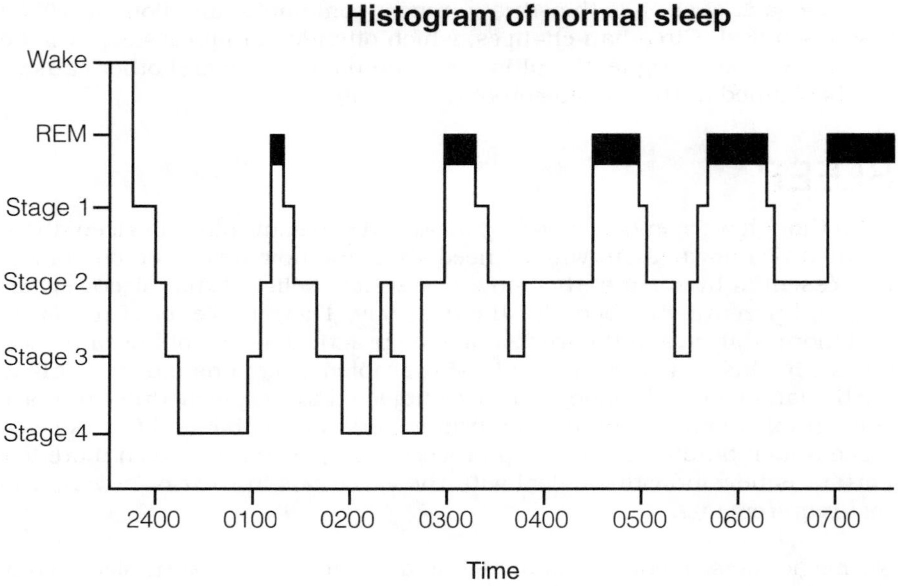

Histogram of normal sleep

Wake
REM
Stage 1
Stage 2
Stage 3
Stage 4
2400 0100 0200 0300 0400 0500 0600 0700
Time

11-1 Over the course of a typical night, NREM and REM sleep occur in a cycle, with about 60 minutes of NREM sleep followed by about 30 minutes of REM sleep. This 90-minute cycle repeats itself throughout a typical sleep period; however, most deep sleep (i.e., NREM stages 3 and 4) occurs in the first third of the night, and REM periods are shorter early in the night and then become longer and occur more regularly later in the sleep period. Overall, about 25 percent of sleep time is spent in REM sleep and about 50 percent is spent in NREM stage 2.
National Aeronautics and Space Administration.

gun its first cycle and then as the person returns to the first stage of sleep and just before going back into the next cycle. These four stages are called NREM sleep.

Stage 1 This is the time between wakefulness and when drowsiness takes over. The brain waves begin to slow down, the body relaxes, and body temperature, respirations, and pulse rate decrease. Some people notice (or see in others) a quick jerking movement of the muscles, which is an indication of a change in the brain activity. This might awaken the sleeper, but she quickly returns to sleep. This stage lasts only about 1–10 minutes, depending on how fast a person normally falls asleep.

Stage 2 Brain activity increases for a time with many short bursts of thoughts and memories. Muscles continue to relax, body functions continue to slow down, and breathing is even and steady. If a person were awakened during this stage, he might deny that he was asleep, though in fact he was. This stage lasts about 10 minutes.

Stage 3 Continued relaxation occurs with even slower respirations and a slower pulse rate. The body is almost totally at rest and it is difficult to awaken someone in this stage. If awakened, she will not feel rested and might actually feel more tired. Some say this is the entry to "delta" sleep, whereas the first stage is often called the "alpha" sleep, both relating to the type of brain waves on an *electroencephalogram* (EEG), which is the measurement of electrical impulses in the brain. Stage 3 is also sometimes called a *transitional stage* and lasts for about 5 minutes.

Stage 4 This is the deepest part of sleep. The body is completely relaxed and the brain activity is slow. A person could be considered unconscious and would be very difficult to arouse. This stage lasts for about 30–45 minutes and is the longest period of the cycle and the most important in getting quality sleep; however, most deep sleep occurs in the first third of the night, and REM sleep periods are shorter early in the night and become longer later on.

It is this fourth stage that gets shorter and less deep as we get older. Furthermore, if this stage of sleep is broken for any reason, and a person is awakened while well into the fourth stage, he or she would be very groggy (called *sleep inertia*), probably incapacitated for 10–15 minutes, and end up with a "disturbed" sleep cycle. Naps must be timed so as not to last too long into the fourth stage of sleep. Usually, depending on the time of night, being awakened before 40 minutes of sleep should not cause a problem of sleep inertia.

REM sleep

Concurrent with the four stages of sleep, REM enters about 60 minutes after falling asleep. Pulse rate and blood pressure fluctuate, larger muscles remain relaxed but smaller muscles (fingers and toes) might twitch, and, of course, the eyes are moving rapidly even while asleep. This REM stage coincides with the cycle returning to stage 1 and can last for 10–30 minutes before the cycle returns to stage 2. As each cycle repeats (about every 90 minutes), less time is spent in stages 3 and 4 and more in REM; therefore, most of the deep and productive sleep occurs early in the total sleep session, and REM occupies sleep more during the end of the session.

It is thought that dreams occur during REM sleep. If you are awakened during REM sleep, you might recall what you were dreaming. At other stages, there is little remembered and most dreams tend not to be put into the long-term memory banks of the brain. In most cases during a night's sleep, you will go through REM sleep without awakening.

Anything that interferes with REM sleep (alcohol, drugs, stress, or being wakened during this stage) results in nonproductive sleep. A good night's sleep implies four uninterrupted NREM stages and a complete REM stage. REM sleep might account for 25 percent of total sleep.

As we get older, we tend to awaken more often as we return to the cycle of stage 1 and into REM sleep. For a brief moment we are conscious of being awake and possibly feel that we haven't even been asleep, but we quickly move on to the next cycle of sleep, if nothing keeps us awake. This frequent awakening by the older pilot is not abnormal and is not an indication of true insomnia; however, some might think that there is a problem, and thinking about that concern now interferes with returning to sleep. In the younger pilot, frequent awakening might be an indication of insomnia and needs to be evaluated by a doctor.

Overall, about 25 percent of sleep time is spent in REM and about 50 percent in NREM stage 2 during the night, with no stage 4 later. How long we actually spend in REM and NREM sleep will determine our quality of sleep. Any sleep disturbance that interferes with feeling rested upon awakening needs to be evaluated. Middle age changes of occupational stresses, family transitions, and financial pressures could eventually lead to a significant sleep disturbance. It should be noted that sleep is a very dynamic process, changing during the night, affected by age, stress, and other factors, and probably different each time, especially when not following a consistent sleeping habit.

Variations to normal sleep

There are individual changes to a typical sleep cycle. Some people are able to feel rested and function normally with as little as 3–5 hours per day. Others require more than 8 hours. The 8-hour norm of sleep is not universal, and each individual must determine what his or her optimum amount of restful sleep is. The indicator of a restful sleep is how you feel; if you are rested in the morning and remain so during the day under normal conditions, then you are getting enough sleep. When you don't get enough sleep, you become sleepy. That's not a startling revelation; however, we tend to take this symptom lightly as opposed to the symptoms of lack of fluids (thirst) and lack of food (hunger). Feeling sleepy is a physiological signal of a sleep debt and no less important a signal than thirst and hunger.

Some people can fall asleep easily if just given the opportunity, even in noisy, uncomfortable locations. They are able to take quick "cat naps" and keep ahead of their sleep needs. Others find it very difficult to get to sleep and then stay asleep, even in ideal conditions. Again, this is a normal variation and not something for you to be overly concerned about and try to change. Avoid quick fixes that attempt to change a sleep habit that is normal.

Everyone's sleep cycle is determined by a personal circadian rhythm, which will be subsequently described in this chapter. In fact, the cycle of sleep is the most important rhythm, and it can be easily disrupted if that rhythm is changed, as with jet lag. Some people are "larks," or morning people. They awaken early, ready to go and alert to the world around them, but they also tend to begin to lose that energy about midafternoon. The "owl," or night person, on the other hand, takes hours to get going after awakening, with his or her full energy available about midafternoon and lasting well into the evening. Larks go to bed earlier than owls, who stay up much later.

Most people know whether or not they are larks or owls. Their challenge is to fit in with the rest of the world or work with people and cockpit crews who are opposite to their cycles. It is important to recognize this normal variation in other people's mental alertness, energy, and motivation at different times of the day and take into account that they have to deal with your variations as well. Tolerance, understanding, and patience are the keys for owls and larks to effectively and safely work and fly together.

Sleep disorders

When sleep is disrupted or inadequate (not restful), sleeplessness occurs. This is also called *insomnia*. There are three different degrees of insomnia: transient, acute, and chronic. The most common and least significant is the transient kind, which lasts only a few nights and is a result of excitement (just received a major salary increase), the "jitters" (anticipating an upgrade checkride you don't feel ready for), and other reasons we have all experienced that keep us from getting to sleep. The mind just won't let us relax. When the situation causing this form of insomnia is resolved, sleep quickly returns to normal.

Acute or short-term insomnia results from longer lasting stress (your company might be sold) or illness (a bad flu). This type of insomnia usually disappears in a few days, and the sleep cycles return to normal. Little sleep is actually lost and is easily recovered after several good nights of restful sleep; however, depending on the amount of sleep lost, health and age of the person, and other factors, the recovery might take longer.

Chronic insomnia is more serious and comes from continued sleep deprivation. This can last for days, weeks, or even months. The most common cause in the general public is serious unresolved stress or illness. For the pilot, it's a combination of several factors, including stress. Sleep researchers have observed that it's not so much the lack of sleep as much as *when* the sleep is allowed to occur in a 24-hour period. Furthermore, it is now known that the quality of sleep is more important than length or when.

Long-haul trips, flight instructing day after day, and frequent night flights are all inherent to aviation. Selected people can adapt; others can't. If you can't and you feel you are never fully rested, then changes must be made. Insomnia is a symptom; therefore, take care of the cause first without focusing treatment of the symptom. Chronic insomnia usually needs the help of a doctor.

Alertness management

In the early 1990s, NASA introduced its program called "Alertness Management," which was a product of the agency's fatigue countermeasures studies. NASA's approach to sleep and how it affects safety, especially in long-haul operations, is a clear view of the real world of sleep. This program is a major accomplishment because for the first time the physiology of sleep and how the lack of quality sleep can affect performance and safety has been scientifically defined and proven. This lends credibility beyond anecdotal information that aviators have known for years: Lack of restful sleep is unsafe, and no amount of experience or motivation can overcome the body's need for adequate amounts of quality sleep.

Taking the physiology of sleep as described earlier one step further in the real world, all pilots can relate to three observations.

Sleep is vital

Like human requirements for food and water, sleep is absolutely essential. We have signals from the brain when there is a need for each of these three vital entities: hunger, thirst, and sleepiness. Sleepiness is no less an important signal than the others, yet it is common to try to overcome this signal through various means and minimize the only true solution, getting adequate sleep. Unfortunately, it is easier to satisfy hunger and thirst: not so with sleepiness.

When the body is deprived of sleep (acute or chronic), the human brain can spontaneously shift from wakefulness to sleep in an *uncontrolled* fashion in order to meet its physiological need for sleep.

The sleepier the pilot, the more rapid and frequent are these intrusions of sleep in wakefulness. These episodes can be very short, such as *microsleeps* lasting only a few seconds and unknown to the pilot. Others are extended, lasting for several minutes. With sleep loss, these uncontrollable sleep episodes can occur while standing, sitting, and operating aircraft. Pilots do fall asleep while at the controls, especially on long flights. The concern: The pilot is not aware that sleep has occurred nor is he or she able to determine when it started.

Sleepiness, as a signal from the brain, is often dismissed as a minimal nuisance or easily overcome. Aside from being asleep, the sleepiness, like so many other symptoms in flight that have already been described, can potentially degrade most aspects of human performance.

Sleep debt

Continued sleep loss accumulates into a sleep debt, and is cumulative over days. If you need 8 hours of sleep per day and you get only 6 hours, then you are in sleep debt. This will continue over subsequent days of sleep loss, and eventually the body will *require* sleep to continue functioning, hence the need for additional sleep on weekends or sleeping late when the opportunity arises. Furthermore, it is the quality of sleep, not just the quantity, that is important. In other words, getting a deeper sleep for a shorter time will be more effective than a light sleep for a longer time. An indication of sleep debt is the feeling of sleepiness and someone telling you that you were asleep.

Physiological and subjective sleepiness

Sleepiness, as simple as it might seem, is differentiated into two distinct and critical components. The first is physiological, which has been just described. Sleep debt results in sleepiness, which forces a pilot to sleep in order to meet the physiological need. This will happen spontaneously. *Subjective* sleepiness is the pilot's introspective self-report on how sleepy he or she is. That subjective feeling can be affected by many factors such as caffeine, physical activity, stimulating environment (a rough ride through weather) or a stimulating conversation. As a result, the pilot will claim to be more alert than is physiologically possible. In other words, the stimulated subjective sleepiness (perceived as being minimal) will mask or conceal the physiological status of need. The pilot might feel a high level of alertness but actually be functioning under an accumulated sleep debt. This apathy can be an operationally significant dilemma.

Factors affecting sleep quality

Most pilots know what interferes with a good sleep: noisy rooms or sleeping quarters, uncomfortable beds, hot or cold temperatures, too much light, and the like. Wearing ear plugs reduces noise levels, putting a blanket over a window shuts out the light, and extra blankets make a bed more comfortable.

Other controllable factors are self-imposed stresses, such as too much caffeine or nicotine. Alcohol is a major disturbance because it interferes with REM sleep, and many believe that if REM sleep is disrupted by alcohol, restful sleep is compromised. This effect can last for several hours after the last drink, and that includes wine and beer if taken to excess.

Other factors include naps that are either too long or too frequent, which disrupt your normal sleep circadian rhythm and cycle. Too much sleep during the day prevents a good night's sleep. Sleeping in late on days off creates problems with sleep later on in the day, especially for older pilots.

Hypoxia is not often considered as another source of problems. Sleep is difficult if a person is not acclimated to an altitude at which she is sleeping. Coming from a sea-level airport and staying overnight in Denver after drinking a nightcap will not allow a restful and quality sleep. And the following morning will be unpleasant as well. Hangovers tend to be more severe when you are hypoxic. Remember, you can be hypoxic without being in an airplane.

Circadian changes (*desynchronosus*) can affect sleep, especially when you are trying to sleep at times your body is not ready to sleep. A variety of techniques, many anecdotal, that are conducive to sleep will be discussed under circadian rhythms in this chapter. For now, one should expect disruption of sleep when crossing time zones. In addition to the pure fatigue of the long trip, sleep will be compromised when trying to sleep. The continuous flow of adrenaline during the trip (caused by waiting for something wrong to happen, flying around weather, etc.) can keep the pilot from getting to sleep after the trip. That same pilot will no doubt mentally "relive" those flight situations, and it will be difficult to get to sleep. By anticipating the upcoming need for a good night's rest, the pilot can try to avoid or resolve those situations that can lead to problems.

Symptoms of insomnia

Insomnia symptoms are similar to those associated with fatigue and will be subsequently detailed in this chapter. Generally there is a pronounced impairment of performance, especially in decision making and fine muscle skills. Motivation to do a good job is diminished, and similarly to hypoxia, you don't really care about your performance, even taking risks you wouldn't normally take. There will be lapses in alertness, a tendency to slip into microsleep, and no awareness that you were asleep. Small errors and mistakes will be exaggerated. A major problem is that insomnia can be a vicious circle: Stress from not sleeping well coupled with the fear of being an insomniac when you go to bed lead to poor sleep.

How to sleep

You can take several measures to improve your chances of getting a restful and "productive" sleep. Try to maintain the same sleep schedule, especially when at home. Staying up to watch the late movie and sleeping in late on weekends only confuses the body's rhythms and disrupts sleep cycles. Exercise and keeping in shape is a major factor that allows the body to relax more efficiently and to tolerate those factors that could keep you awake. At home and in the hotel, sleep in a soundproof and lightproof room in comfortable conditions.

Stress and thinking about the stress often interferes with sleep. It's difficult to mentally turn off the concern and avoid lying awake considering all the bad things that can happen. At night, alone with your thoughts, these stresses are often out of proportion to reality. Imagination runs rampant. Getting up and writing down your thoughts with some ideas of how to resolve the concerns will sometimes help break the thinking cycle and allow sleep to take over.

Diet plays a role in sleeping. Carbohydrates tend to make a person drowsy and therefore can be consumed before retiring. Some feel that L-tryptophan, an amino acid found in many carbohydrates, is the sleep inducer. (See chapter 9 for a discussion of OTC sleep medications.) Consequently, the old-fashioned warm milk before bedtime has a physiological basis. Foods to avoid (in addition to caffeine and alcohol) are high in protein and animal fat.

For some pilots, prescription medications can help, but only in situations where sleep is otherwise impossible and there is adequate time for the drug to be out of your system before you fly. Your doctor can advise you on these medications, some of which are quick to act and are generally out of your body in a few hours (half life of 3–4 hours); however, be wary of depending on these crutches because tolerance can be built up, and eventually higher doses will be required to accomplish the same objective. Also, there can be a rebound effect where it becomes more difficult to sleep without medication.

Over-the-counter sleeping pills are almost always antihistamines. (See chapter 9 for a discussion of OTC sleep medications.) Antihistamines have significant side effects in flight (their half life is in excess of 12–18 hours), in addition to sedation, and certain antihistamines might interfere with REM sleep. Read labels, and you'll see that the same kind of chemicals are used in allergy medications as are used in sleeping pills.

Staying awake

Staying awake or alert is the other side of sleeping. In fact, "alertness management" and not sleep management is the term used by NASA's fatigue countermeasures program, as previously described in this chapter. Using reverse reasoning, the above measures that interfere with sleep can be used to keep from falling asleep. Caffeine is the traditional drug, but timing is important so as not to interfere with sleep when you are ready to sleep. Avoid carbohydrates and sugary foods. In addition to the drowsiness brought on by carbohydrates, hypoglycemia is a common result. Eat small meals, high in protein. Peanuts would be a good stop-gap measure. Keep active. Avoid boring or nonmotivating tasks; save them for later when you want to sleep. Drink lots of water, and change positions frequently, even getting up and walking around.

Crew duty time

Crew duty time is related to *crew rest*. This is a subject on which physiologically there will probably be little new discovered over the years. Lack of adequate rest along with poorly managed rest periods results in sleepy and fatigued pilots. Furthermore, pilots are often tired and in a sleep debt when they report for duty and are already behind the power curve of rest, just adding to the sleepiness caused by the long flight. Much research is being done detailing and confirming the unsafe and dangerous situations that can result from a crew that has not had adequate crew rest. A classic example is the Guantánamo Bay accident of August 1993 where for the first time the NTSB stated that fatigue from lack of sleep was the cause and not just contributory factor.

NASA's fatigue countermeasures research program recognized and proved that pilots on long-haul trips do fall asleep on the flight deck and are impaired at the most crucial phases of flight: approach and landing. When one pilot is allowed to sleep during a planned and controlled rest period, even on the flight deck (less than 40 minutes, so as not to be awakened during the deep-sleep phase during which it takes longer to fully awaken), it is much safer than the situation where entire crews are known to fall asleep without their knowledge.

This is the "NASA Nap," and it is anticipated that this will be allowed by the FAA under certain conditions. There are many considerations regarding this program, and further information about NASA's research and development of its alertness management is beyond the scope of this text but is recommended. The point is that sleep is essential and a pilot must get that sleep, even while in flight.

However, translating known insight into regulations and policy is another matter and becomes more of a corporate, financial, union, and regulatory challenge than a matter based upon the realities of the impaired pilot who hasn't gotten enough sleep. The other side of the coin is that some pilots tend to (and willingly) disregard the fatigue from tiring trips because the pay or job security is higher. There are significant differences between crew-duty limits for military and civilian flying and also between countries. Corporate flying and air-carrier flying have FAA-prescribed duty times, but these still don't address the ultimate concern of ensuring an adequately rested crew *during* flight.

Companies and flight crews cannot rely on outside agencies to govern duty time because there are too many variables. It will still be necessary for the pilot and members of the crew to maintain their own levels of adequate sleep, primarily through applying the knowledge of sleep physiology to their own habits and lifestyles. An understanding for "alertness management" by management is essential in respecting the potential financial loss and liability from impaired performance of a fatigued crew.

CIRCADIAN RHYTHMS

Closely related to sleep and alertness management is how the pilot is affected by jet lag or disruption of biological rhythms. Biological rhythms of living things have been recognized for hundreds of years. The Chinese have been carefully recording for a thousand years the blooming of a bamboo that occurs every 120 years. The American cicada hatches every 17 years. Hibernation of several kinds of animals goes on for several months and is carefully controlled through physiological rhythms.

It is also recognized that the human body has internal periodic physiological rhythms, at least 300 or more, that are relative to the revolution of the Earth and to the rising of the Sun. *Circadian* literally means "about a day" and usually represents a 24-hour day. Another term is *diurnal*, which means that a cycle occurs within a day.

Human rhythms

Studies have shown that if you take a person and put him or her completely out of touch with any outside source of light or time, the body's rhythms settle around a cycle of about 25–26 hours. We reset our internal clocks (which establish our biological rhythms) every day by getting up at the same time, reporting to work at a given and repetitive time, and, more important, by being exposed to sunlight at these times.

During long winters, where daylight is present for shorter periods of time, it is more difficult to adapt. There is even a disorder called *seasonal affective disorder* (SAD), which is a form of depression and decreased motivation that is believed to be related to the lack of enough sunshine to strengthen our circadian rhythms. Unrestricted sunlight, by the way, is about 100 times brighter than artificial lighting within even a well-lit building or office. But the Sun, coupled with our own schedule, is still the key to maintaining our circadian rhythms at 24 hours and maintaining an active and productive day.

Many studies have been accomplished to determine if there is a single "timekeeper," or body clock. The German word for "time giver" is *zeitgeber*, which is a term commonly used to refer to this internal control. It's also called a *pacemaker* or *oscillator*. Whatever the term, it is now believed that there is, indeed, a pacemaker, possibly two, and that they are located in the hypothalamus (among other potential locations) of the brain. Future research is now geared to confirm these findings. Then we will learn how to manipulate this pacemaker so that we can change our circadian rhythms at will, without compromising our health or safe flight, hence making long-distance travel more tolerable. There is increasing research regarding chemicals than can adjust our pacemaker to a local time. Some are natural hormones within the body, such as *melatonin*, but there is still much research to be done to officially recommend their use as a safe method to adjust these rhythms. Controlling sleep and wakefulness is a *very* complex process, and there will not be a simple resolution to controlling these rhythms.

It is also recognized that in addition to an internal pacemaker, there are also zeitgebers in our environment that concurrently affect the cycle of our internal rhythms. For example, body temperature variation is a specifically known circadian rhythm; however, this internal temperature is also affected by environmental temperatures.

Another term commonly heard is *biorhythms*. This is an unscientific claim that how a person feels, does business, makes decisions, and should act during a day is based upon her or his biorhythm for that day, based primarily upon birthdate. This is comparable to astrology. There *are* biological rhythms, which is what we are discussing, but there is no scientific basis for biorhythmic predictions.

Circadian rhythms are true cyclical rhythms. There are hourly variations for each function of our body and mind: highs, lows, and steady states. A curve can be plotted for each function that has a rhythm.

Common human circadian rhythms

Most of the 300-plus rhythms relate to the complex metabolic process within our body, and they all indirectly or directly affect how we function. The more obvious functions are also related to a 24-hour period.

Probably the most important is our sleep cycle, which we have just discussed. Many people are able to awaken without the help of a clock, and others are able to estimate time without a watch. We feel tired and ready for bed at a specific time, and we tend to follow that urge. If that cycle is not followed, the body responds with the same symptoms recognized for sleep deprivation and for fatigue in general. In aviation, this is a major problem because time zones are frequently crossed (trans meridian), which means the Sun and the people at the destination can be several hours off our own rhythms for sleep.

Weekends at home base can result in a miniature change in sleep circadian rhythm. Friday, Saturday, and possibly Sunday night are usually spent staying up long after the usual bedtime. We add to this by sleeping in late on Saturday and possibly on Sunday. By Monday morning the body thinks it is on the West Coast when you are really getting up at 7 a.m. on the East Coast; it's the Monday-morning blues, or weekend jet lag.

In addition to telling the body when to sleep and awake, this circadian rhythm curve has a dip about late afternoon (between 4 and 6 p.m., and again between 4 and 6 a.m.) when it suggests that a nap would be in order. We are less alert and have some of the symptoms of lack of sleep. We recover for a few hours, but then the curve begins to bend downward until it reaches the time to go to bed.

Fluctuating throughout the day, temperature is another powerful rhythm that affects performance. Temperature reaches a peak about midday and then begins to drop off. Owls, or night people, peak toward the end of the day; Larks peak earlier. Increased circadian temperatures indicate when our level of activity, feelings, and alertness can change.

Our appetite is diurnal. The body expects food at certain times based on how the body is programmed by our own habits. If we don't eat at these expected times, our appetite goes away, and we might not eat. Or we might eat a large meal when the body isn't quite ready, subjecting it to the additional stress of unexpected metabolic energy. The body also expects to have a bowel movement at a certain time; if the movement doesn't happen because we are off schedule, the urge goes away. Over a period of a few days, constipation occurs, and we might not be aware that we haven't had a bowel movement. Being dehydrated and failing to eat enough roughage as a result of a long trip complicates this problem.

Other rhythms relate to alertness, urine output, and changes in the levels of hormones and body fluid. Heart rate and blood pressure increase as the day progresses, then they return to a lower level as night approaches.

If all the curves were on one graph, it would appear as a jumbled mess; however, these are finely controlled cycles, essential for health and a productive day. The body and mind, attempting to work efficiently, expect these rhythms to go on without interruption or disruption. Like a symphony orchestra, if each instrument is played independently, the sound would be unpleasant; however, "orchestrated" to play "in concert" (like keeping our biological rhythms in synch), the resulting sound is more than pleasant.

Symptoms of jet lag

When circadian rhythms are disrupted, desynchronosus (jet lag) occurs. When traveling north and south, there are no time zone changes, yet there are symptoms similar to those of traditional jet lag. A major source of those symptoms is the length of the trip, which is fatiguing in itself. The same added problems of dehydration, hypoxia, sitting for long periods of time, and poor nutrition complicate the symptoms no matter which direction you go.

There are, obviously, symptoms that are directly related to trans meridian travel but only because the circadian changes conflict with the destination's sunlight and activities. These symptoms include headache, poor sleep, constipation, disrupted eating habits, vertigo, poor short-term memory, depression, and others. Note that these are similar to the symptoms associated with fatigue alone. So fatigue is essentially the most significant symptom of desynchronosus but from different causes.

It is also recognized that traveling west is easier for most people than traveling east. Part of the reason is that when heading west, one is traveling with the Sun. Since the real "day" for the body is 25 hours (previously noted in this chapter), there is less of a problem; however, you still have to come back toward the east, and it's this continuous changing that is particularly fatiguing. NASA's studies indicate that the problems and symptoms are cumulative. There is also the fact that as we get older, most of us are less tolerant to these circadian changes and disruptions.

Coping with jet lag

Two kinds of circadian changes must be identified. If we are discussing how to adapt to a distant location where we plan on staying for several days or weeks, then we compare this to shift work, such as when a factory worker now goes on the graveyard shift (working nights) after working days. Completely changing our circadian rhythm (resetting our body clock) so that it is in synch with the new location takes about one day for each hour difference. Obviously we can function during this transition, but we are apt to be less than efficient and proficient in our mental and physical performance. Studies show that the greatest chance for error occurs between 4 and 6 a.m. on our current rhythm (Fig. 11-2). If you travel to a location where you are expected to perform safely but it is in the middle of *their* day and 5 a.m. according to *your* body time, then you are at high risk for making errors. That risk decreases as you begin to adapt to that local time.

Pilots, however, rarely stay for very long in one place during a trip, no matter how far away they fly from their home; therefore, in addition to recognizing that they are going to be affected, there are measures that might help them cope with this change in body rhythms. NASA's studies indicate that layovers longer than 24 hours are not recommended so that crews can try to maintain their home base rhythms.

For stops longer than a few days, some suggestions are to get into the sunlight as soon as possible relative to your destination's time. Become active and involved with your destination's time-zone activities. Avoid all alcohol and caffeine before and during the trip. In fact, alcohol on arrival will interrupt your sleep when you want to sleep. Eat lightly, keeping in mind the sedative effects of carbohydrates. "Jet-lag diets" are of questionable value according to most circadian experts, but there are some basics that are helpful, such as the type of foods to eat and the other suggestions already mentioned in this chapter.

The better your health and fitness, the better you can tolerate the change and the quicker you can adapt. Increase your fluid intake, especially water, and minimize the diuretics of alcohol and caffeine. Eat plenty of high-fiber foods, such as raw vegetables and bran.

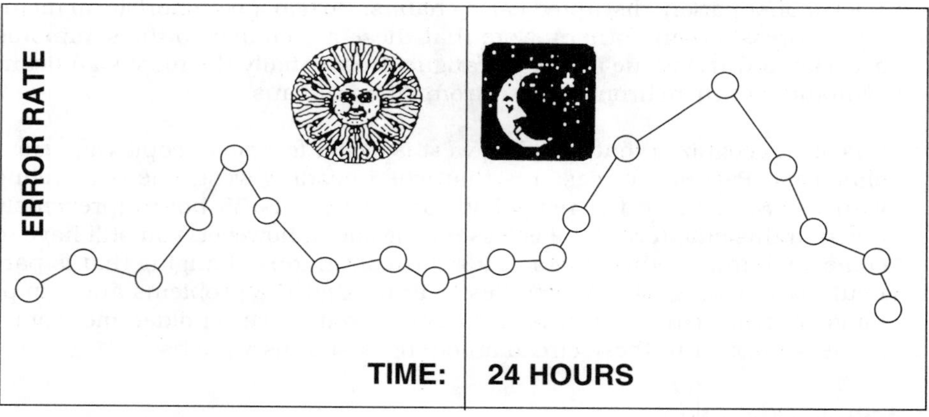

All human functions ebb and flow predictably: the ability to solve math problems, temperature, hormone levels, sensitivity to drugs and pain, oxygen consumption, heart rate and blood pressure and, of course, sleeping and waking. Changing your daily schedule can disrupt htis pattern of cycles. Once disrupted, body rhythms readjust at different rates.

At the time of day when your body temperature falls mental alertness is also depressed. During an all-night trip, your temperatuve low and mental dip will probably occur between 4 a.m. and 6 a.m.

This graph shows how the likelihoood of making an error—from simple lapses of attention to falling asleep—varies through the day. At 4 a.m. the chance of error is 60% higher than average. Between 8 and 11 a.m., people are about 25% less likely to err.

11-2 Jet lag has an adverse effect on performance.

Try to rest before the trip; otherwise, you are fatigued even before you get there. Wiser long-haul pilots watch with knowing glances as their younger crewmates enjoy the nightlife in an exotic new city only to suffer from fatigue the next day on the leg back. It's tempting to enjoy the layover, but the price paid the next day might turn out to be too expensive.

More studies are coming out from NASA and the FAA relative to better defining how desynchronosus affects pilots and how the aviation industry can cope and still run a profitable business and have realistic regulations. This information will be very visible in the aviation press because desynchronosus is a major human factor in aviation. This text can only introduce you to the issue and alert you to pursuing more information on these topics as they are published. Pilots are urged to study this information as it becomes available.

FATIGUE

Fatigue is the common denominator of virtually all symptoms related to deprived sleep and desynchronosus as well as many other physiological stresses unique to flying (as has been stated throughout this chapter). This subsection will review the many causes of fatigue and concentrate on the various symptoms related to fatigue. It is important to recognize that many of the causes are occurring at the same time as the symptoms are present. There can be a quick accumulation of causes and effects, often creeping up on the unsuspecting pilot without her or him being aware of it. Like hypoxia and alcohol, fatigue is an insidious deterrent to safety because one of its main symptoms is a lack of awareness and a feeling of indifference to its effects.

Two general kinds of fatigue are acute and chronic. Like sleep, the acute kind is related to current events and activities, and the pilot will recover when the situation resolves itself and he or she can rest. The chronic kind is more serious in that there is a cumulative effect over several months, creating more fatigue without the benefit of adequate recovery. Such fatigue makes the body vulnerable to illness and increased stress and is generally unhealthy.

Causes of fatigue

The following are specific causes of fatigue of which the pilot must be continually aware. The pilot then maintains a self-monitoring of judgment and performance when these conditions are present and is suspicious of deteriorating performance. Watching for these symptoms in crewmates is essential.

Lack of restful sleep The key is restful. It doesn't matter how long or where the sleep is achieved. If it is not restful, fatigue is the result. It often becomes a vicious circle. The pilot is so fatigued that she cannot relax because she is still mentally rehashing the day's activities and anticipating the next leg of a trip. She is unable to allow sleep to take over, and she becomes more fatigued.

Strenuous activities Like with sleep, working hard is fatiguing and the symptoms are the body's way of expressing that you used up a great deal of energy. One can expect to be more fatigued during a trip if you show up already tired from pushing your physical activities the day before.

Dehydration Often overlooked as a cause of fatigue, dehydration is a common occurrence on the flight deck. Being dehydrated results in fatigue that might not be noticed. The solution is simple—drink water. Awareness of the need for adequate fluids in flight should alert crewmembers to maintain a high intake of fluids. By the time a person feels thirsty, dehydration is already present.

Caffeine Although considered a stimulant and often used to stay awake (an acceptable practice when used under the right conditions), ingesting too much caffeine keeps the body in a high degree of alertness, even to the point of being tense. That, coupled with any trip that might require a lot of mental and physical energy (and adrenaline flow), leaves the pilot fatigued when the destination is reached. This is a form of withdrawal and is a significant factor if he is not rested enough for the next flight, whether it departs in a few hours or the next day. The letdown of not having the stimulation of caffeine leads to fatigue.

Noise and vibration Most prop-driven airplanes and all helicopters are noisy. Noise is a form of energy that is not well tolerated by the body for long periods of time. This is frequently proven when pilots begin to wear earplugs and noise-attenuating headsets and realize that they are not as tired as expected after the trip. Because you can't avoid noise in most aviation situations, fatigue will be a factor unless adequate protection is worn. Vibration is similar to noise (except the frequencies are lower) and still causes fatigue. Little can be done to prevent vibration, but the combination of earplugs and noise-attenuating headsets, along with adequate padding in the aircraft seats, will help.

Illness The body fights illness through its own metabolism. Colds and flus are common, and some pilots go ahead and fly while they are ill. These situations are fatiguing because of the increased energy needed by the body and the fatiguing effects of the illness itself. Recall how tired and miserable you felt the last time you had the flu or a cold.

Over-the-counter medications Many medications have antihistamines, which have sedation as a side effect. The combination with caffeine (often used in these medications to overcome the effects of drowsiness) is fatiguing. Now there are the cumulative fatiguing effects of the treatment coupled with the fatiguing effects of the illness one is treating.

Hypoglycemia Fatigue is the most serious effect of this commonly self-imposed stress. Proper nutrition and prevention is the best way to deal with hypoglycemia. If the pilot is prone to these symptoms, carrying some quick energy in the form of snacks, peanuts, or candy will help. While not being the most nutritious meal, such high-energy foods are important under these circumstances. Fatigue from hypoglycemia is slow to develop, and the pilot will probably not relate how he or she is feeling to being hypoglycemic.

Hypoxia To many, fatigue is their first sign of being hypoxic. This is a very subjective sign, and the pilot might think his or her feeling of fatigue is the result of something else. Working at altitude, whether in the plane or on the ground, will cause fatigue. Those who have been hypoxic and know they were hypoxic (as in a chamber ride) realize that they are fatigued even after they have returned to their normal altitude.

Impaired vision Pilots do not like wearing glasses, especially as they get older. Not wearing glasses—and hoping the eyes and brain will sort the out-of-focus images—is fatiguing. Most have heard of eye strain, and that is what the eyes are doing; they are being strained. Unless you have unforced 20/20 uncorrected vision, the obvious solution is to wear glasses with adequate lighting all the time, especially for reading.

Thermal Excessive heat and cold are common sources of fatigue. The body is working to maintain its core temperature and requires calories and energy to accomplish this task. The result is fatigue, even after temperature control is achieved.

Flicker Fatigue is often noticed when helicopter pilots are exposed to the flicker effects of the rotor blades.

Boredom One needs to be physically and mentally active to keep from being bored. Although not directly fatiguing, the feeling is perceived as fatiguing and many of the same symptoms are present.

Circadian change Fatigue is the most common complaint with long-haul trips across time zones. *Fatigue is caused by the combination of trip length and desynchronosus.*

Skill fatigue This little-recognized source of fatigue is a result of demands of persistent concentration with using a high degree of skills. This is an operational problem related to the pilot's motivation to continue pushing himself beyond his capabilities. There is a conflict between the pilot's ego and his good judgment. The result is fatigue, which is not noticed until after the trip and is especially dangerous if there is another trip (or leg) soon afterward. This is common in emergency medical evacuation where the mission becomes all-important, or when any pilot is flying in instrument meteorological conditions for hours with a lot of traffic, or when practicing for a checkride. The "can-do" altitude can be dangerous.

Unresolved stress This is common for everyone. It's been said that nothing is more fatiguing than the eternal hanging on of an uncompleted task. Personal stress that is not resolved drains the body of energy and is often associated with depression, which also is fatiguing or gives the perception of fatigue.

Symptoms of fatigue

Various symptoms are directly related to fatigue; other symptoms are also associated with other situations. Hypoxia, for example, has several symptoms similar to those caused by fatigue; therefore, there is a cumulative effect of various causes resulting in a more pronounced and significant symptom.

Probably the most important symptom of fatigue, as with hypoxia and lack of sleep, is the feeling of indifference, of "settling for less" in performance. The pilot is unaware she is fatigued and allows her personal acceptable tolerances and limits for performance to expand, which allows for mistakes or substandard skills. Taking unnecessary risks is common.

A key element in crew flying is to watch your crewmate. Maintain a high index of suspicion if he or she isn't performing as expected or responding to checklists or ATC appropriately. Because of an increasing unawareness of increasing discrepancy in the fatigued pilot, he is probably the last to know he is impaired. In single-pilot operations, she must be continually monitoring her own performance, analyzing how she feels, and then taking action.

Unfortunately, the mind has minimal long-term memory for fatigue. In other words, we do not really recall how impaired we were when we were fatigued. The mind tries to protect us from being overwhelmed by such bad experiences. The same is true for pain. We know we hurt but can't really define how incapacitating it was after the event. The next time that you are fatigued and you know it, make mental or written notes regarding how you really feel, whether or not you could fly an ILS at the end of the day in turbulence, and whether or not you really care. It's like hypoxia. We go to chamber rides primarily to remind us what our personal symptoms are for hypoxia. Fatigue is no different. If conditions are ripe for being fatigued, recall how you felt the last time.

Other specific fatigue symptoms unique to a pilot in flight include the following:

Increased reaction time The pilot takes longer to react to a change or emergency. Automatic and instinctual response to any abnormal flight is slowed, such as recovery from a tight turn that is losing altitude or recognizing a stall. It has the potential to turn minor problems into major accidents, both in the air and on the ground.

Channelized thought processes The pilot tends to concentrate on one thought or activity at a time, rather than on several, which is essential in flying. Concentration is also narrowed as the pilot focuses only on what the weather is ahead and does not plan his approach. He might be focused on issues outside the cockpit, such as an argument at home or the unknown results of his medical evaluation. Flying requires a multitude of thoughts, all processed by the brain in a sequence or sometimes at the same time. The number of separate thoughts that can be processed decreases as fatigue gets worse.

Fixation Here the pilot fixates on a single instrument, staring at a blinking warning light on the panel or listening intently to a strange sound that is not a factor of the flight. The instrument scan becomes less efficient and traffic is missed because the pilot is concentrating on some object below.

Short-term memory loss The pilot quickly forgets what ATC's last clearance was, and she didn't think to write it down. Or she forgot to change the radio and is calling departure on tower's frequency. While she is occupied with other activities, she hears a crewmate call for flaps, but she forgets to respond to the call.

Impaired judgment and decision making Dumb mistakes are made, and, while they might be insignificant, they might just as easily become another link in a growing chain of events leading to an incident or accident. The pilot takes chances and shortcuts rather than following acceptable procedures (most accidents could be prevented if procedures were followed); he really knows better but just doesn't know he's messing up. Risk assessment is minimized.

Easily distracted The pilot notices a map falling from the panel and reaches for it rather than holding back. A sound or other visual cue away from a scan becomes more important than the scan for a brief moment. She will miss an item on the checklist because she is distracted by conversation that is irrelevant to the flight. (The sterile-cockpit policy helps prevent this.)

Sloppy flying Fine motor skills are compromised, and the pilot finds himself holding onto the yoke with his hand rather than his fingers. He loses the fine-tuning of turns, climbs, and power-setting changes. More importantly, he has to think about what to do rather than instinctive and "automatic" flying using well-learned skills. He has to think too long about which direction to turn to change heading, or in what order controls are changed to begin a descent. His instrument flying is not precise and there are errors in timing. Much of our flying is automatic, a habit. Anything that interferes with these skills impairs the pilot and might not be noticed.

Decreased visual perception It takes longer to focus (accommodate) from a distant object to a near object. This becomes especially crucial while flying an instrument approach to landing. The scan takes longer because the pilot isn't focusing well, or he or she might just skip a part of the scan.

Loss of initiative The pilot doesn't really care anymore and is unwilling to maintain a high level of skill or accuracy. He or she becomes passive about his or her own high standards of flying. He or she lets someone else make an error and won't say anything.

Personality change He becomes more irritable and is easily depressed with minor problems. There is an increase of intolerance to small nagging and minor irritants of a crewmate's habits.

Attitude It bears repeating that the pilot is willing to settle for less and is not aware of his change in attitude and performance.

Depression Even though the pilot might ordinarily have a positive outlook on life, fatigue can generate a rich imagination of things that can go wrong. Depression, although short-term, can result in a gloomy perception of life. Fatigue and depression are a poor combination of feelings to be used in flight.

Coping with fatigue

Unfortunately, there is no "magic pill" that prevents fatigue or makes it go away. Only through education and awareness can the safe pilot make a professional decision as to whether or not to fly. There is often a fine line between acceptable fatigue and fatigue that could seriously impair the pilot. The challenge is that most often significant fatigue occurs well into the flight, just prior to a descent and approach to a tricky landing.

Someone once said that "takeoffs are optional, landings are not." Before taking off, the pilot must make a careful assessment of what condition he or she is in and what can be expected during the flight or during the trip (outbound, layover, and inbound). All these factors must be considered as cumulative, and if there is doubt, either turn the plane over to a crewmate or forgo the takeoff.

Research studies concerning fatigue are always in progress. There is a lot to learn about fatigue and its effects on flight, but the basics are already well known. The problem is that the policymakers are the ones who ultimately determine when the pilot should fly, and only the scientific proof from research will validate and mandate changes in policy. Obviously, policy and regulatory changes to protect the pilot from fatigue are driven by budget and "cost effectiveness"; there would have to be direct proof that fatigue caused a specific incident or accident, which is not an easy thing to prove. Flight crews must take it upon themselves to regulate themselves and to deal with causes from a preventative position.

It is ultimately up to the pilot to make the final decision to fly. Prevention of fatigue is in the same category of enhancing sleep. Being healthy and physically fit and avoiding the causes that have been discussed in this chapter are the only practical means of assisting the pilot in deciding if he or she is safe.

To put the topic of fatigue in perspective, the following is a possible scenario of what could happen on a typical flight. The obvious continuous thread of concern is the pilot's fatigue. The author wrote this for commercial pilots in *Business & Commercial Aviation Magazine* in 1991.

> Let's begin. It's already been a long day with several stops and waiting for several hours at two of the stopovers. This is now your last leg and you're flying.
>
> You are cleared to taxi and, en route to the active, you go through the rest of your "before takeoff" checklist. You are transfixed by an emergency vehicle heading across the tarmac to the terminal and as a result of this minor *distraction* you miss the command to set flaps. Your copilot has to remind you later on into the list.
>
> At the last minute, tower tells you to "hold your position for landing traffic" that seems a long way off on final. You let tower know about your dissatisfaction of having to wait for traffic that, in your opinion, is a safe distance. You are usually patient in these conditions, but right now you are *more irritable and easily annoyed.*
>
> You are now cleared for takeoff on 23 Right and to fly a heading of 210°. You forget the clearance and maintain runway heading until ATC reminds you of the 210° heading. You call departure on the tower frequency, another lapse in *short-term memory.*
>
> You climb out into the clouds as you *fixate your concentration* on a flashing yellow light on the annunciator panel, nothing significant, but you also are sliding into a turn as you stare at the light. Your copilot gets you back on your scan after tactfully saying "heading 210."
>
> You finally reach cruising altitude when ATC requests you to turn 30° to the right for traffic, your copilot is asking where the approach plates are, you are trying to trim the altitude hold, and the waypoint light is flashing. Your mental focus is really on the radar, which is showing thunderstorms on the flight path. Only after ATC has to remind you of your heading change do you realize that your *thoughts were channelized* on only one situation, the radar, when you should have been alert to everything.

The rest of the cruise was uneventful—boring. Your mind is wandering as you go through your scan. You begin to think about the unopened letter from the IRS sitting on your desk, your FAA medical that is due next week, and your borderline blood pressure. Your teenager got his or her first speeding ticket yesterday. You imagine many undesirable scenarios for each of these situations, and you are beginning to scare yourself; you fear you are *becoming depressed* even though nothing serious has actually happened.

Finally, you're "cleared to descend to 4,000 at pilot's discretion" while still 75 miles out. You begin an immediate descent. Your copilot questions your *judgment* and wonders aloud if your *decision* to go down so soon is premature, knowing it's bumpy lower and your passengers are finishing their lunch. "What difference does it make? Besides, who is in charge anyway!" you snap back.

ATIS reports 1,000 and 1. You choose to fly the ILS by hand. You find yourself chasing the needle, driving with a firm grip on the yoke instead of flying the airplane with a gentle touch. Then your first bounce is hard, and your second floats to a mushy touchdown. You think to yourself that you haven't experienced this kind of *sloppy flying* since you were a student. Your final act of frustration is to yell at your copilot for not alerting you to your altitude.

Most of us have been in one or more of these situations. Fatigue is insidious, and there is no way to avoid its symptoms. The degree of impairment, however, is going to be a result of the pilot's awareness of and insight into the whole subject of fatigue.

12
Acceleration

The last thing Karen remembered was looking straight up and seeing nothing but blue sky. She had started the pullup at 10,000 feet and a constant 3 Gs. About halfway through the loop, her vision started to tunnel and turn gray. By the time she had returned to what she thought was her starting position, she had lost all vision and was literally flying by the "seat of her pants." After several seconds, her vision returned, and the cockpit instruments showed that she was at 10,300 feet and in a slight descent. It was an uncomfortable situation, and she was glad that she had put a lot of space between her aircraft and the ground.

ACCELERATION IS A PART OF FLYING and it can uniquely affect the body in every axis. In most cases, other than in the high-performance fighter aircraft of the military, there is little chance for high-G maneuvers, except in aerobatic aircraft or extreme unusual attitudes. Because such extreme situations are uncommon, the pilot can usually tolerate these forces that are present without impairment; however, it is important that the pilot be familiar with acceleration forces because newer high-tech aircraft are capable of pulling Gs often in unexpected situations. Furthermore, the physical forces and physiological response to any accelerative situations are the same and will, sometime in your flying, affect your performance in an unpredictable way.

Before discussing the forces of acceleration, certain terms must be reviewed. *Speed* describes the rate of movement of an object and is expressed as distance covered in a unit of time (miles per hour). *Velocity* describes both the magnitude and direction of motion and is measured in distance per unit of time in a particular direction. The velocity of a body changes if it changes direction, speed, or both. *Acceleration* is a change of velocity in magnitude or direction and is generally expressed in distance per second per second.

TYPES OF ACCELERATION

Acceleration is the rate of change in velocity and is measured in G units. As a point of reference, the pull of gravity is considered 1 G. Several forms of acceleration are distinguished by their impact on speed and direction:

- *Linear acceleration* is a change in speed without change in direction, such as an increase in thrust in straight-and-level flight.

- *Radial* (or *centripetal*) *acceleration* can occur in any change of direction without change in speed—for example, when executing a turn or when pulling out of a dive.
- *Angular acceleration* occurs when both speed and direction are changed, as in a tight spin. For practical purposes, this type is not common, but can play a role in disorientation.

G force in aviation is usually referring to the force exerted on the long (or vertical) axis of the body, but it can also affect the horizontal and transverse axis. How these G forces impair the pilot is determined by the intensity and duration as well as the direction. For example, jumping off a table that is 4 feet high can result in a force of 12–15 Gs; however, a pilot in a 12-G turn would be unconscious in about 2 seconds.

In general, therefore, the body can usually tolerate about 5 Gs for up to about 5 seconds if the pilot isn't otherwise protected and adapted to tolerating higher Gs. Beyond that point, the pilot can experience a variety of symptoms, some incapacitating, most distracting or uncomfortable, and often disorienting.

Positive G forces occur when the body is accelerated in the headward direction and the body is forced downward into the seat. Negative G force is in the footward direction, and the body is lifted out of the seat. The body is less tolerant of negative G forces than positive G forces (Fig. 12-1).

COMMON SITUATIONS IN CIVILIAN FLYING

The study of G forces is far more important in military aviation. For civilian purposes, the discussion will be limited to predictable effects on the unprotected body and how the pilot can decrease his or her tolerance to unexpected G forces.

During any maneuver that produces positive Gs, the weight of the body is increased in direct proportion to the magnitude of the force. The 200-pound pilot on the ground will weigh 600 pounds under 3 Gs. More important is how the body organs are affected and how the physiology of the body is disrupted during this exposure.

The skeleton and soft tissues of the body can tolerate significant G forces without problems; however, the circulatory system, similar to any hydraulic system, is greatly affected by changes in G forces. This occurs because there is a direct effect on the hydraulic pressures of blood in the body, especially as the blood tries to get to the brain in a continuous flow (Fig. 12-2): therefore, in tight turns, rapid accelerations or decelerations during flight, and unexpected changes in flight resulting in accelerative forces, there will be physiological changes that can range from an awareness of pressure into the seat to various levels of consciousness and visual disturbances.

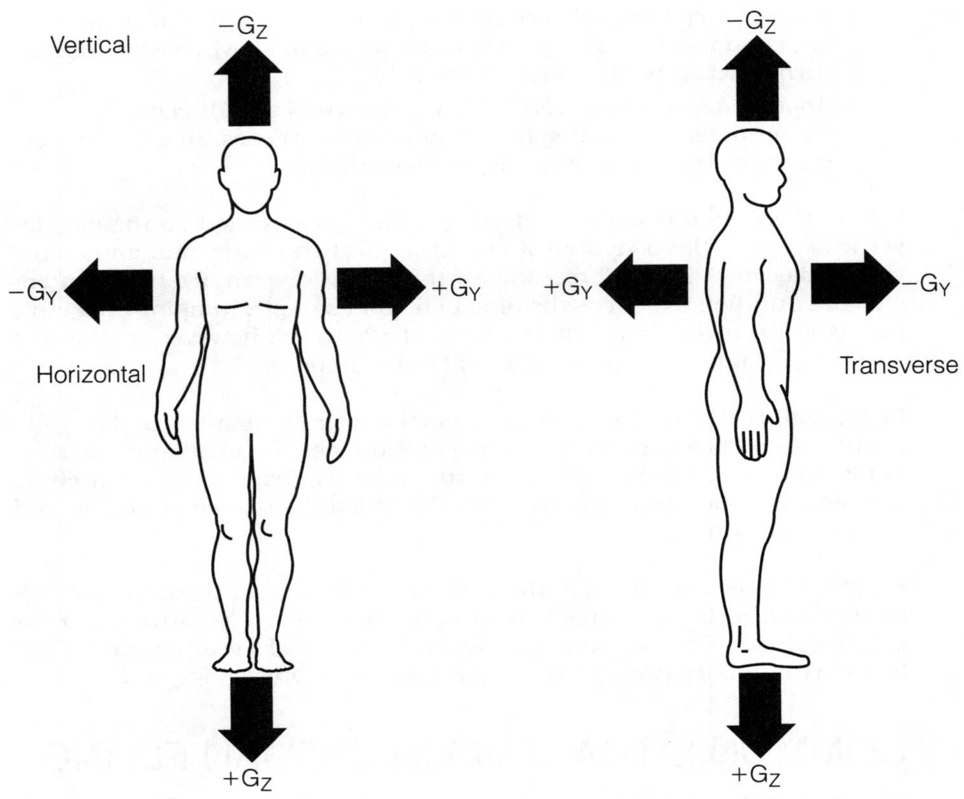

12-1 The three axes of G forces that act on the body are *horizontal*, *vertical*, and *transverse*.

SYMPTOMS OF G FORCES

Most problems associated with G forces are related to the inability of the blood to reach crucial organs of body functions, specifically the heart and the brain. If G forces are significant, the heart is unable to get blood from various parts of the body (such as parts distant from the heart like the lower abdomen and the legs) returned through the lungs and to the brain. The brain must have an adequate blood supply at all times. A lapse of even a few seconds can compromise oxygen levels to brain cells, and make impairment imminent. Since the eyes are an extension of the brain, visual changes will be the first symptoms noticed (Fig. 12-3).

Closely following visual changes will be a change in consciousness, which is obviously much more incapacitating. This is important to keep in mind because after the vision begins to change, various levels of unconsciousness are quick to follow. Of course, the pilot is unaware that she is approaching unconsciousness, and her fate now is dependent on how soon and how much the G forces are reduced.

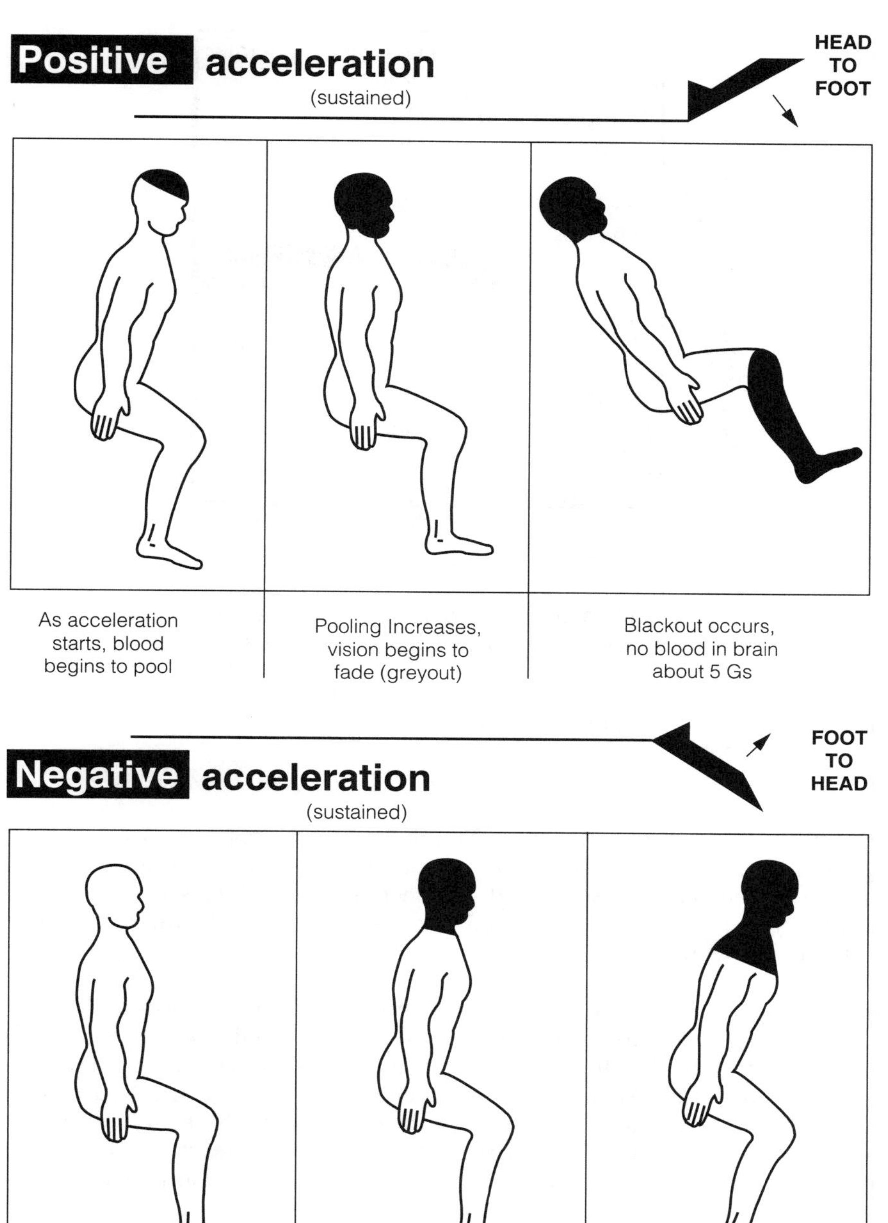
Positive acceleration
(sustained)
HEAD
TO
FOOT
As acceleration
starts, blood
begins to pool
Pooling Increases,
vision begins to
fade (greyout)
Blackout occurs,
no blood in brain
about 5 Gs
FOOT
TO
HEAD
Negative acceleration
(sustained)
Blood pools in
the head, face
feels flushed
Effect continues,
vision begins
to redden
"REDOUT" occurs,
feeling of eyes
"popping out"
Maximum G factor in negative acceleration for pilots is 3
Effect on pilot—5 seconds

12-2 Acceleration forces to the body (sustained).

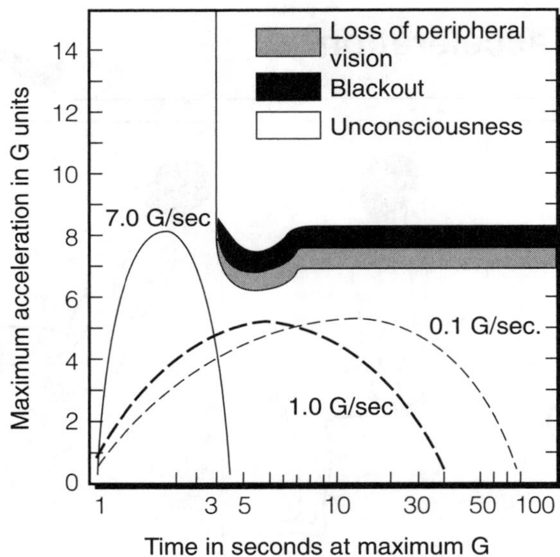

12-3 Acceleration and time at maximum G required to produce visual symptoms and unconsciousness. Curves showing different rates of G development are given to show the importance of this parameter for the occurrence of peripheral vision and blackout.

To complicate matters, even after the pilot regains consciousness, there is a period of amnesia where the pilot doesn't know what happened. This whole sequence can take up to 20–30 seconds, which doesn't sound very long, but it is plenty of time for a pilot to lose control of the aircraft and be unable to recover.

Some of the more specific symptoms include the feeling of pressure in the early stage of acceleration stress. The increased weight of the body's organs tends to pull the diaphragm down and thus interferes with respiration. As G forces increase or persist, there is a change in the hydrostatic pressure in the circulatory system. With a force of 5 Gs, this hydrostatic pressure is higher than the pressure produced by the heart pumping, and blood no longer gets to the brain. At this point, unconsciousness occurs rapidly.

Prior to this final event, there is a blackout period, at about 4 Gs. During this time, the pilot has a complete loss of vision but is still conscious. Tunnel vision might precede this stage and would be the first indication of impending impairment. Since hypoxia (lack of oxygen getting to the brain cells) is the symptom and decreased blood pressure is the cause, the combination results in a period of disorientation and loss of memory that can last a minute or more.

TOLERANCE TO G FORCES

Tolerance limits are usually fairly constant between individuals; however, various factors can increase or decrease one's tolerance to intensifying G forces. The most obvious is anything that interferes with the efficiency of the cardiovascular system.

The body, however, is able to increase its tolerance through a number of other factors. This includes being in excellent physical condition. Another is to use what the military calls the M-1 or L-1 maneuver, or "grunt," which involves tightening all skeletal muscles and straining hard while holding your breath. (An L-1 maneuver included some exhalation during the straining.) It's somewhat related to straining when constipated or trying to lift a heavy weight. Then take a quick breath, and do it again about every 4–5 seconds until the G forces have diminished. This technique raises arterial blood pressure enough so as to endure an additional 1.5-G force or more.

NEGATIVE G FORCES

Experiencing negative G forces is a very uncomfortable feeling that is not often experienced except in aerobatic aircraft. It's similar to going over the top on a Ferris wheel ride. Here, hydrostatic blood pressure is the reverse of positive Gs, with the blood being forced to the head and into the brain tissues. The symptoms are commonly referred to as a "redout," compared to "blackout" in positive Gs. A force of only 3 negative Gs is considered the limit for most humans.

Most G-force stresses are short lived in civilian aviation (except for aerobatics), and the body usually tolerates these forces well. As with most physiological events, it's the recognition beforehand that such things can happen to the body in flight that keeps the pilot out of trouble. The susceptibility of the pilot to any G forces is related to the physical condition of the pilot as well as the level of hypoxia and fatigue.

13 Crew resource management (CRM)

First officer Nelson sits in his seat on the last leg of a two-day "quick turn." In addition to fatigue, there are some personality conflicts that have broken down the lines of communication in the cockpit. The crew has been given an altitude-crossing restriction by ATC, which FO Nelson understood to be 11,000, but the captain has set the altitude capture for 10,000 feet. FO Nelson would normally speak up when he is sure he is correct, but at this point in the trip he is kind of glad to see the captain making a mistake because the captain had been correcting him ever since the trip started. At about 10,300, ATC queries the crew about their altitude, at which time the captain corrects back to 11,000 feet. He then brings it to FO Nelson's attention that he should have been backing the captain up. He is right. FO Nelson has let his personality as well as the difficulties encountered when flying with rotating crewmembers interfere with the normal and safe operation of the flight. Missing an altitude reflects badly on the entire crew and violates standard operating procedures as well as the safety of the flight, all as a result of a petty personality dispute.

CREW RESOURCE MANAGEMENT, or CRM (at one time called *cockpit resource management*), is now a common aviation term, especially in commercial aviation. This concept of improving the safety of flight through viewing the crew as a management team just like any business is only a few years old. Some say it originated in a major airline with the help of human factors specialists in several government agencies when they recognized that the inherently independent personality of a pilot often interfered with a safe and productive crew working together effectively. Their studies and reviews of incidents and accidents clearly indicated a high percentage of mishaps were a result of some pilot "error" in the form of not following procedures and not effectively communicating with other members of the crew. This was coupled with poor judgment and decision making.

The intent of this chapter is not to give comprehensive insight into how to accomplish the goal of improving the management of cockpit and crew resources. In fact, this whole concept is still new enough beyond the major airlines that there is no clear and consistent method of achieving CRM's goals for a given company or flight department. New information is coming out all the time from the experiences of companies that are generating their own CRM programs, often under different titles. Articles, books, and seminars are becoming more available as more is learned about the optimum way to develop, teach, and reinforce the principles of good CRM. Because any CRM program must be tailored to the culture of the company, there are many variables that interfere with any description of a generic program.

Therefore, the purpose of this chapter is to acquaint the reader with the concept of CRM and how psychology of flight (including stress, attitude, personality, and other personal traits) fits into the program. The role of judgment and decision making is also considered. Equally important is the impact of the pilot's medical airworthiness—health and well-being—on the effectiveness of management techniques.

The pilot can be legal and well trained in CRM, as well as experienced and proficient in flying performance; however, if the pilot is not medically airworthy and does not respect the issues discussed in this text, then that pilot is not safe because learned skills in CRM will be compromised. Every pilot has been subjected to some of the disabling situations discussed in flight physiology. It should be apparent that even under the best of conditions, performance can be eroded by fatigue, hypoxia, illness, stress, and other "abnormal" conditions, and the pilot is less safe despite knowing what should be done.

A long-term NASA research program and other studies have identified and shown that nearly 80 percent of air carrier (and other flying) accidents and incidents are related to some form of human factors. A common characteristic is that the problems encountered by aircrew are often associated with poor group decision making, inadequate leadership, ineffective communications, and poor management of available resources. It was also noted that training programs place strong emphasis on the technical aspects of flying, often at the expense or even lack of training in the various types of crew management strategies and techniques that have been found to be increasingly essential to safer flight operations.

Publications dealing specifically with the issues of CRM, stress, and other "nonmedical" aspects of safe flight are readily available, and it is recommended that these additional resources be used to further educate the pilot in CRM concepts and techniques.

DEFINITIONS IN CRM

The basic meaning of CRM is crew coordination of the use of all resources in all phases of flight. This can be further defined, using traditional terminology common in any situation dealing with people, personalities, tasks, and management skills.

The cockpit crew

The cockpit crew was once considered the sole source of skills and expertise concerning flight and management performance; however, it is now recognized that the cabin crew (flight attendants and available trained passengers) is also essential in a total team, as well as ATC and company personnel on the ground. The increasing usage of terms such as "crew resource management" or "crew leadership resources" reflects the role of other people who can have a direct assistance to safe flight, especially in a flight that is getting out of control.

Resources

The term *resources* pertains to any kind of resource that can be used to safely fly an aircraft, especially under less than ideal or expected circumstances. The crew, meaning all usable people on board, is the first to be considered. Ground crew, such as dispatchers, meteorologists, mechanics, doctors, air traffic control, company operations, and other associated personnel, are also available resources. Mechanical and electronic assistance must play a role in the total picture and includes the wealth of information from instruments, monitoring gauges, radios, radar, computers, and many other sources of information crucial to any flight or emergency.

Even single-pilot operations need to know CRM, and the pilot must be aware of the vast resources available before, during, and after any flight. Incidents and accidents don't happen for lack of resources or guidance. But a crew's reluctance or inability to access those resources leads to serious consequences. For example, one of the most significant causes of mishaps is the failure to follow established procedures. Is this pilot error, incompetence, complacency, arrogance, indifference, or just not doing the job?

Management

The foundation of any successful business is management of available resources. The best resources, even in abundant supply, are worthless if not managed properly. Companies often fail because of poor management of the resources of people, equipment, and facilities made available to their managers. The same is true onboard an aircraft. There is a chain of command—levels and boundaries of responsibility—as with any business or organization. But historically this chain of command has either not truly existed or has been abused in the sense of overbearing pilots working with unassertive crew. Frequently in the past, a flight proceeded from A to B without employing basic yet essential management skills, often with unacceptable results. CRM is now recognized in aviation as the way to improve this situation.

Another way of considering the definition of management is that it represents the clear understanding of "who's in charge," and "who's responsible," and "who's available" to manage and use the resources to perform essential tasks plus a mutual respect for each of these entities.

SKILLS LEARNED IN CRM

Many crew resource management skills were once thought to be "untrainable." Yet research and experienced trainers have shown that these techniques can be learned. The industry is still learning, especially in ways to effectively train pilots and other crewmembers. Guidelines are becoming more evident as the management skills known for years in business are incorporated into flying airplanes safely.

This list can be long when considering specific components of these skills. For the purposes of this chapter, however, the following are the key elements of any CRM program, especially as expected by the FAA. The agency has several regulations and advisory circulars explaining usage of CRM, with AC 120-51B, *Crew Resource Management Training*, being the most informative. A training program consists of three phases:

- Initial indoctrination/awareness
- Recurrent practice and feedback
- Continuing reinforcement

The following skills should become as automatic as latching the seat belt and shoulder harness in a car or aircraft.

Decision-making skills

Decision-making skills are necessary for all crewmembers in all situations. This means taking into account the information from the various resources and making an intelligent and professional decision at that moment and in a timely manner. Such decisions cannot be effective if poor CRM skills are practiced. A pilot must be confident of the choices she makes and be able to competently communicate her decisions. That can only happen after considering all options and accessing all resources.

Leadership skills

Leadership techniques are very individualized, but everyone can be taught basic qualities of being a leader as well as being a follower or team player. Such insight is not just for leading others but to also understand and respect the challenges to any and all leaders during all phases of flight. Part of the team is a mutual respect for the leader and a clear understanding of who is ultimately responsible. Additionally, the leader must respect the opinions and observations of the rest of the team by setting an example for all to follow.

Communication skills

Communication is probably the most important factor in any relationship. Without knowing and comprehending what the other is telling you, there is chaos. Failing to communicate is the age-old problem for everyone. Many studies have shown that most people are poor communicators and even worse listeners. If a situation is not clearly understood or heard by all parties involved, this is a setup for an accident. Active questioning is essential if any issue is not clearly understood.

Judgment skills

The use of good judgment is difficult to teach, but it can be done. It must be associated with real-world situations and real-world practice. Attitude, personality, intelligence, and knowledge are key elements, much of which is learned only from experience and maturity along with continued reinforcement of the skills that can be taught. Judgment also comes from applying lessons learned in situations that went wrong. That means keeping an open mind and being willing to change old habits.

Crew coordination skills

Crew coordination comes from applying all these skills and working as an effective team to accomplish a common objective: getting the aircraft off the ground, to its destination, and back on the ground with the passengers thinking your job is simple. This becomes extremely crucial in an emergency situation, where everyone must depend on the resources and skills of each other.

Other skills that accompany those just defined are learned in most CRM courses. The skills include stress management, respecting situation awareness, problem solving, team management, and other interpersonal skills. All these skills must be monitored and tested on a periodic basis to ensure that they remain current. One way to track how a crew manages its resources and works within CRM parameters is through simulator training and a program called *line-oriented flight training* (LOFT) where actual "chock-to-chock" trips are planned with scenarios of several situations that require crew coordination and management of resources to salvage a flight from disaster. CRM LOFT uses the same facilities but incorporates CRM challenges. Instructors then review how the individuals functioned as a crew, often using a videotape of the ride to enhance instructional effect.

TRAINING IN CRM

Part of CRM training includes how to monitor crewmates and how to communicate notable situations or substandard performance observations, real or anticipated. Furthermore, the pilot must be able to communicate to other members in a way that is understood and respected by all. The terms *crew concept*, *teamwork*, *working together*, and *CRM* all mean the same thing. The days of flying "solo" in a vacuum of communications and performance skills are over. Today we depend on many people and many sources of information to make flying safe, enjoyable, and more dependable. Pilots are known for their independent and "in-control" personalities, but it is recognized that this attitude no longer fits in the "team concept" of modern flight. CRM is a program and a philosophy, not just a course. It's a way of life that is introduced at the beginning of flight training and reinforced throughout the flying career.

A good training program in CRM should include at least three distinct phases:

- An awareness of the many CRM issues
- A practice and feedback of lessons learned using CRM techniques
- A continual reinforcement of CRM principles

CRM, therefore, is defined various ways. The FAA has a good example:

- It is a comprehensive system for improving crew performance.
- It is designed for the entire crew population.
- It can be extended to all forms of aircrew training.
- It concentrates on crewmember attitudes and behaviors and their impact on safety.
- It provides an opportunity for individuals to examine their own behavior and make individual decisions on how to improve cockpit teamwork.
- It uses the crew as the unit of training.
- It is a training program that requires the active participation of all cockpit crewmembers.

DETERRENTS TO GOOD CRM

Probably the greatest challenge to effective management techniques is our individual personality. We can be taught how to act and respond to others, how to communicate, how to manage people, and the like. But if our own personality traits, good and bad, interfere with how we act, and if the other person doesn't know how to cope with our traits, the whole management concept erodes.

We can't significantly change personalities for any extended period of time; therefore, we must learn how to deal with these differences and still be able to interact and communicate. A breakdown in many of the skills mentioned earlier is attributed to an individual's inability to practice the skills and a lack of reinforcement of the principles of good CRM. The results are many and varied and thus are often a cause of a breakdown in safe flight procedures.

Furthermore, if the pilot is not physically fit, psychologically stable, or "medically airworthy" as described throughout this textbook, these management principles and techniques become less effective and even detrimental to safety. For example, some of the examples that follow could be the result of the effects of fatigue, disorientation, self medication, stress, or all of these together. We all recognize that we can be less than productive when we are not feeling well; therefore, when considering CRM and its skills, keep in mind that these skills can be eroded by how we feel with a simple *head cold.*

Poor cockpit procedures

A simple review of a checklist is sometimes interrupted by a distraction, and takeoff flap settings, for example, are forgotten. Focusing on unnecessary conversations in the cockpit results in a missed clearance from ATC.

Lack of assertiveness

When the copilot/first officer notices a problem or a situation that could lead to a problem, the captain must be notified, no matter what kind of a person he or she is. If others on a crew are intimidated and are not assertive, then the chain of events that lead to an accident will not be broken. This happens, for example, when fuel levels are allowed to deteriorate without apparent recognition by the crew and/or no one informs the captain.

Lack of effective leadership

The captain is in charge and ultimately responsible; however, if he doesn't delegate authority, prioritize duties, or set a good example, there will be a breakdown in effective resource management.

Poor group decision making

The entire crew must be a part of the flight. Anyone noticing a change in expected performance of a crewmember, the aircraft components, or flight path needs to make the correction or alert the crew so they all can decide the course of action. An approach to minimums requires everyone doing his or her job, and if there is a change, such as a heading deviation, everyone has to be aware of the change.

These and more are the conditions that can lead to unsafe situations, incidents, and accidents; however, on the positive side, there is clear evidence that CRM works and, indeed, has saved crisis situations from turning into disasters. This was the case when a cargo door was lost on an airliner over water, and because the crew worked as a team, they recovered. Another example is the survival of more than 100 passengers after a wide-body aircraft lost all hydraulic fluid. Both cases are examples of where CRM and its techniques and skills have paid off. There is no doubt in the minds of these crews that CRM training saved the day.

With changing regulations regarding training, especially when added to requirements for most of commercial aviation, CRM will play an increasing role. In fact, even though human factors includes more than just CRM and continues to be vitally important, evolving CRM expectations will include familiarization of fatigue, stress management, and situational awareness.

CRM is an evolving process as we all learn what works best and what doesn't. There will always be skeptics and those who fear that the threat of intervention by others will remove their authority. CRM should be the ultimate program for all pilots and is most effective when all share their insights and observations throughout the learning process. Indeed, convincing upper management that CRM is essential is often a hard sell, similar to showing that a safety program is cost effective. Without the full support of incorporating CRM into a flight department's program, CRM is doomed to fail.

STRESS MANAGEMENT

Stress is probably discussed and written about more than any other medical or psychological topic. This topic can be found in daily newspapers, weekly magazines, department-store bookshelves, and video programs. Everyone experiences stress to varying degrees throughout their lives, some more than others. In fact, stress is essential to keep us from becoming complacent. Someone once said that stress is the price paid to be a racehorse instead of a plow horse. Stress by itself is not the problem. How we cope with stress can become the problem.

Flying airplanes can be stressful. There is much to learn and much to remember, and there is the challenge of many crucial things happening at the same time. We learn something new every time we fly. We are being continually monitored and checked. Many pilots have high expectations of themselves, often becoming perfectionists. These often unrealistic expectations increase the amount of stress with which the pilot must deal, especially if that pilot perceives that he or she is not meeting those expectations or is losing control over a problem. Another common cause of stress is taking on too many responsibilities and activities, not necessarily related to flying. Nothing can be more fatiguing or stressful as an uncompleted task over which there is a sense of lack of control and not enough time. Even good stressors such as marriage, buying a new house, and upgrading to a new airplane or to the left seat are still stressors.

If stress becomes unmanageable, a variety of symptoms develop, such as anxiety and unhappiness and sometimes panic. More subtle forms of impairment become apparent to others but often unnoticed by the pilot. No one can be free of stress. If we can learn to cope with stress effectively, we can become better pilots. Unresolved stress disrupts the body's immune system, making it more susceptible to disease and intolerant to environmental problems. Sleep is disrupted by the continuous rehashing of the day's stresses.

Stress that is not dealt with effectively leads to *dis*tress. The distressed body will respond with a variety of symptoms, such as upset stomach, headaches, muscle cramps, and diarrhea. The most common symptom is a feeling of fatigue; therefore, the variety of symptoms, especially fatigue, are cumulative with those from other sources as described in the other chapters.

Some situations that are common to pilots under stress are distraction from tasks, inattention to flight activities, preoccupation with insignificant problems, judgment errors, a "mind set" on only one way to do something, and technical errors. The responsible pilot who begins to notice these changes in himself now adds these deficiencies to the underlying stress causing the changes. A circle of stressful and often unrealistic perceptions develops that leads to an impaired pilot.

The stressed person often has a rich imagination of how badly things can turn out, even though nothing serious has actually happened. Stress causes people to continually ruminate about their predicament and become obsessed with their status. They go to bed thinking about their problems and wake up where they left off, thinking about their problems.

The only way to deal with stress is to first recognize that stress is present, which is sometimes difficult for the pilot's independent personality. The pilot, even after sensing there might be too much unresolved stress, will often feel that he or she can deal with it himself or herself. This belief has been proven many times to be a great deterrent to resolving a stressful life. These stresses can be identified and dealt with only through "talking out" the problems with someone who can be unbiased. Sometimes professional help is necessary to break through the denial that stress is affecting performance and help is needed.

One technique is to write down all the stressors, good and bad, along with responsibilities and activities that take up your time and mental attention. Then prioritize them in order of importance and those that concern you. Often it is helpful to have someone who knows you review that list. The objective is to recognize what's a part of your life and sort out those items that aren't worth the stress.

There are a multitude of good publications on stress, its recognition, and how to cope. Since stress is present for everyone, it often is better for the individual to do personal research into the topic and become familiar with the pertinent alternatives. Then when symptoms develop that could be secondary to stress, appropriate action can be taken.

FAA rules require pilots to disclose any treatment by a medical professional on the FAA medical application form. This includes therapists counseling someone with significant stress. If it's relatively mild stress and doesn't involve alcohol or drugs, it doesn't even have to be reported. Getting counseling is not a cause for concern and certainly not a reason to avoid help. The FAA, indeed everyone, would prefer the pilot to seek help, resolve the stress, and press on. The FAA will not question or deny the pilot who does this, unless the stress has been allowed to get so severe that it is a compromise to safety or it has gotten to the point where medication is required.

14
Human factors of automation

This time we made it down 500 feet too low due to a high degree of descent engendered by a crossing restriction given too close in by ATC. My fault for assuming that the F/O had it under control. His fault for assuming the autopilot had it.

ASRS

AUTOMATION IS NO LONGER "NICE TO HAVE." It's an absolutely essential part of flying. Writing just a short chapter on cockpit automation and the human factors that are related doesn't come close to the amount of information accumulating in the industry. Entire books, articles, seminars, and training modules are devoted solely to automation and an equal number discuss the human factors of automation. As with chapter 4 on situation awareness, this chapter is meant only to raise the student's awareness of this issue and to respect the subtle intrusion of flight physiology into how one performs in the glass cockpit.

The industry is in a transition mode, and will be for years, when it comes to putting pilots of different training backgrounds and experience into an airplane that to one age group is alien to how they are accustomed to flying. The other end of that spectrum is the younger generation where computers instead of slide rules and word processors are an expected part of their lifestyle, both professional and recreational. Our entire individual perception of where and how automation fits into flying and how well the pilot adapts and utilizes yet maintains control is being constantly challenged—that with the pilot who is unimpaired from any physiological event.

Even back in the late 1980s, studies from the Flight Safety Foundation were showing that "there are numerous instances where the inherent limitations of humans in monitoring automated systems have led to accidents or near accidents, particularly when combined with the effects of fatigue or circadian dysrhythmia." For example, a Boeing 747 over the ocean rolled inverted and dove 31,000 feet before recovering. The event occurred at about 4 a.m. and was the last leg of the flight (sound like a familiar time for an incident?). The crew, according to the investigation, showed overreliance on the automated systems of the aircraft and loss of situation awareness when the aircraft began to deviate (undetected by the crew) until the aircraft was in a dive.

Another example is a jumbo jet losing cabin pressurization over a 40-minute period without any member of the crew noticing the problem, which was being accurately displayed on their instruments throughout the incident. This is a case of being distracted or not concentrating. Yet another example is reflected in these Aviation Safety Reporting System (ASRS) comments: "While in navigation mode, the aircraft turned the wrong way over a checkpoint (a misprogrammed FMS). Although the wrong turn was immediately noticed, the aircraft turned more than 45 degrees before the pilot took action." Here, complacency was present with the assumption by the pilot that the computer would take over and correct the problem.

While the active event gets your attention, the underlying related "passive" human factors can be easily minimized and often forgotten, especially if there was recovery. Who would want to admit physiological or psychological impairment?

The author described the relationship between human factors and the automated (glass) cockpit in a presentation to the Flight Safety Foundation and in an article written for *Business & Commercial Aviation* magazine. (Portions of both have been revised and included in this chapter.)

This rapidly advancing technology of our aircraft and how we fly them will always be a challenge. What used to be stick-and-rudder control of a flying machine—monitored by needle, ball, and airspeed—has evolved into aircraft that can't be flown *without* the presence of computers and cathode-ray tubes (CRTs) and liquid crystal displays (LCDs).

A statement of dismay often heard in the cockpit used to be "Where are we?" This led to "What's it doing now?" "Why is it doing that?" and then the worrisome "What's it going to do next?" That "it" is what we used to call the "black box." Now it's called EFIS or "electronic flight instrumentation system." "IT" is cockpit automation—the glass cockpit.

Automation is nothing new and some form of it has been in use since before World War I. The objectives, however, remain the same: how to optimize flight performance, decrease workload, increase safety, and reduce pilot error. The new technology has virtually outgrown our ability to incorporate new programs into our systems. But a pilot's *goal* remains the same—get the aircraft safely from A to B. The pilot's *role* is still to navigate, communicate, and operate. During all this, the pilot must continue to monitor, plan, supervise, and make decisions. Automation is meant to assist the pilot in these duties, and to that end, the industry has adopted several classes of automated systems; however, the ultimate result from these systems is no better than how that human functions. Aging and the physiology of flight will affect those functions.

COCKPIT AUTOMATION DEFINED

The primary type of cockpit automation is aircraft control, such as the autopilot and fly-by-wire systems. The next is aircraft performance monitoring, which includes flight management systems (FMS) and alert and alarm signals. It logically follows that there needs to be a ready source of operational information that leads to information access such as flight plans, checklists, radar, and GPS. We now have a form of automation that assists in communication, such as DataLink.

These black boxes were initially installed as a backup to traditional aircraft control and monitoring. This increase of boxes has led to one of the *new* problems associated with automation: overreliance on these systems so that they become "de facto" *primary* devices, whereby the pilot tends not to confirm what is happening in a real-world situation. Curt Graeber, who was once with NASA in its fatigue and sleep studies, and who is now with Boeing in its human factors department, stated that "The great challenge has become information management: what's available, how and when it should be available, and how it should be displayed. We no longer have the 'touch-feel' feedback of what's going on with the aircraft." Curt's comment illustrates why automation can be both a help and a hindrance.

In addition to having to manage all this information, there is the increased use of CRTs and LCDs with their many variations of alphanumerics, colors, and symbology with smaller-typeface displays. The cockpit has become a *visual* environment with less need for manual control. A generalization, therefore, is that automation tends to decrease our hands-on flying and increases monitoring. The challenge is how to effectively use automation, recognizing there are many human factors involved.

When considering automation, one must understand the evolution to our present status. As the number and use of black boxes increases, the pilot becomes more distant from the actual control of the aircraft. The pilot is becoming part of the periphery, spending more time on programming and monitoring than using his or her flying skills. A frequent observation is that the pilot feels "out of the loop." One pilot stated: "It's not so much being out of the loop as (it's) flying alongside it."

Visual displays are now a required part of these boxes in the glass cockpit, and they can do wondrous things. Up to 256 different colors and hues can be used for the symbols and backgrounds to determine changes in aircraft performance. Size and shape of the characters has no end for the creative engineer. The number of items and pages is limitless. For example, there are over 200 items on one checklist of the A320. Remember also that computers do exactly as they are told, and if garbage goes in, garbage comes out. How each of us adapts to this evolution is the same challenge faced when a person is programming a VCR: mystification, intimidation, effortlessness, or frustration that leads to giving up.

For the professional pilot, however, safety necessitates adapting to automation as well as learning how to mentally process the information it provides.

INFORMATION PROCESSING

This process begins by taking in sensory inputs, mentally processing this data, coming to a decision, and then taking an appropriate action with the controls. That action creates new information to be processed. This is a closed loop of mental processing, with each action affecting the others. Furthermore, several loops are usually occurring at the same time.

The role of the pilot employing automation can range from ignoring the boxes to thinking that the boxes can act as the copilot. For example, the pilot might feel that the box can think by itself and do more tasks than it's programmed to do, and therefore he or she relies upon automation rather than his or her information processing. On the other hand, automation might intimidate the pilot because of the increased number of system components. Furthermore, there can be a distrust in any piece of equipment where reliability is suspect. Early TCAS is a good example where one false conflict alert led to pilots ignoring subsequent alerts.

The single greatest ongoing concern, therefore, is identifying where the pilot has the greatest risk of breaking this loop of information processing, often with surprising and embarrassing results. At best, it's a quick and easy way to become complacent and lose situation awareness. Most errors or disconnects in this loop are the result of a particular physiological or psychological impairment.

This concern is made worse by the fact that humans are inherently poor monitors. We are poor monitors because we are easily distracted; our concentration is often broken. We become bored easily unless we are active. Our minds can analyze, extrapolate, and seek novel ideas, but we can only interpret and think of one thing at a time. When impaired, we often have difficulty discriminating and prioritizing different types of data, which has fostered various kinds of symbology displays. Furthermore, we cannot efficiently assimilate very large amounts of raw information in short periods of time nor can we effectively handle complex tasks under tight time constraints.

As a result, as the use of automation is increased, mental workload frequently reaches overload, while concurrently, memory reaches saturation during crucial phases of flight. Mental "underload" happens during cruise, leading to boredom and a decreased awareness of aircraft performance and position. All this results in spending more time monitoring the visual displays.

Remember why we have a sterile cockpit? It's to avoid distraction from unnecessary conversation. Now we have another distraction. It's an increase in "heads down," especially when flying in an ATC high-density environment (perhaps in a terminal area) with many changes to the flight track. A major airline's human factors training calls this "eyeball sucking." Everyone is watching or working the automated systems instead of focusing on the results of their actions.

PHYSIOLOGICAL IMPAIRMENTS

What about our attention to sensory inputs that we use when working in this information-processing loop? These senses can be impaired by a lack of medical airworthiness. As stated earlier in this text, some might call this flight physiology, but it's easier to accept and respect these topics as we relate them to the airworthiness of an aircraft. Pilots, like an aircraft, must be airworthy to function safely.

For example, because we are working in a highly visual environment, our visual senses are primary. Now we must consider and recognize potential physiological impairment. A case in point is that many of the high-tech aircraft are flown by older pilots who need reading glasses that are too often in a coat or shirtpocket. Since the FAA doesn't require a pilot to wear reading glasses no matter how defective the near vision is, the glasses go on only when the pilot feels there is a need. For example, consider this ASRS comment: "The aircraft began to rotate, but it just didn't feel right. The numbers looked good, but I wasn't sure. By the time I took my reading glasses out of my pocket, the copilot was reaching for the controls. I had misread the display on the CRT and was 20 knots too slow for my takeoff roll."

Our focal lengths for clear acuity change with age. Our normal focused reading distance is between 14 and 16 inches. As we go beyond age 40 that distance increases in order to keep images in focus. That means a necessity for reading glasses. Furthermore, pilots often will move their seats closer to the controls during takeoffs and landings. In cruise, they will push their seats back for comfort. Moving the seat back and forth further interferes with the acceptable limits for sharp focus. And, of course, flying at night increases any visual impairment already experienced during the day.

Deficient color vision can be another unrecognized factor and should be one of increasing concern as crucial visual displays increase in number. Many color-deficient pilots have valid waivers from the FAA based upon identifying colored light signals from the tower or the fact that they have flown safely for a number of years. Yet, CRTs and LCDs have many different colors and hues, all meant to detect change in the aircraft's performance. Hypoxia can further impair color discrimination. Now add sunglasses with various colored lenses, such as the blue-blockers or amber Serengetis, and you end up with additional color distortion. Consider this ASRS report: "While checking out on a new radar, the magenta wedge to test the turbulence mode was pointed out. I observed no magenta, it all looked red to me. At the time I was wearing blue-blocker sunglasses. When blue is filtered from magenta, you get red."

Let's go back to information processing and add some of the other physiological and aging impairments that go with flying, with fatigue being the most important. This was described in chapter 11, but needs repeating.

We all know and have experienced the effects of fatigue, insomnia, and jet lag. Yet we often minimize its effects until tiredness overwhelms us. Our brain tends to push back in our memory anything that feels bad, such as with pain. This suppression of what the mind chooses not to remember is the same for how we feel when becoming fatigued. We tend to minimize its effects when we are rested or think we are. Then, by the time we begin to feel fatigued, we are already impaired. And while fatigued, we have trouble concentrating, monitoring, and analyzing data from the displays. Tolerance to fatigue diminishes and reserve energy is decreased as we grow older.

A suggestion: The next time you are really fatigued, make a mental note to remember how impaired you felt and whether or not you could mentally keep track of everything happening in the cockpit. When you know that you are fatigued and must fly in the future, be especially wary of your cross-checks, scans, and program changes into any automated systems. In a crew mode, tell your colleagues that you are tired and request double-checks of crucial activities.

The same problem holds true for hypoxia. It is now common to spend many hours at cabin altitudes of 5,000 feet or more. This by definition is hypoxia. As covered in an earlier chapter, hypoxia's first effects are on vision, especially at night. Once again, our tolerance to these effects lessens with age.

Other examples of how we can be less than medically airworthy are a hangover, even though your last drink was well over 12 hours ago, or when suffering from a cold or flu. We then treat these symptoms with over-the-counter medications that also have side effects. As we get older, there are more aches and pains, and we will notice more significant symptoms with common colds and flus. Each of these situations will affect your ability to process information in the automated cockpit, often concurrently.

What about those effects that are directly and solely related to aging? We might be great pilots as we get older and we are more experienced. But this does not overcome the impairment from age-related changes. For example, as noted earlier, our near vision deteriorates, and accommodation, or focusing from near to far, is slower. And, although not publicly stated, we know that our reaction time is decreased, and we frequently feel somewhat behind the power curve when dealing with multiple tasks.

Physiologically speaking, therefore, even if we are competent in flying skills and masters of the glass cockpit, if we are not medically airworthy, we might miss a clearance change, be easily distracted by mental and visual overload, misprogram the FMS, and lose situation awareness. Unfortunately, as we age, our ability to maintain the same medical airworthiness we are accustomed to has deteriorated, and although we are unwilling to admit that we are affected, we are impaired.

QUANTIFYING IMPAIRMENT

Most of these physiological issues are difficult to quantify. How do we truly ensure that a pilot is medically airworthy and capable of performing safely in a glass cockpit, regardless of age? There is no straight answer. Only through training can we maintain awareness of these stresses of human factors.

Despite regulatory requirements, it's still a hard sell because physiological issues like fatigue are not found as a cause of accidents. The Guantánamo Bay accident of August 1993 is the only exception where fatigue was singled out as causative and not just contributory. We minimize fatigue and other physiological issues because we feel OK and don't expect to be impaired.

Having recognized the potential physiological impairment of our ability to safely process information in a glass cockpit, how do we cope? Where are the problems?

John Lauber, formerly with the NTSB and now with Delta, once said: "I am not entirely convinced that many so-called automation problems really are uniquely associated with automation technology. The real problem is that we, once again, have failed to appreciate the support requirements for the humans who end up having to use the systems. Training, performance monitoring and evaluation, [and] operating procedures all need to be satisfactorily addressed."

Essentially three issues are related to this challenge. The first is to maintain an awareness of the many human factors that can affect performance in an automated environment. The second objective is to change our corporate philosophy about automation so that automation is not so intimidating that the pilot is kept in the loop and the company makes proper and realistic use of the benefits. The third point is that training needs to be modified and expanded to incorporate at all levels (plus management) the most efficient and safe use of automation, including the effect of human factors and flight physiology.

Primarily, we must all recognize and respect that the pilot is still the one in charge. Dr. Charles Billings, a retired NASA senior scientist in human factors and now at Ohio State University, states that automation must be human-centered, not technology-centered.

Gone are the days of just discussing the pilot and the aircraft. We speak now of system components. As Dr. Billings says: "Computers are fine assistants, but they are lousy commanders because they lack flexibility and intelligence. They can't innovate when new circumstances require it. If human pilots and controllers are to remain responsible for system safety, they must remain in command of the systems and they must retain the authority necessary to command them." It follows that the human must be performing without physiological or psychological impairment to perform those functions.

There has been the tendency for some companies to say "We bought it, you use it." This pressure reinforces the perception that automation must be a primary resource inflight simply because it's there. To counter this perception, many companies effectively set corporate policy similar to the following upper management statement. "Automation should be used at the level most appropriate to enhance the priorities of safety," and the like. In other words, if you aren't comfortable the way the plane is flying under the computer's control, then take over manually. That also means that pilots must determine if they can safely perform in an automated cockpit. If not, informing crewmates or turning the flying over to someone else have to be the options.

It's important to remember that we are still pilots, not just system managers. There is increasing training in resolving the lack of mode awareness, especially in highly automated aircraft. It cannot be assumed that an experienced pilot will not encounter mode confusion. Only through continuing training in flight physiology and associated human factors can the pilot and crew keep ahead of not only the airplane but the automation that keeps it flying and on course.

Pilots must respect the human factors of automation. The pilot must be aware of how he or she can be affected by the level of medical airworthiness. The best-trained pilot even in CRM will not be safe if there is impairment of information processing and subsequent actions.

15
Inflight medical emergencies

It was an uneventful trip so far. The two-man crew reached cruising altitude, put their charts down, and began to discuss the events of the day. During the conversation, the captain interrupted and stated that he suddenly had a pain in the back that was severe enough that he didn't think he could fly. The copilot took over the controls and asked the captain what he thought was happening. "I passed a kidney stone many years ago, and it really hurt. I think I'm passing another one!" By this time, the captain was in real pain and said that they better get down at the nearest airport. The copilot became so concerned about the captain that he couldn't concentrate on his flying. He didn't even consider declaring an emergency and kept asking the captain how he was doing and what to do. Finally, the captain, in spite of his pain, said: "Look, get back on course, forget about me, call ATC and tell them you have a sick person on board and need to be vectored to the nearest airport with medical facilities, and then fly the airplane—please!"

IT'S ONE THING FOR THE PILOT TO BE AIRWORTHY and follow all the advice given in this textbook and another to fly only when he or she is fit to fly. There is only a minimal chance that the pilot will experience any medical problem in flight. This is not true, however, for passengers and crewmembers who are not following a good health maintenance plan or developing a medical problem that they have been warned about. And a problem can occur completely unexpectedly. Medical emergencies do occur in flight and, although uncommon, can be more than just distracting. If someone becomes ill or is injured, there is little anyone can do in the confines of an aircraft with limited medical aids or insight. There are, however, some basic techniques that can help.

The greatest challenge to the pilot is determining what the problem is and how serious it is. Then it becomes necessary to decide what to do and how quickly. No matter what is happening to others in the plane, there is one activity that takes precedence over all others—flying the plane. This sounds simplistic, but dealing with the sick or injured is, at best, distracting. Focusing on someone's expression of pain or abnormal actions, even for a moment, is a sure way to lose control of flight.

Most medical emergencies can wait until you get the plane to the ground. Declaring an emergency to ATC might require some paperwork afterward, but the controllers are trained to get you to the airport with the best facilities. They will want to know the extent of the problem, and you will have to tell them. It is better to err on the side of severity, letting the doctor decide later if you might have overreacted.

Often the medical emergency occurs with someone who can tell you what's going on. If the emergency involves someone unconscious, then basic CPR, if indicated, is about all that can be done, if even that. Remember, the primary role of the pilot is to fly the plane. If possible, ask someone else on board to help the patient, and then decide the next step. Either press on or head for an emergency airport.

It is said that nobody dies in flight, and legally that is true in most cases. The reason is that only a medical doctor can pronounce someone dead, which means a doctor, usually on the ground, must see and examine the ill (or dead?) person. It also means that, short of first aid, most medical emergencies ultimately have to be seen by medical professionals anyway; therefore, knowing basic first aid is essential during your inflight emergency diversion to medical personnel. In larger commercial aircraft, more extensive care can be given, depending on the resources available.

Another medical emergency is survival after a crash in a remote area. In most situations, because of sophisticated search-and-rescue resources throughout the world, help is not far away. But you should understand fundamental techniques of survival in case rescue is delayed by weather, terrain, or other factors.

FIRST AID

Most injuries requiring first aid include lacerations (cuts), bumps (contusions), sprains, and broken bones (fractures) suspected or real. Most bleeding can be controlled with pressure over the laceration. Don't waste too much time looking for "pressure points" unless you know exactly where they are and there is extensive or uncontrollable bleeding. Pressure must be maintained for a minimum of 5 minutes for minor cuts, longer for extensive lacerations.

A bloody nose can be controlled by simply holding your nose for 5 minutes. Techniques such as placing a cold compress behind the neck or holding a tissue under the nose have little effect, if any. The point is to put pressure on the bleeding vessel(s) inside the nose. Compressing the fleshy part of the nose between your thumb and forefinger will control the bleeding.

Sprains, contusions, and suspected fractures are treated the same way, initially. Immobilize the injured part, put ice (if available) over the injury, and wrap it with some protective material. This will be all that is necessary until medical assistance is available.

Burns require only protection until seen by the doctor. Ice or cold water might help. Do not apply ointments or creams. It's probably better just to keep the injured person comfortable until landing.

First aid kits are easy to find in drugstores and usually contain all that is needed. It's easy to make your own kit, with bandages, tape, gauze, scissors, elastic bandages, safety pins, Q-tips, tweezers, and a basic first aid manual. Put everything into an easily accessible plastic bag or small container.

Remember, all you need to do during this period before you can land is to keep things stable and the patient relatively comfortable until you get to the airport or on the ground. This is usually only a matter of minutes.

Fainting

Fainting is a common occurrence and is often self-limiting. That is, when the person is prone (lying down), blood flows to the brain much more efficiently through the vascular system and recovery often takes place spontaneously. Oxygen might help if available, but it's not essential. Try to determine why the person fainted. Was he or she scared, hypoglycemic, hyperventilating, or did something more serious take place (heart attack, stroke, diabetic collapse, etc.)? If the person does not recover in a few minutes and is unable to tell you what happened, consider it a more serious problem. If the passenger is able to talk, he or she might be able to help provide assistance or make a request. Sometimes, like with diabetics, providing some candy or juice will help. If there is no improvement no matter what you try, tell ATC that you have an unconscious passenger that needs immediate medical attention.

Cardiopulmonary resuscitation (CPR)

Selected true medical emergencies require immediate attention, such as a stroke, heart attack, or diabetic collapse. There is, however, little you can do in these cases, especially if the person is unconscious. If oxygen is on board, try to get the mask over the mouth and nose. This might revive the person. If there is still a problem, CPR should be considered but only after you get clearance for an emergency landing at an airport with medical support. Again, fly the airplane, and let someone else work on the passenger. If there is no one else, just fly.

When someone collapses and appears unconscious, there are several actions that must be taken immediately. They are often called the *ABC's of CPR*. Prior to this, it is important to determine the extent of the situation and level of consciousness. Ask the person first, "Are you OK?" If there is no answer, then assume the worst and proceed to the ABCs.

A is to remind you to check the airway, which includes mouth, nose, and throat, to ensure there is no obstruction. If there is, try to remove it.

B represents breathing. Check to see if there is active movement of the chest and movement of air from the mouth or nose. Clear out any obstructions in the mouth.

C indicates circulation. Feel for a pulse on the side of the neck. Listening for the heart might be helpful, but not hearing anything doesn't mean the heart isn't pumping. The ultimate check is to feel for a pulse.

After checking, if there appears to be a lack of breathing and/or heart function, then CPR is initiated. It is even difficult to do CPR in larger aircraft. Once again, if this happens in flight, fly the plane and don't get distracted into trying to help. This requires medical intervention. You might be able to guide another passenger by explaining what to do.

Choking

Choking often occurs with food. The person, often in a hurry, tries to swallow a large firm piece of food, such as meat, and it gets stuck in the pharynx. It's fairly obvious what is happening because choking often happens during a meal and the person is having trouble breathing and is usually holding his or her neck or pointing to the throat. He or she is unable to speak and explain. If you can get the obstruction out by removing from the open mouth, try. The best procedure is the Heimlich Maneuver. Someone else will attempt to force air out of the lungs, pushing the obstruction out of the throat. This is best accomplished by getting behind the person with the obstruction, wrapping your arms around the upper part of the abdomen, and making a fist that fits just below the sternum (breastbone). Then make quick upward-motion thrusts to force an exhalation. More than one thrust might be required to dislodge the blockage.

ONBOARD MEDICAL KITS

There will always be an ongoing controversy regarding what medical supplies and medications should be on commercial airplanes. The problem is not only what should go in the kit but who should use the materials, medications, and equipment. Sometimes there are doctors or nurses aboard, but that does not mean they are familiar with medical emergencies. Not all "doctors" are medical doctors, but they might pass themselves off as being capable of handling the emergency. Also, more can be accomplished by concentrating on getting the plane to an airport, trying to determine the seriousness of the problem, and stabilizing the person. The FAA now requires a medical kit on board for Part 121 operations. The kit is released only by the captain to be used for administering first aid.

Liability is also an issue, especially in the United States. There is a real concern on the part of many health professionals as to how involved they want to be in any treatment on board an aircraft. The real fear is becoming a party to a lawsuit if the outcome of any medical care, even first aid, is not as expected by the patient's family or himself. Occasionally, medical professionals on the airplane will not identify themselves because of this concern.

TRANSPORTING ILL PASSENGERS

The military has transported sick and injured people for years. It has special aircraft with adequate medical supplies and equipment. Medical personnel are onboard, often including doctors. Civilian "air ambulances" are common and often well equipped and staffed. Technology has increased the effectiveness of air evacuation.

Because the flight environment can make a medical condition worse and trigger other problems, it is often unwise to transport known sick people in any aircraft not medically equipped or staffed unless they are accompanied by medical personnel with adequate resources and expertise. Yet this happens frequently in all kinds of aircraft, whether an airliner, corporate jet, or private airplane.

Several situations unique to flight make most aircraft unsuited to transport the ill or injured. The situations include lack of adequate space to work on a patient, vibration, hypoxia, hyperventilation from scared patients, inability to perform basic tasks such as blood pressure readings, noisy conditions that preclude hearing the heart and lungs, difficulty controlling intravenous fluid and medicine, and lack of other medical resources. This becomes a real issue if the patient's condition becomes unexpectedly critical during the trip and there are no means available to assist him or her.

The professional, moral, and ethical liability of transporting passengers who are already ill or injured in a poorly equipped or staffed airplane also becomes a major factor. The best solution would be to check with a doctor and attorney familiar with aviation law and medical liability before allowing any sick passenger to board the aircraft.

BASIC SURVIVAL TECHNIQUES

Basic survival techniques are rarely discussed in the civilian aviation community because the chances are slim of not being found within a few hours after a crash landing. Yet those who fly in and over remote locations, such as mountains and wilderness areas, are wise to consider the real possibility of having to survive for several days before search and rescue can get there. Filing flight plans is an obvious means of protection, but there are many who fail to take this simple step.

The purpose of survival training is not to make campers of pilots. Survival training is common and required in many military flight operations. In fact, the main objective in any actual survival course is to prove to the student that he or she can survive in virtually any condition by employing basic techniques. Instilling this self-confidence is a major step in preparing a pilot to successfully survive an emergency landing.

Coupled with this self-confidence is the knowledge of survival techniques and having basic tools on the aircraft. Many good manuals and pamphlets detail how to "live off the land" and cope with weather extremes. A scout manual also serves the same purpose. The manuals are available in most bookstores, and a good manual is a necessary item in your survival kit.

Elements of survival

Several priorities must be dealt with in any survival situation. The first is tend to injuries, if any, and to ensure that all persons are accounted for. The next step is probably the hardest: Stop and think.

In situations of fear and mild panic that are understandably associated with a forced landing, the worst thing one can do is rush into futile efforts or unimportant tasks. Sit everyone down, try to figure out where you are, check your resources, determine if an emergency signal is going out, and then plan your strategy.

The following are high-priority steps that can be assigned to each person:

- Find shelter, especially if the weather is bad or menacing.
- Get all your supplies into that shelter.
- Look for water.
- Begin to plan for essentials for continued survival. (Food is not important at this stage because the human body can go for days without food.)

Survival tools

Your survival kit need not be elaborate and often can be put together at home. All items must be necessary and practical and compact enough to fit into a small container. For example, a gas heater might be appropriate if flying over the far Northern Hemisphere but unnecessary otherwise. Avoid bulky, heavy, and cumbersome survival kits because they are often dumped as a nuisance before they can be used.

Here are a few suggestions for the basic essentials to include in the survival kit. This, plus your first aid kit, can provide the necessary tools to cope with just about any situation:

- Matches that have been sealed with candle wax
- Small candle
- Safety pins
- Knife
- Thin pliable wire (for snares)
- Aluminum foil
- Small thin plastic tarp
- Fish hooks
- Nylon line
- Plastic whistle
- Small compass
- Balloon (water container)

- Salt
- Bouillon cubes
- Small flashlight (batteries covered to prevent contact) (extra batteries optional)
- Signal mirror
- Water purification tablets

REMINDER

Keep in mind the word **STOP**. It stands for Stop, Think, Observe, and Plan. With some minimal preparation and self-confidence that survival in extreme environments is realistic, your chances for getting out alive are greatly improved.

16
Health maintenance program

George had been passing his FAA physicals without a hitch. Sure, he was overweight and smoked a pack of cigarettes a day for the past 12 years, but he thought he was healthy because he felt good and always passed his flight exam. Then he saw his personal doctor because of a bad cough and the nurse noted his elevated blood pressure. His doctor did a few more tests and then put him on blood pressure medication. Now he had a dilemma. Would the FAA certify him? And why was the blood pressure now a problem? His doctor had said that the weight and cigarettes were a major factor causing his elevated blood pressure, but now he had to be on medication, at least until he got his lifestyle back in order.

HEALTH IS SOMETHING EVERYONE TALKS ABOUT. We learn about nutrition in grade school. Exercise is a challenge to many. You can't read a newspaper, magazine, or watch TV news without getting more information about being healthy. Health is big business and for good reason. No other subject is so commonly discussed and so much on people's minds. Lack of good health is expensive, so programs to protect, improve, and maintain health have become a major effort in industry, government, education, business, and especially aviation. Unfortunately, it's a frustrating situation and a hard sell to get individuals to comply with even the basic concepts of health until something goes wrong, then there is an increased motivation to change a lifestyle. That's a good example of crisis management.

Health in aviation is not just a matter of feeling good or being more productive. Safety now becomes an important expectation by everyone, and the health of the pilot, crew, and support personnel must be considered in any safety program. The unhealthy pilot is a potential safety risk. This chapter is not meant to educate the fine points of health. Plenty of good resources are available, ranging from newspapers and magazines to good (and not so good) books and manuals available in drugstores, supermarkets, and bookstores.

The goal of this chapter is to raise the level of awareness of the individual pilot that health is a daily concern and daily intervention is required to maintain it, which is not an easy task for many. It is a topic, more often than not, that is taken for granted. Those who do not make a conscious effort to maintain good health will be at greater risk of developing a medical problem in the future. Even more significant, the pilot could become less than safe (incapacitated or distracted) and contribute to events that show up in statistics. Often it's a subtle and unexpected impairment. And, of course, the FAA medical certificate is based upon the status of your current and future health.

Therefore, this chapter is meant to alert you to the fundamentals of what leads to poor health and a potentially unsafe pilot. This must become a part of your health maintenance program, a program that only you can follow and accomplish. There is no quick fix or "magic health pill." And no one can compel you to be healthy, except maybe the FAA when it identifies an unhealthy and unsafe medical condition and holds up your medical certification until the problem is resolved.

As noted, this chapter is not meant to be all inclusive. It is strongly recommended that further reading be done to elaborate on the process of the metabolism of foods, exercise, and other health maintenance topics. Continued reinforcement of these basic principles is also a healthy attitude.

NUTRITION AND DIET

Before considering what we should do to maintain health, it's important to review some basics of nutrition. Granted, there is a lot of information in the press about what we should and should not eat, but there is little that has changed in nutritional insight over the past several decades. Whenever looking over the variety of diets, food products, and supplements with which we are surrounded, always keep in mind the basics that you were taught in high school.

Basic food components

All foods are made up of three basic substances: protein, carbohydrates, and fat. Other components are vitamins and minerals, but they are present in adequate amounts in any balanced diet.

Carbohydrates, fats, and proteins yield, upon digestion, glucose, fatty acids, amino acids, and other biochemicals necessary for metabolism. If nonprotein calories are not available for metabolism, energy will come from body storage (primarily fat) or from other resources: glycogen as formed in the liver. Metabolic products from food are used, among others, to regenerate tissue cells and provide energy for organ functions (especially the brain) and muscle use. The body does not need much energy to function at rest, but that level increases significantly when stressed or worked.

Protein is the basic substance of all cells in the body and is required by all living things. Approximately 20 percent of diet should come from protein. It is a major player in the body's metabolic processes and tissue growth. During digestion and metabolism, ingested protein yields peptides and amino acids, which are the basic chemical building blocks of all proteins.

They also provide 4 *kcal*/gm of energy (kcal, or kilocalories, is basic measure of energy related to metabolism per gram of food product).

Proteins are commonly found in animal products: meat, fish, poultry, and dairy products. Fruits, vegetables, and grains are other sources but to a lesser degree and quality of amino acids. Ingested proteins cannot be stored and must be replaced every day.

It is a misunderstanding that ingested protein directly replaces protein lost or absent from the body. All foods are broken down during digestion to simpler components, and then the body forms its own protein in appropriate locations in the body; therefore, it is not true that eating more proteins will build muscle mass, for example. Proteins in the diet provide the elements for the body to enhance muscle tissue. Calories from proteins are also necessary, and protein already in the body mass is sometimes used as a source of calories. The body's proteins, however, can be spared by ingesting more fats and carbohydrates. In other words, if sufficient nonprotein calories are not available, then the body's proteins become part of the metabolic process to generate calories.

Fats are very concentrated, and therefore are the primary long-term source of energy: 9 kcal/gm. Only ethanol (the alcohol in beverages) is close at 7 kcal/gm. Carbohydrates have only 4 kcal/gm. Because the body needs this energy at various times during any day, unused sources are stored in fat cells and tissue. Each person has his or her own rate of use of this energy and ability to store energy in fat. Therein lies the challenge and dilemma of weight control. Only controlling calorie intake is not going to be successful in all people. It's a factor of individually inherited metabolic efficiency.

After digestion, fats provide the body with fatty acids and glycerol, which are used in metabolism. Fats also carry fat-soluble vitamins throughout the body for use during digestion and metabolism. Although many people try to avoid fatty foods, fat in the diet is essential. As with the other food components, ingested fat does not go directly into body fat. Fat reaches the fat stores via the metabolic route, and any unused calories are stored for future use at varying rates. Approximately 25 percent of diet should be fat.

Fats are often classified into three different categories: saturated, monounsaturated, and polyunsaturated. As mentioned in chapter 2's subsection on the anatomy of the heart, the coronary arteries bring essential blood to heart muscle and can be blocked or narrowed by cholesterol deposits (another "form" of fat). It is important, therefore, to recognize that fats are saturated because they are considered to be a major source of "bad" fats, leading to these cholesterol obstructions. Most saturated fats are solid at room temperature and are found in animal products, especially beef and pork. This doesn't mean they are also high in cholesterol since fat and cholesterol are two separate and different materials. The body makes its own cholesterol in addition to that ingested.

It is generally accepted that the cholesterol in the blood (as tested in the doctor's office) is related to that found in tissues, but not directly. That means that some people with low blood cholesterol will still have unhealthy cholesterol deposits in their arteries. Other components that affect this process are called *lipoproteins. Low-density lipoproteins* (LDL) are thought to transport cholesterol throughout the body via the blood. *High-density lipoproteins* (HDL) are thought to protect the tissues from accepting cholesterol from the blood. Thus, HDL is a "good" fat or cholesterol. By the way, cholesterol is a normal and required metabolic component. It's the levels of cholesterol in our blood and within our tissues that become the problem.

It should be apparent that this subject is complex and requires more elaboration, something that is not possible in this text. Suffice to say, it is important and you must understand and respect these crucial components of metabolism and health.

Carbohydrates are broken down and metabolized to *glucose* in the body (also called *blood sugar*). The function of glucose is to provide immediate sources of energy, being more accessible than fat or protein. It is, therefore, an important body "fuel" to be used by muscles and especially the brain. The majority of dietary intake should be carbohydrate, approximately 55 percent.

In addition to these calories being stored in the body's fat tissue, some is stored in the liver as glycogen as a backup if adequate blood sugar is not available. It is, however, depleted within a few hours. If all ingested and stored energy plus all the glycogen is used up, then the body, through a process called *gloconeogenesis*, will use up skeletal muscle tissue to maintain adequate blood glucose.

Carbohydrates are no more fattening than fat or proteins. This misconception is common even though fat is recognized as having more calories per weight than carbohydrates. Granted, the body metabolizes each component in a different way and at a variable rate, but you cannot maintain an ideal weight solely by avoiding carbohydrates, nor should you avoid fats and proteins. In fact, that would be unhealthy.

Carbohydrates are found in grains, potatoes, beans, and peas. Other vegetables and fruits have smaller amounts of carbohydrates in the form of starches and sugars. Plain sugar, by the way, is an "empty calorie" because it does not have other nutrients or vitamins or minerals associated with those calories.

An important carbohydrate is fiber. *Fiber* is essentially carbohydrates that the body cannot digest or break down into usable materials. It is now recognized that fiber is important in a balanced diet even though it is not directly metabolized into other components for the body's use. Fiber assists in helping move ingested foods and by-products through the gastrointestinal system. Lack of fiber leads to constipation.

Vitamins and minerals

Vitamins and minerals are adequately supplied to the body with a basic balanced diet. Vitamins, by the way, provide no nutritional or caloric value. They act like a catalyst, assisting in the biochemical metabolism of the body. Some vitamins are excreted in the urine and therefore need to be replaced. This is adequately accomplished by following a nutritious diet. Taking supplemental vitamins does not harm and in fact might help in professional working conditions that do not always provide opportunities for the best food; however, all vitamins are the same, no matter who manufactures, markets, or sells them.

Nothing is gained by taking "megadoses," although this is somewhat controversial in some scientific and medical communities. Because excess vitamins are excreted in the urine and expensive vitamins are no better than the cheapest, you are literally wasting good money with your expensive urine. There is also some indication that "megadoses" can accumulate and become detrimental rather than helpful. Minerals are in the same category, most being provided in basic balanced diets.

Hydration is worth mentioning here because water is an integral part of a good diet. Most foods have a high percentage of water and provide enough fluids in many cases, along with the usual nonalcoholic beverages (milk, soft drinks, etc.); however, when the body is subjected to extremes of exertion such as heavy work or exercise, especially in hot conditions, the body obviously needs more replacement fluids.

When you realize that you are thirsty, you are already behind in fluid replacement. Supplemental water is essential to prevent digestive problems as well as physiological situations such as heat stress. It is always a good idea to drink at least one glass of water with each meal and often between meals. Additional salt, replacing that lost through perspiration, can be obtained by simply salting your food. This is generally unnecessary because most lifestyles are already high in salted food products.

Calories

As mentioned in the above definitions, calories are a part of every food that we eat. Calories that are not used by the body are either stored in fat tissues or excreted. But this varies with individuals because some have a metabolic system that easily gets rid of excess calories. Others have a difficult time losing weight because their metabolism is such that it is very efficient in using available calories and storing the rest. Nevertheless, our caloric input is still what decides our weight. Most overweight people are simply taking in more calories than their body needs to function.

Diets are big business, with the public spending billions every year to find the "secret recipe" that takes away the excess weight and keeps it off. Diets often fail over time because of the absence of a commitment to a maintenance program that keeps caloric intake under control after the initial loss. Counting calories taken in and used by the body—similar to basic fuel management principles with the airplane—is the best way to keep weight under control. Furthermore, rather than being successful over the long run, many diets are quick fixes at the expense of the body's metabolic process and generally fail after a few months. Then another diet is tried.

Most bodies with usual levels of activity need less than 2,000 calories per day to function, often less than 1,500; therefore, by counting calories, keeping track of the intake in a diary, and staying around 1,000–1,200 calories per day, a dieter will usually accomplish weight loss. As you approach your optimum weight, more calories can be added. This is very simple and effective, but most people need more motivation to stick to calorie counting. That's one reason there are so many different variations and creative methods to weight-loss diets—they are just another unique way to count.

Other "medical" diets accomplish different goals: a reduced-salt (sodium) diet controls blood pressure; a diabetic diet controls elevated blood sugar; a high-fiber diet prevents constipation, a low-cholesterol and low-fat diet helps prevent heart disease, and so on. Each diet has a definite purpose, and the dieter must have a clear understanding of why a special diet is recommended. The dieter must learn which foods to avoid and which ones to emphasize.

HYPOGLYCEMIA

By definition, hypo (low) glycemia (blood sugar) means that the body has inadequate required levels of blood sugar to function, resulting in impairment. To a doctor, hypoglycemia usually refers to a major problem associated with diabetes out of control. To the pilot, it means there isn't enough immediately available energy, and it is taking too long for the body to access its stores. It is closely related to poor diet and missed meals and leads to poor efficiency, weakness, headache, and sometimes fainting. The body metabolizes ingested and digested food by changing it to energy using insulin. Without insulin, glucose cannot be used by the body in its metabolism. Too little insulin means diabetes. Too much insulin results in glucose being quickly used up (hypoglycemia).

Most people are someplace in between: not enough insulin when eating or too much insulin when there isn't enough food to digest. One circadian rhythm occurs when hungry and the body expects food to enter. Insulin is often increased in anticipation of expected food. If a good breakfast is expected, as it usually is for that person, blood insulin levels begin to increase as soon as any food is ingested. Then the body is fooled by only getting a candy bar and a cup of coffee. Now the only calories to digest are instantly used up and blood sugars quickly go down: hypoglycemia. That's why most people experience hypoglycemia midway through the morning. It's a rebound effect. The treatment is to get sugar back into the digestive system with another candy bar or fruit juices. Not a nutritious cure, but adequate until a balanced meal is eaten. The prevention of hypoglycemia, therefore, is to eat food that is readily available, yet takes longer to metabolize. Proteins are the best, even if it means bacon and eggs for breakfast.

The short-term prevention of hypoglycemia is more important than the long-term but minimal risk of developing heart problems.

EXERCISE

Exercise is as important as a diet in any health maintenance program. Exercise means different things to different people. The main objectives are to increase the efficiency of the heart and lungs (aerobic), to keep the body's muscles, tendons, and bones in tone, and to maintain an increased sense of well-being. Various means accomplish these goals, and once again, many good publications help define which are best for you and which are realistic to follow under the conditions of your flying schedules.

The basis of any aerobic exercise is to gradually bring your pulse rate up to a target rate, often defined as 220 minus your age for a maximum heart rate. (Actually, getting up to 85 percent of the maximum level is still beneficial.) Then the object is to maintain that rate for no fewer than 20 minutes, after which you slowly return to your resting pulse. This can be accomplished by one of several methods: jogging, fast walking, biking, cross-country skiing, and the like.

While jogging might be the most productive in a shorter period of time, it is not enjoyed by everyone and can be very hard on hips and knees over a long period of time. It is important to begin some form of exercise program and stick with it. A brisk walk will accomplish the same beneficial effect. Many indoor exercise machines (stationary bike, treadmills, Nordic Track, etc.) will also meet the objectives of an exercise program. Some aerobic-dancing exercise is good, but certain aerobic programs are actually too strenuous.

For best results, exercise three times a week or no more than three days between exercise sessions. Working out to achieve more than your maximum heart rate might not be safe nor will it gain any more conditioning than when reaching the 85-percent level. Furthermore, the more you exercise and the more competitive you get, the greater the chance of injury, such as an ankle or knee sprain. That injury could keep you from flying during healing.

THE MEDICAL EXAMINATION

It's one thing to recognize the need for good health maintenance. It's another thing to eat nutritious meals, participate in an active exercise program, and pursue a prudent lifestyle of adequate rest and relaxation. Unfortunately, while pursuing good health maintenance is absolutely essential in protecting your quality of life, it does not guarantee that you will experience good health, free of disease or injury. This is another area where we can compare the body to an aircraft.

As stated throughout this textbook, the measurements of the body's medical and health status is expressed *in ranges, not absolutes*. Aircraft performance and standards *can be measured in absolutes*, and if you fly the aircraft "by the numbers," it should perform as stated. For humans and aircraft, there are only two ways to determine if something is going wrong or about to fail or fall apart. Either you wait for something to happen (or change) and then take corrective action (crisis management), or you test the "machine" periodically to deliberately try to identify a problem, hopefully in the early stages of failure. Once identified, you treat the problem before it gets out of hand. For mechanical machines, this strategy of early testing is called a *preventive maintenance program*. For the human machine, it's the medical evaluation or physical examination in a health maintenance program.

The FAA requires all pilots to have a periodic medical exam: the flight physical. The same is true in the military. Without this requirement, it is doubtful that many pilots would seek out their own exam. The reason, of course, is the potential threat that something will be found that could ground the pilot. And it happens. Many pilots reason that as long as they feel good, there really is no need to have further tests to see if anything is wrong. Some pilots feel that a flight medical evaluation is unnecessary and that the pilot can determine if he or she is fit to fly. This is not the way they would view keeping ahead of their airplane's airworthy status where an evaluation by an aircraft mechanic is sought out. The difference is the perceived lack of control the pilot has over his or her medical evaluation as opposed to the known control over the maintenance check of the aircraft.

It's important to recognize what happens in a typical exam. By being aware of this, the threat of losing control should be minimized. It might seem like a trade-off—looking for something wrong and finding something wrong—or finding something wrong in time to correct it before the problem becomes unacceptable. The other choice, of course, is waiting for something to go wrong (if you really can detect that something is going wrong!) and then hopefully being able to correct the problem; therefore, the key to health maintenance is a combination of what was explained earlier—commonsense lifestyles—and the preventative medical evaluation.

There are varying levels of medical exams, ranging from a basic overview like an FAA or insurance physical to a comprehensive medical evaluation (an "executive-type" exam) that includes several tests and evaluations. While even a basic exam is important, compared to the rest of the population who rarely have any physicals, only a comprehensive physical will really tell you and your doctor very much that can help you. Here is a description of the typical medical exams available to you as a pilot. All exams consist of a medical history, a physical exam, a series of tests, and a review of the results. Basic exams usually are shorter in each part and do not often have many tests. Comprehensive exams include a more elaborate history, a more thorough exam, and many different physical, blood, and urine tests, plus X-rays and EKGs.

The medical history

Your medical history is the most important part of any medical exam, especially for your doctor (Fig. 16-1). How you feel, what has changed in your

1. Name of Proposed Insured. Print in full:

2. Date of Birth: Month: Day: Year:

3. Have you ever received disability benefits because of injury or illness?

4. Have you ever applied or been examined for life, accident or health insurance which was declined, postponed or modified as to rate or amount?

5. Have you applied or been examined for life insurance within the past six months? Give name of company and results.

6. HAVE YOU EVER HAD OR BEEN TOLD BY A MEDICAL PRACTITIONER YOU HAD: Yes No

a. Epilepsy, Alzheimer's or Parkinson's Diseases, Paralysis, or ANY Brain, Nervous or Mental Disorder? .... ☐ ☐

b. Pneumonia, Emphysema, Asthma, Chronic Cough, Tuberculosis or ANY Respiratory or Lung Disorder? .... ☐ ☐

c. Chronic Diarrhea or Indigestion, Ulcer, Liver Disease, Colitis, Rectal Disease, or ANY Abdominal Disorder? .... ☐ ☐

d. Kidney Stone, Albumin or Blood in urine, Kidney, Bladder, Prostate, or ANY Genito-urinary Disorder? .... ☐ ☐

e. Chest Pain, Heart Attack, Stroke, Heart Disease, Murmur, High Blood Pressure or ANY Heart or Blood Vessel Disorder? .... ☐ ☐

f. Anemia, High Cholesterol, Sugar in your urine or Diabetes? .... ☐ ☐

g. Rheumatic Fever, Arthritis, Gout, Back trouble, or ANY Bone or Joint Disorder? .... ☐ ☐

h. ANY Sexually Transmitted Disease? .... ☐ ☐

i. Cancer, Tumor, Goiter, or ANY Blood, Gland, Spleen or Skin Disorder? .... ☐ ☐

j. Immune Disorder? .... ☐ ☐

k. Were you ever treated for use of alcohol or drugs and have you ever used narcotics or hallucinogen drugs (except under a physician's care)? .... ☐ ☐

l. Enlarged lymph nodes, unexplained weight loss, Kaposi's sarcoma? .... ☐ ☐

m. Herpes, Candida, Epstein-Barr virus? .... ☐ ☐

n. ANY injury, operation, medical attention or special diagnostic tests (EKG, X-ray, Blood, etc.) not stated above? .... ☐ ☐

7. Have you taken prescription drugs during the last 12 months? ☐ ☐

8. Other than above, any examination or treatment by a doctor, practitioner or hospital in the past five years? ☐ Yes ☐ No

9. Have you ever smoked cigarettes? ☐ Yes ☐ No
If so, number of months since you last smoked ________
Do you use any other tobacco products? ☐ Yes ☐ No

10. Family Record (List parents or siblings that died before reaching age 65)

Relationship	Age	Cause of Death
1		
2		
3		

11. Name & Address of personal physician.

Date and reason last seen.

12. What is your: Weight: ____ lb.; Height: ____ ft. ____ in.
Have you gained or lost weight in the past year?
Gain ______ lb. Loss ______ lb. Reason for change?

Please give DETAILS of all "YES" answers.
Date—Durations—Results—Doctors' names and addresses

16-1 Typical medical history form.

life, your habits, your family's medical status—all are important clues to your current and future health. But to the pilot, this is the one area where there is a reluctance to be completely candid and thorough. It's that "threat" thing again and denial that a simple symptom could be meaningful or helpful to the doctor. Why mention anything unless you think it's significant?

The items of a good history should always be the same. Next time you take a physical, whether FAA, employment, insurance, or for your own use, check out the similarity of the questions asked. You need to be familiar with these questions in order to have consistent answers and explanations. For example, if you had surgery as a child, you need to know the circumstances, what happened, when, and how you did afterward. Many people have little knowledge of what happened to them when they were children. It's the same for family history. You must know (and have on file in your own records) the specifics of your relatives' deaths, illnesses, and medications, and their ages when these events occurred.

Although some medical evaluations have only a brief review of medical history, you need to know everything that has happened to you medically, and you should keep this information for your own record. There will be a time in your flying days when you will need to report that information to someone, if only to compare data over a period of time to detect any changes or trends.

The frequent question: "Have you *ever* had a history of...?" That means, have you ever had "X" disease, injury, surgery at *anytime* in your life? Even if it was decades ago that you had the problem, broken bone, surgery, or hospitalization, it's important to you and your doctor.

It would be wise to have a copy of a typical history form that you have filled out completely, checking with family members to confirm dates, ages, and events. Check with your personal doctor to make sure that everything is adequately evaluated and explained. Keep this in your personal files for future use.

Another part of the medical history is a "review of systems." This is a review of how you feel, what symptoms you have, and how you perceive your body is functioning or not functioning. The doctor is interested in any significant pain, discomfort, congestion, aching, dizziness, and any other symptom that is not normal for you.

Equally important is a perceived change in one of these feelings—is it better or worse, occurring more often or less? It's important to realize that what you think is insignificant might be important to the doctor as she considers all the data from the rest of the evaluation. Let the doctor decide what to do and also let her elaborate if she thinks it is important.

You should have your shot records (immunizations) in your file and bring the dates to your doctor. You will also be asked about your habits: smoking, alcohol consumption, coffee intake, exercise, diet, and the like. She is, by the way, not interested solely in what you have been doing the last week. She wants to know that information, but she also wants to know what you have done in the past, even years ago.

The physical examination

Once again, like the history, the physical will be as extensive as the doctor feels it ought to be, depending on why you are having a physical. For example, a doctor could spend 15 minutes listening to your heart but often only a few minutes is necessary if there are no abnormal findings. Basically, the doctor is looking, feeling, listening, and smelling for anything that isn't normal or expected.

Usually the doctor will start from the head and work down during the physical exam (Fig. 16-2). That means, in a comprehensive exam, checking the scalp, eyes, ears, nose, mouth, and the like. He or she will look for lymph nodes (small swelling of glands) in your neck, check your thyroid (in the lower front part of your neck), and evaluate other parts of the upper anatomy.

Next is the chest, listening for breathing sounds, watching how you breathe, examining the heart and its rhythm, sounds, and size. Lymph nodes are sometimes checked out in your armpits, a common place for abnormal lumps. This is the first major area of skin to be examined.

The abdomen has several organs to be examined, including the liver under the right rib cage and the spleen under the left side. He or she might ask you to take a deep breath, which pushes these organs from under the ribs for easier palpation. He or she will poke deeply, looking for masses that shouldn't be there. Listening to the bowel sounds with a stethoscope will give an idea of what should be going on with the digestive system.

Next comes the check for hernias (by the way, the reason for turning your head when you males are checked is so that you don't cough into the doc's face—a little known medical secret). Pelvic exams for females and testicular exams for males are also important. A rectal exam, although not the most pleasant part of any exam, is very important for everyone, especially after age 40. Hemorrhoids are quite common in professional pilots at any age. The prostate is also examined during the rectal examination.

The extremities—arms, legs, joints, hands, feet, backbone, and spine—are also examined for swelling, pain, range of motion, and distortion. Your circulation is also checked by feeling pulses in your neck, groin, wrists, and feet.

Although this text is not meant to be a medical "how-to" book, a word about high blood pressure is in order because this part of the exam results in most concerns with pilots. Blood pressure as measured by medical people is the "hydraulic" pressure within the arterial circulatory system. Normal blood pressure is about 120/80 mm Hg (millimeters of mercury). The top number (*systolic*) of this reading is the highest pressure exerted by the heart during a contraction. The lower number (*diastolic*) is the "resting pressure," or the lowest pressure reached in the artery before the next contraction. Remember that it is the elasticity of arteries that can maintain that resting pressure between each contraction.

Blood pressure is supposed to be high under conditions of extra work, illness, stress, and, of course, poor health. Blood pressure that remains elevated despite control of these factors is called *hypertension* and requires treatment. Often, a doctor will consider medication if blood pressure consistently stays above 140/90. Obviously, for the pilot, every measure must be taken to keep blood pressure under 140/90 by controlling known causes—the very issues discussed in this text. And the only way you can keep ahead of blood pressure problems, as well as other medical problems, is with a good health maintenance program. If your doctor feels that your high blood pressure needs treatment with medication, the FAA will still

MEDICAL EXAMINER'S CONFIDENTIAL REPORT

11. How long have you known the applicant? ______ Are you related? ______ Are you his/her physician? ______

12. Height ____ ft. ____ in. Chest, full expiration ____ in. Chest, full inspiration ____ in. Abdomen, at umbilicus ____ in.
Weight ____ lb.

	YES	NO
Did you weigh?	☐	☐
Did you measure?	☐	☐

Please comment below on any significant gain or loss of weight in past five years.

13. Does inquiry (history) or examination (operative scars, etc.) indicate any past or present disease, function impairment or abnormality of the:
Nervous System? ______ Abdominal Organs? ______ Cardiovascular system? ______
Respiratory System? ______ Genito-urinary System? ______ Glands, Skin, Joints? ______

14. Pulse: Rate per minute: ______ Rythm: ______ If over 90 or irregular complete #19 below.

15. Blood Pressure:
Systolic 1 ____ 2 ____ 3 ____
Diastolic (5th Phase, end of sound) ____ ____ ____
If over 140 or 90 report several readings and complete #19 below.

16. Is there evident arteriosclerosis? ______
Is there a heart murmur? ______
Is there any hypertrophy? ______
Is there cyanosis, dyspnea, edema? ______
If any "YES" answers complete #19 below.

17. Is general appearance healthy? ______
Is appearance older than given age? ______
Any eye or ear disease or function impairment? ______
Are there any abnormal reflexes? ______
Is there any deformity or physical defect? ______
Any disorder of prostate? ______
Are there any varicosities? Any hernias? ______

18. Urinalysis: Please send the urinalysis specimen in the container provided.

Please give DETAILS and YOUR DIAGNOSTIC OPINION of any "YES" answers

19. HEART SECTION: Please give DETAILS and DIAGNOSTIC OPINION.

RIBS ARE NUMBERED
1
2
3
4
5
6
7
MIDSTERNAL LINE
MID-CLAVICULAR LINE

PLEASE MARK ON ABOVE DIAGRAM:
X = Apex
O = Maximum intensity of murmur
⁘ = Area over which murmur is heard
➤ = Direction of murmur transmission

A. Heart Murmurs:
1. Report Intensity as Grade I to Grade VI.

	YES	NO
2. Location?		
Apical Area:	☐	☐
Aortic Area:	☐	☐
Pulmonic Area:	☐	☐
Other	☐	☐
3. Timing?		
Systolic:	☐	☐
Presystolic:	☐	☐
Diastolic:	☐	☐
4. Transmission:		
Axilla:	☐	☐
Neck:	☐	☐
Scapula:	☐	☐

5. Constant?
6. Effect of exercise?
7. Effect of recumbency

B. Hypertrophy?
☐ None ☐ Moderate
☐ Slight ☐ Marked

C. Apex is located in the ______ intercostal space ______ inches to left of the midsternal line.

D. Exercise test: If not done, i.e., contraindicated, please state why. Have applicant do at least 50 vigorous hops or, preferably, 15 ascents on an ordinary chair in one minute in order to secure an adequate exercise response, i.e., an increase of more than 20 beats per minute.

Exercise Test	Pulse Rate	Irregularities Number per min.	Blood Pressure	Murmurs
a. At rest before exercise				
b. After exercise				
c. 3 min. after exercise				
d. 5 min. after exercise (p.r.n.)				

this ______ day of ______ 19 ____ A.M. P.M. Signature ______
Examining Physician

Agent ______ Address ______

16-2 Typical medical exam form.

certify you but will want some additional medical information that your doctor can get about you.

This is only an overview of what the doctor does. He or she is trained to detect subtle findings during the examination. By the way, you should know the difference between doctors. Medical doctors, the kind you should look to for most medical evaluations dealing with your health and medical certification, are physicians: M.D. (medical doctor) or D.O. (doctor of osteopathy). All others who call themselves doctors are not medical physicians, and you should be cautious as to their expertise and training, especially in matters that directly or indirectly affect your health and ultimately your medical certificate and fitness to fly.

Blood and urine tests

Medical technology has advanced to the point where a multitude of test results can be achieved from a few tubes of blood and a small sample of urine. The cost is reasonable enough to warrant full blood profiles, which are a series of 25 or more blood-test results (Fig. 16-3). The other side of that benefit is the identification of test results of little value in a screening situation (as opposed to diagnostic). At times, these results of minimal worth are not within acceptable (normal) limits. (The next subsection of this chapter examines "normal.")

The doctor is now faced with the dilemma of what to do next. More tests could be considered even though, realistically, there is little clinical significance to a profile's single test result. Or she could choose to minimize the medical significance to the result because the rest of the medical evaluation is normal. In any case, to the pilot, test results in the "not-normal" range could be considered by the flight surgeon and the FAA as being abnormal or unacceptable until proven otherwise, which is another reason to be sure your medical and certification needs are met by competent aeromedical physicians.

Key blood-test results include blood sugar, liver function, kidney function, cholesterol, HDL, LDL, and triglycerides, as well as others relative to metabolism and heart functions. When you look at the computer print-out of blood-test results, you will notice that absolute single-number normals are nonexistent. Normal *ranges* are stated. (Fig. 16-3).

These ranges, by the way, will change as the medical profession gains more knowledge as to the significance of the results relative to your health. A good example is that the upper limits for acceptable total cholesterol in 1987 was considered to be 290. Now that level has been reduced to 240 or even 200 in some situations.

Normal, abnormal, not normal

It's appropriate now to define the terms *normal, abnormal,* and *not normal.* They are used commonly in any medical discussion, yet they frequently have a different meaning to a doctor than they do to a pilot.

REPORT STATUS FINAL	TEST	RESULT IN RANGE	RESULT OUT OF RANGE	UNITS	REFERENCE RANGE	SITE CODE
CHEMZYME PLUS						
CHEMZYME						ML
GLUCOSE		107		MG/DL	70-125	
UREA NITROGEN (BUN)		22		MG/DL	7-25	
CREATININE		1.3		MG/DL	0.7-1.4	
BUN/CREATININE RATIO		17		(CALC)	6-25	
SODIUM		140		MEQ/L	135-146	
POTASSIUM		4.5		MEQ/L	3.5-5.3	
CHLORIDE		105		MEQ/L	95-108	
MAGNESIUM		1.8		MEQ/L	1.2-2.0	
CALCIUM		9.2		MG/DL	8.5-10.3	
PHOSPHORUS, INORGANIC		3.3		MG/DL	2.5-4.5	
PROTEIN, TOTAL		6.7		G/DL	6.0-8.5	
ALBUMIN		4.1		G/DL	3.2-5.0	
GLOBULIN		2.6		G/DL (CALC)	2.2-4.2	
ALBUMIN/GLOBULIN RATIO		1.6		(CALC)	0.8-2.0	
BILIRUBIN, TOTAL		0.6		MG/DL	0.0-1.3	
ALKALINE PHOSPHATASE		46		U/L	20-125	
LDH, TOTAL		112		U/L	0-250	
GGT		15		U/L	0-65	
AST (SGOT)		12		U/L	0-42	
ALT (SGPT)		15		U/L	0-48	
URIC ACID		4.5		MG/DL	4.0-8.5	
IRON, TOTAL		74		MCG/DL	25-170	
IRON BINDING CAPACITY		353		MCG/DL	200-450	
% SATURATION		21		% (CALC)	12-57	
TRIGLYCERIDES		127		MG/DL	<200	
CHOLESTEROL, TOTAL		193		MG/DL	<200	
HDL-CHOLESTEROL		35		MG/DL	35 OR GREATER	ML
LDL-CHOLESTEROL			133 H	MG/DL (CALC)	0-130	ML
CHOL/HDLC RATIO			5.5 H	RATIO (CALC)	<4.98	ML

>> END OF REPORT <<

16-3 Results from blood tests give the doctor insight into how well the body is functioning.

There is little doubt about the meaning of *normal*; it is not only an acceptable result, it also implies that the result is perfect, or at least near perfect for that person. What do people commonly call something that isn't normal? Abnormal. But that could be misleading to the pilot and to the doctor.

Abnormal has a varying definition. Abnormal, especially to a doctor, is not only less than normal, but unacceptable, a result or finding that requires further evaluation and/or treatment. Yet, it is impossible to be normal in every parameter that is used to define good health. Something that is abnormal is not necessarily unacceptable in the eyes of people other than the medical profession.

For example, using total cholesterol again, a value of 200 would be considered normal (as would be blood pressure of 120/80 or vision of 20/20). That doesn't mean that any value above those just stated (or above the normal range) are abnormal or unacceptable, although a doctor might put that value in an abnormal category until it is proven to be acceptable.

In other words, to a doctor's mind, something abnormal is already unacceptable. To a nonmedical person, such as a pilot or the FAA, abnormal might be OK when additional tests are done to prove that the result is safe. Maybe another term should be used by pilots when discussing the results of their exams—*not-normal.*

Not-normal should mean that the value might not be perfect but it is not really abnormal—unacceptable—either. Knowing that the rest of the medical evaluation is within normal limits now places the less-than-normal value in the acceptable range.

Another concept to keep in mind is the bell-shaped curve. We know that such a curve represents how a population of people, healthy and not so healthy, statistically compares to others for a specific parameter. Take the results for blood sugar; a curve from the laboratory will show a wide range of perhaps 70–120+, which will be considered "normal." Values on either side of the curve might still be acceptable but out of the lab's "normal" range. It's up to the doctor to determine the meaning of results that are beyond the "normal" curve's ranges.

Virtually everything about our body, the literally thousands of parameters that define our specifications (weight, cholesterol, hair color, urine output, sleep requirements, etc.) are part of some bell-shaped curve. And to complicate matters, there is an interrelationship with many of these parameters; therefore, if asking if something is "normal," think of where that something fits on your bell-shaped curve.

Yes, this is making an issue of a minor point. But note how people react when they say their cholesterol is 242, or their vision is 20/30, or their blood pressure is 135/85. Some will get very excited and concerned. They are used to relating to absolute terms of aircraft performance. If a test result is above (or below) a value that is perceived as being normal, then they reason that it is unacceptable: not true. And this is all the more reason to have periodic exams to rule out anything that really is acceptable but might not be normal.

You will find this attitude in virtually all aspects of medical tests: weight, sugar in the urine, pulse rate, EKG rhythm, murmurs, and the like. It is important, therefore, to fully understand the significance of all medical test results relative to your health and not be hung up on absolute values. Furthermore, test results such as blood pressure and blood cholesterol often will vary from day to day. Monitoring trends over time becomes an important part of a health maintenance program.

The electrocardiogram (EKG)

The *electrocardiogram* (EKG) test measures the internal conduction of a physiological electrical current through your heart muscle. The EKG does not put the current into your body; it measures the current that is made by the cells in the body. The readings come from 12 leads (connections) attached by suction cups or patches to the chest wall and to your arms and legs. The machine then measures the current and reports the results as tracings on a piece of paper.

From this, the doctor can recognize rhythm disturbances, poor oxygen supply to heart muscle, electrical-conduction defects (like heart blocks), heart size, and other results about the heart muscle and its conduction system. It will not reveal much information about murmurs or the status of the heart's coronary arteries.

The EKG is a valuable screening aid, but like many medical tests, it is not a completely accurate, definitive, or totally reliable diagnostic tool. In other words, the tracings can be interpreted by different doctors as being normal or abnormal. At times, it is a judgment call, subject to interpretation, and necessitating other opinions. If the reading is considered abnormal, other tests can be done (a stress-test exercise EKG, echo-cardiograms, thallium scans, etc.) to rule out any pathology that could be the source or cause of the abnormal results. This added data, if "normal" can make the abnormal result acceptable. The initial test result was not normal, but it's OK.

The spirometry test

The *spirometry*, or *pulmonary function test*, is a measurement of your lung's ability to function: air capacity and exhalation rate. This is tested by blowing into the machine as quickly and forcefully as you can and then for as long as you can. Some people cannot do this test well because they don't understand how to blow into the machine. Here's a tip: Practice blowing out a match that is held in front of your wide opened mouth (do not purse your lips). Continue to try blowing out the match as you increase the distance of the match from the mouth.

X-rays

X-rays, particularly of the chest, are common and important, if just for a baseline for diagnostic purposes in the future. In addition to visualizing the lungs, the heart and major vessels are also seen as well as the bony structures of the ribs and spine of the chest; however, the trend in medicine is not to do routine screening X-rays.

Vision and hearing tests

Eye and ear checks are common to all pilots, and the significance of these results is explained in their respective chapters. It is worth repeating that because these test results are vital to your safe flight and medical certification, it is imperative that you not take shortcuts. A good eye exam by an ophthalmologist (an M.D.) should be done early in your flying followed with periodic exams as you get older. This is in addition to those screening tests with your AME. Hearing is equally important, but the shortcut of using the whispered voice test or conversation (voice) test tells you very little in terms of a quantitative evaluation of your hearing.

Audiograms using an audiometer are preferred, at least once every one or two years.

Drug testing

Urine and blood can be tested for just about any chemical or substance. High-tech equipment can detect minute amounts; therefore, a company or government agency can check for anything it wants to check for and even determine the level or amount that is considered significant. This includes medication, nicotine, carbon monoxide, chemicals, alcohol, and drugs.

DOT and FAA in 1989 mandated that certain professional pilots be tested for drugs: cocaine, marijuana, amphetamines, PCP, and opiates. The urine is tested for the presence of these drugs (or their metabolites) as well as the level or amount in the urine.

This is a two-step process. The first test is for screening the urine for the presence of drugs. Several materials and even foods can result in a "false positive" for a specific drug. In addition, the amount of that drug is also measured and must be at a certain level before noted as positive. If that level is reached, then a confirmatory test is done—such as gas chromatography/mass spectrography and other such test—which is virtually 100 percent accurate. These tests define a footprint for every known chemical in a series of curves. No other chemical matches that curve, which means that if the test is positive, the chemical is present; therefore, there are no false positives in the final report.

Then the level or amount present is once again determined. That test result must be above a certain predetermined level before a positive result is reported. That level is often fairly high, which means that once a test result is reported, not only is the drug present in the urine, but it is there at a level that confirms usage. If the testing program is to be faulted, it is, in fact, that a true user will not be reported unless that threshold level is reached.

As of January 1995, alcohol is also tested using a breathalyzer and for the same reasons as for illicit drugs. Blood testing for alcohol is not used in usual situations.

Time will tell how the drug-testing program will evolve. The biggest challenge is the administration of the program, what drug levels are considered significant, and what to do with the results. No doubt that this testing will evolve and change over the years as experience is gained in testing and the number of users actually found.

FAA AND COMPANY FLIGHT PHYSICALS

The FAA exam is as comprehensive as the aviation medical examiner (AME) wants to make it. The FAA really doesn't require many tests, leaving the AME to decide what evaluations to do to meet the medical standards and the intent of the regulations; therefore, you can find very simple exams given by some doctors and more-meaningful exams given by more-thorough doctors.

It is common knowledge that many pilots will seek out the easiest examiner so as to avoid finding potential abnormal results that might ground them (see chapter 17 regarding medical standards). Passing an FAA medical certification exam does not mean that all medical problems are ruled out. Because it is a basic exam, many medical and especially psychological conditions, some significant, will be missed.

Some companies (especially those with larger flight departments) require periodic comprehensive medical examinations for their flightcrews. The problem sometimes arises where the results of this exam are not dealt with from an aeromedical perspective. Some pilots feel so threatened by these company exams that they go to great measures, often through their union, to keep company exams from being required. The fact remains that the only way a pilot can be determined to be medically safe is with an adequate evaluation, just as you would expect for your airplane.

The threat of possibly losing your medical certificate is real, but that doesn't minimize the necessity of getting a proper evaluation on a periodic basis. Appropriate health maintenance includes comprehensive checkups, the results of which should be reviewed by an aeromedical doctor. This type of doctor—flight surgeon—is crucial in protecting your health and medical certificate as well as being in a position of interpreting the significance with safety of all test results. By understanding the procedures for certification, there is no professional or responsible reason to avoid the very program that keeps a pilot safe. Furthermore, by understanding what to expect in a complete physical and a health maintenance program, the pilot is ahead of the process of maintaining a valid and meaningful medical certificate.

17

Medical standards, regulations, and certification

Frank never did think much of the required FAA medical exam. He thought it was a waste of his time because he felt he was in a better position to tell if he was safe, based on how he felt, than an annual medical. As a result, he sought out and went to the AME in town who had a reputation for giving easy exams without making waves. Then he got a good job with a regional airline and began looking forward to going to work. His medical was due by the end of August, and he found himself in a distant town on the 31st of August. He wouldn't be legal on his return flight tomorrow. He checked the Yellow Pages, found an AME, and explained his plight. The AME had a cancellation and was able to see him later that day. But this examiner did a more complete exam, which included an audiogram. Frank couldn't pass this test and didn't know he had a hearing loss because he never had one before. And his blood pressure was up. The examiner was too busy and didn't desire to get involved with the FAA. Frank wasn't certified, and the AME sent in the form to the FAA. "I'm sure the FAA will certify you, but I don't have the time to help you out. Besides, your own AME should be able to help." Frank had heard that it could take weeks before the FAA responds, but now what could he do. He had to call his company and try to explain.

NOTHING IS MORE INTIMIDATING to most pilots than the required FAA flight physical. The fear of losing a medical because of the AME and FAA denying the pilot is often more intense than the fear of busting a checkride. The main reason for this perception is the fact that it happens and the pilot feels out of control.

However, this fear is fueled by many well embellished and often erroneous war stories of pilots denied their medical certification either for a minor reason or because the FAA took months and/or years to reach a conclusion. It's no wonder that pilots dread this ordeal and seek out the most lenient AME so as not to "rock the boat." In addition, the concerned pilot will often deny or minimize medical problems when asked specific questions on the application form or by medical personnel.

Much of the threat comes from a poor understanding of how the certification process works, relying more on crew lounge counsel than on facts. In the pilot's defense, however, there is little credible information available to which the pilot has access that explains how the process is supposed to work. There is a lot of misinformation and poor counsel that tends to lead the pilot into traps, delays, and red tape.

FAA medical certification is a good example of crisis management by the pilot. As long as the pilot receives his or her flight physical, there is little motivation to learn how the FAA works until there is a problem picked up during the exam. This attitude is a source of great stress to the pilot, one more stress added to the many that have already been identified elsewhere in the text; therefore, knowing how the certification process works is essential to remove one more source of distraction in the cockpit (worrying about the FAA medical due next week).

Another area of concern for pilots is the company physical. The pilot might be required to take a comprehensive medical evaluation, similar to what is given to the company executives. The reasoning is that if the exam is required for top employees, it should be adequate for the pilots. This is only partially true. Granted, the physical is usually very thorough, but the problem lies in the fact that the results are interpreted by company doctors not necessarily familiar with FAA requirements or what equates to a pilot shown to be fit to fly. As noted in previous chapters, there are medical conditions that are safe on the ground but become a risk to safe performance in flight.

Furthermore, some nonaviation doctors might expect unrealistically strict medical standards for their pilots; therefore, a slightly abnormal EKG or blood test found during the CEO's exam will not keep this company officer from going back to work. The same findings with a pilot might have entirely different consequences. If the company doctor is a "white knuckle flier," he or she won't allow you to fly because of his or her perception of the potential risk. On the other hand, a doctor familiar with aviation medicine and FAA aeromedical standards would consider other tests, possibly allowing the pilot to fly in the meantime if he or she doesn't feel there is any immediate risk to safety. Furthermore, the AME or flight surgeon can identify a medical problem that could impair a pilot in flight that would otherwise be minimized by other doctors.

DEFINITIONS

Many of the following terms are used in pilot conversations dealing with medical certification; however, as noted before, there is a lot of confusion as to who really does what. Here are the key players in the FAA medical certification process.

Aviation medical examiner (AME)

Aviation medical examiners (AMEs) are medical doctors designated by the FAA to examine pilots. They are not employed by the FAA. They can have any medical specialty, probably are not pilots, and often fit you into their schedules with the rest of their patients. The FAA recommends that AMEs be pilots, have a background of military aviation medicine, and be active in aeromedical organizations; however, many do not have this background. An AME has no authority to certify a pilot unless everything is normal or meets standards in the examiner's eyes. Anything suspected as being abnormal has to be reviewed by FAA doctors before the pilot can be certified. The good news is that there is more authority being given to AMEs, providing they strictly follow FAA guidelines for evaluations.

Flight surgeon

All aeromedical doctors in the military are called flight surgeons. They have been trained in an extensive aviation medicine course and have thus earned their professional wings. In comparison to many AMEs who have only minimal knowledge and interest in aviation medicine and certification, some civilian "flight surgeons" choose to be more knowledgeable, involved, and effective. An AME-flight surgeon is someone who specializes in working with pilots, the aviation community and the FAA, while other AMEs view themselves as only examiners without taking on additional roles or involvement expected by the pilot. The only way the pilot can determine what role a particular AME chooses to play is by talking to other pilots and even the FAA. They know which AMEs are willing to work with the pilot in complicated cases.

Oklahoma City (Oke City)

All applications for certification (Form 8500-8) go to the FAA's Mike Monroney Aeronautical Center in Oklahoma City for review by FAA-employed aeromedical doctors and their staff. They receive in excess of 2,000 forms per day. The results are fed into a computer. If anything is abnormal or incorrectly filled out, the computer rejects the form, which means it must now be reviewed by a medical person (not necessarily a doctor). If there is doubt as to the significance of the results of the exam, one of the doctors gets involved.

All EKGs for first-class certification must be transmitted electronically to Oke City. Some AMEs can now transmit the Form 8500 electronically. Both of these procedures have greatly reduced the number of errors and omissions. Despite some critic's fears, there is no adverse effect on the accuracy of the data when it arrives in Oke City.

The Oke City level has much more authority to certify and deny than in years past. In fact, that staff now can process "special issuance" requests for otherwise deniable situations, instead of having to send the request to the Federal Air Surgeon in Washington, D.C. This office will also direct the pilot (not the AME) to complete further tests, if indicated, to determine certifiability.

Regional flight surgeon

The FAA has several regions around the country. Within each region is a flight surgeon who has more authority than the AME and can often guide the pilot and AME in problem cases. These doctors still are responsible to Oke City but can often provide an OK to fly while other medical tests are being done.

Federal air surgeon

The federal air surgeon is the chief flight surgeon responsible for all civilian aviation medicine issues, including medical certification. Although most of the evaluations and decisions come out of Oke City, the federal air surgeon has the final say in very complex cases.

The federal air surgeon and support staff are those who are responsible for ensuring that the public is protected from unsafe pilots. Not an easy role to play, considering they must abide by federal regulations, policies, and laws, in addition to their own medical insight. There is little room for discretionary judgments, such as those enjoyed by other doctors who treat you; however, with adequate documentation of what your true health is, the FAA will try to certify all pilots if they meet specified standards and can prove their medical condition is an acceptable risk in flight. In fact, it creates much more work for the FAA to deny you, so there is no incentive to issue frivolous denials, contrary to what some critics will say.

Part 67

This refers to Part 67 of the Federal Aviation Regulations, which outlines the medical regulations that specify aeromedical standards for certification. These standards are usually quite general as opposed to military medical standards, which are comprehensively spelled out. The FARs are adequate because there is plenty of room for interpretation of the standards, allowing for exceptions and waivers based upon thorough medical evaluations. The FAA will always have the authority to certify any medical situation if the situation is proven safe.

FAR 61.53

This is a specific regulation, within FAR Part 61 (which refers to all pilots, commercial and general). It is noted on the back of a pilot's medical certificate as a regulation by which the pilot is governed. Specifically it states: "OPERATIONS DURING A MEDICAL DEFICIENCY. No person may act as pilot in command, or in any other capacity as required pilot flight crewmember, while he has a known medical deficiency, or increase of a known medical deficiency that would make him unable to meet the requirements for his current medical certificate." In other words, the pilot is not legal if he or she has any problem that would make him or her unsafe and unable to pass an FAA flight physical just before the flight. This often places the burden on the pilot to determine if he or she is legal and safe for each flight. This regulation might be expanded in future changes, but the intent will be the same.

Form 8500

All forms dealing with the application and evaluation for medical certification are labeled as 8500s, or FAA Form 8500. The next number refers to the specific form, whether it's the medical certificate (8500-9), the waiver (8500-15), eye evaluation form (8500-7), or the application for medical certification. It's this application form number—8500-8—that is usually being referred to when someone states simply "8500" (Fig. 17-1). Some companies, when considering disability (loss of license) insurance for a pilot receiving a final FAA denial, require a formal denial from the FAA in addition to the denial by an AME. Both forms are 8500s, but they have different suffixes; therefore, one must be certain to which form someone else is referring if only "8500" is used.

Limitations

To be legal, the pilot must follow any specific condition or conditions stated on the medical certificate within the "limitations" area. For example, the most common limitation is the need to wear or possess corrective lenses for vision. If the pilot with the limitation is not wearing corrective lenses (glasses or contact lenses) for distant vision, or does not have glasses for near vision in the cockpit during flight, he or she is in violation of the regulations.

Waiver

This term has a double meaning. One is a "Statement of Demonstrated Ability" or SODA, which means that the pilot has demonstrated his ability to fly safely despite the fact that he doesn't meet absolute standards, such as single-eye (monocular) vision, deficient color vision, missing limbs, hearing loss, and the like. Such a waiver is granted after special practical evaluations or a "medical" checkride and then becomes a separate document that supplements the medical certificate.

The other meaning of a waiver is the usual terminology expressed when discussing medical problems (illnesses and diseases and their treatments) and whether or not they are certifiable. If a disease initially considered to be unsafe can be proven to be safe in flight, then the standards are waived. It's like an exception to the rule because it can be proven that the rule doesn't apply to the pilot who has had a complete medical workup. This definition is not a separate document. Instead, it is a letter stating that after review of the medical file, the pilot meets the intent of federal regulations. This is one reason why medical FARs should not be too specific because the lack of specificity allows the FAA to use its professional discretion as to whether or not the pilot is proven safe and an acceptable risk despite the stated standards. This letter is not something that you carry with you. The SODA, however, does have to be carried with you and shown to an examiner if requested.

This Page (Except For Shaded Areas)
PLEASE PRINT

UNITED STATES OF AMERICA
Department of Transportation
Federal Aviation Administration
MEDICAL CERTIFICATE VOID CLASS

This certifies that (*Full name and address*):

Date of Birth	Height	Weight	Hair	Eyes	Sex

has met the medical standards prescribed in Part 67, Federal Aviation Regulations for this class of Medical Certificate.

Limitations

Date of Examination | Examiner's Serial No.

Examiner: Signature | Typed Name

AIRMAN'S SIGNATURE VOID

Sample

1. Application For: ☐ Airman Medical Certificate ☐ Airman Medical and Student Pilot Certificate

2. Class of Medical Certificate Applied For: ☐ 1st ☐ 2nd ☐ 3rd

3. Last Name | First Name | Middle Name

4. Social Security Number

5. Address: Number/Street | Telephone Number ( ) | City | State/Country | Zip Code

6. Date of Birth (M M D D Y Y) | 7. Color of Hair | 8. Color of Eyes | 9. Sex

10. Type of Airman Certificate(s) Held:
☐ None ☐ ATC Specialist ☐ Flight Instructor ☐ Other
☐ Airline Transport ☐ Flight Engineer ☐ Private
☐ Commercial ☐ Flight Navigator ☐ Student

11. Occupation | 12. Employer

13. Has Your FAA Airman Medical Certificate Ever Been Denied, Suspended, or Revoked?
☐ Yes ☐ No If yes, give date (M M Y Y)

Total Pilot Time (Civilian only): 14. To Date | 15. Past 6 months

16. Date of Last FAA Medical Application (M M Y Y) ☐ No Prior Application

17. Do You Currently Use Any Medication (Prescription or Nonprescription)?
☐ Yes ☐ No If yes, give name, purpose, dosage, and frequency.

18. MEDICAL HISTORY — Have you ever had or have you now any of the following? Answer "yes" for every condition you have ever had in your life. In the EXPLANATION box below, you may note "PREVIOUSLY REPORTED, NO CHANGE" only if the explanation of the condition was reported on a prior application for an airman medical certificate and there has been no change in your condition. See Instructions Page

	Yes	No	Condition		Yes	No	Condition		Yes	No	Condition		Yes	No	Condition
a.	☐	☐	Frequent or severe headaches	g.	☐	☐	Heart or vascular trouble	m.	☐	☐	Mental disorders of any sort; depression, anxiety, etc.	r.	☐	☐	Military medical discharge
b.	☐	☐	Dizziness or fainting spell	h.	☐	☐	High or low blood pressure	n.	☐	☐	Substance dependence or failed a drug test ever; or substance abuse or use of illegal substance in the last 5 years.	s.	☐	☐	Medical rejection by military service
c.	☐	☐	Unconsciousness for any reason	i.	☐	☐	Stomach, liver, or intestinal trouble					t.	☐	☐	Rejection for life or health insurance
d.	☐	☐	Eye or vision trouble except glasses	j.	☐	☐	Kidney stone or blood in urine	o.	☐	☐	Alcohol dependence or abuse	u.	☐	☐	Admission to hospital
e.	☐	☐	Hay fever or allergy	k.	☐	☐	Diabetes	p.	☐	☐	Suicide attempt				See v. & w. Below
f.	☐	☐	Asthma or lung disease	l.	☐	☐	Neurological disorders; epilepsy, seizures, stroke, paralysis, etc.	q.	☐	☐	Motion sickness requiring medication	x.	☐	☐	Other illness, disability, or surgery

Conviction and/or Administrative Action History — See Instructions Page

v. Yes ☐ No ☐ History of (1) any conviction(s) involving driving while intoxicated by, while impaired by, or while under the influence of alcohol or a drug; or (2) history of any conviction(s) or administrative action(s) involving an offense(s) which resulted in the denial, suspension, cancellation, or revocation of driving privileges or which resulted in attendance at an educational or a rehabilitation program.

w. Yes ☐ No ☐ History of other conviction(s) (misdemeanors or felonies).

EXPLANATIONS: See Instructions Page | For FAA Use

19. Visits to health professional within last 3 years. ☐ Yes (explain below) ☐ No See Instructions Page

Date	Name, Address, and type of Health Professional Consulted	Reason

- NOTICE -

Whoever in any matter within the jurisdiction of any department or agency of the United States knowingly and willfully falsifies, conceals or covers up by any trick, scheme, or device a material fact, or who makes any false, fictitious or fraudulent statements or representations, or entry, may be fined up to $250,000 or imprisoned not more than 5 years, or both. (*18 U.S. Code Secs. 1001; 3571*).

20. APPLICANT'S DECLARATION

I hereby authorize the National Driver Register (NDR), through a designated State Department of Motor Vehicles, to furnish to the FAA information pertaining to my driving record. This consent constitutes authorization for a single access to the information contained in the NDR to verify information provided in this application. Upon my request, the FAA shall make the information received from the NDR, if any, available for my review and written comment. Authority: 23 U.S. Code 401, Note. (All persons using this form must sign it. **However, NDR consent does not apply unless this form is used as an application for Airman Medical or Airman Medical and Student Pilot Certificate.**)

I hereby certify that all statements and answers provided by me on this application form are complete and true to the best of my knowledge, and I agree that they are to be considered part of the basis for issuance of any FAA certificate to me. I have also read and understand the Privacy Act statement that accompanies this form.

Signature of Applicant VOID | Date (M M D D Y Y)

FAA Form 8500-8 (1-91) Supersedes Previous Edition | OMB Approval No. 2120-0034

17-1 FAA medical application form (front).

TYPICAL CERTIFICATION SCENARIO

The intent of the FAA medical certification process is to identify any medical, mental, physical, or psychological condition that would make the pilot impaired and unsafe. Certification also anticipates any abnormal conditions that might develop between the time of the exam and the next required medical evaluation. Problems with this process occur when unexpected abnormalities are found during the exam. If the AME doesn't know how to deal with the finding by working directly with the FAA, the entire application is sent to the FAA for its review.

During this review, which could take a minimum of six weeks, the pilot is grounded. It's this lack of control on the part of the pilot that is the most threatening to him. He perceives his certificate and his livelihood as being judged by some bureaucratic doctors who are making their decision only upon an impersonal file of medical reports, and he has nothing to say about it. This just isn't the case.

Recall that some pilots will seek the most lenient AME, one who also is less knowledgeable about the certification process and is unwilling to do more than fill the squares, certify only those that he or she feels are legal, and send the rest to the FAA to decide. Some AMEs tend to discourage pilots who have a medical problem that the AME feels won't be certified, often stating that the pilot will never fly again. The fact is that the FAA certifies many conditions often thought to be a permanent grounding. These conditions include hypertension using medication, bypass heart surgery, heart attacks, kidney stones, alcoholism, and many more. In the hands of a competent and knowledgeable AME, anything is potentially certifiable if it can be proven to the FAA's satisfaction that the condition will not impair the pilot's performance or lead to unexpected incapacitation, now or in the near future.

Two of the controllable problems leading to delays or even denials are failing to answer all questions ("filling in all the squares") on the front of the application form (FAA Form 8500-8) and not explaining all new affirmative answers or changes from previous exams (Fig. 17-2). The 8500-8 (revised in 1991) has some additional questions and a different format. Your AME should be able to explain what is required of you. On the back of the form, the AME is supposed to clarify all the responses the pilot made in the affirmative, fill in his or her portion dealing with his or her findings with the examination (including test results), and then state his or her action: certify, deny, or defer (Fig. 17-3). Here is where some AMEs will not take the time or effort to further evaluate and explain to the FAA a medical situation that could be questioned later. It's simpler, in the doctor's mind (especially when he or she is busy and doesn't want to get involved), to deny you and let the FAA take over the recertification process (Fig. 17-4).

This is why the FAA will correspond with the pilot (not the AME) when there is a problem noted or there is an incomplete file. The FAA expects the pilot to initiate and coordinate the requests necessary to complete the process. This "passing the buck" to the FAA becomes a very time-consuming process and certainly is not in the best interests of the pilot. It need not happen that way.

Instructions for Completion of the Application for Airman Medical Certificate or Airman Medical and Student Pilot Certificate, FAA Form 8500-8

Applicant must fill in completely numbers 1 through 20 of the application using a ballpoint pen. Exert sufficient pressure to make legible copies. The following numbered instructions apply to the numbered headings on the application form that follows this page.

NOTICE — Intentional falsification may result in federal criminal prosecution. Intentional falsification may also result in suspension or revocation of all airman, ground instructor, and medical certificates and ratings held by you, as well as denial of this application for medical certification.

1. APPLICATION FOR — Check the appropriate box.

2. CLASS OF AIRMAN MEDICAL CERTIFICATE APPLIED FOR — Check the appropriate box for the class of airman medical certificate for which you are making application.

3. FULL NAME — If your name has changed for any reason, list current name on the application and list any former name(s) in the EXPLANATIONS box of number 18 on the application.

4. SOCIAL SECURITY NUMBER — The social security number is optional; however, its use as a unique identifier does eliminate mistakes.

5. ADDRESS — Give permanent mailing address and country. Include your complete nine digit ZIP code if known. Provide your current area code and telephone number.

6. DATE OF BIRTH — Specify month (MM), day (DD), and year (YY) in numerals; e.g., 01/31/50.

7. COLOR OF HAIR — Specify as brown, black, blond, gray, or red. If bald, so state. Do not abbreviate.

8. COLOR OF EYES — Specify actual eye color as brown, black, blue, hazel, gray, or green. Do not abbreviate.

9. SEX — Indicate male or female.

10. TYPE OF AIRMAN CERTIFICATE(S) HELD — Check applicable block(s). If "Other" is checked, provide name of certificate.

11. OCCUPATION — Indicate major employment. "Pilot" will be used only for those gaining their livelihood by flying.

12. EMPLOYER — Provide your employer's full name. If self-employed, so state.

13. HAS YOUR FAA AIRMAN MEDICAL CERTIFICATE EVER BEEN DENIED, SUSPENDED, OR REVOKED —If "yes" is checked, give month and year of action in numerals.

14. TOTAL PILOT TIME TO DATE — Give total number of civilian flight hours. Indicate whether logged or estimated. Abbreviate as Log. or Est.

15. TOTAL PILOT TIME PAST 6 MONTHS — Give number of civilian flight hours in the 6-month period immediately preceding date of this application. Indicate whether logged or estimated. Abbreviate as Log. or Est.

16. MONTH AND YEAR OF LAST FAA MEDICAL EXAMINATION —Give month and year in numerals. If none, so state.

17. DO YOU CURRENTLY USE ANY MEDICATION (Prescription or Nonprescription) — Check "yes" or "no." If "yes" is checked, give name of medication(s), purpose, dosage, and frequency (e.g., daily, twice daily, as needed, etc.). See **NOTE** below.

18. MEDICAL HISTORY — Each item under this heading must be checked either "yes" or "no." You must answer "yes" for every condition you have ever had in your life and describe the condition and approximate date in the EXPLANATIONS box.

If information has been reported on a previous application for airman medical certificate and there has been no change in your condition, you may note "PREVIOUSLY REPORTED, NO CHANGE" in the EXPLANATIONS box, but you must still check "yes" to the condition. Do not report occasional common illnesses such as colds or sore throats.

"Substance dependence" is defined by any of the following: increased tolerance; withdrawal symptoms; impaired control of use; or continued use despite damage to health or impairment of social, personal, or occupational functioning. "Substance abuse" includes the following: use of an illegal substance; use of a substance or substances in situations in which such use is physically hazardous; or misuse of a substance when such misuse has impaired health or social or occupational functioning. "Substances" include alcohol, PCP, marijuana, cocaine, amphetamines, barbiturates, opiates, and other psychoactive chemicals.

Conviction and/or Administrative Action History — Letter (v) of this subheading asks if you have ever been: (1) convicted (which may include paying a fine, or forfeiting bond or collateral) of an offense involving driving while intoxicated by, while impaired by, or while under the influence of alcohol or a drug; or (2) convicted or subject to an administrative action by a state or other jurisdiction for an offense for which your license was denied, suspended, cancelled, or revoked or for which you were required to attend an educational or rehabilitation program. Individual traffic convictions are not required to be reported if they did not involve: alcohol or a drug; suspension, revocation, cancellation, or denial of driving privileges; or attendance at an educational or rehabilitation program. If "yes" is checked, a description of the conviction(s) and/or administrative action(s) must be given in the EXPLANATIONS box. The description must include: (1) the alcohol or drug offense for which you were convicted or the type of administrative action involved (e.g., attendance at an alcohol treatment program in lieu of conviction; license denial, suspension, cancellation, or revocation for refusal to be tested; educational safe driving program for multiple speeding convictions; etc.); (2) the name of the state or other jurisdiction involved; and (3) the date of the conviction and/or administrative action. The FAA may check state motor vehicle driver licensing records to verify your responses. Letter (w) of this subheading asks if you have ever had any other (nontraffic) convictions (e.g., assault, battery, public intoxication, robbery, etc.). If so, name the charge for which you were convicted and the date of conviction in the EXPLANATIONS box. See **NOTE** below.

19. VISITS TO HEALTH PROFESSIONAL WITHIN LAST 3 YEARS— List all visits in the last 3 years to a physician, physician assistant, nurse practitioner, psychologist, clinical social worker, or substance abuse specialist for treatment, examination, evaluation, or counseling. Give date, name, address, and type of health professional consulted, and briefly state reason for consultation. Multiple visits to one health professional for the same condition may be aggregated on one line. Routine dental, eye, and FAA periodic medical examinations may be excluded. See **NOTE** below.

20. APPLICANT'S DECLARATION — Two declarations are contained under this heading. The first authorizes the National Driver Register to release adverse driver history information, if any, about the applicant to the FAA. The second certifies the completeness and truthfulness of the applicant's responses on the medical application. The declaration section must be signed and dated by the applicant after the applicant has read it.

NOTE: If more space is required to respond to "yes" answers for numbers 17, 18, or 19, use a plain sheet of paper bearing the information, your signature, and the date signed.

Applicant—Please Tear Off This Sheet After Completing The Application Form.

FAA Form 8500-8 (1-91) Supersedes FAA Form 8500-12

Applicant Must Complete
This Page (Except For Shaded Areas)

17-2 Instructions for completion of the application form.

NOTE: FAA's Copy of the Report of Medical Examination Must be TYPED.

REPORT OF MEDICAL EXAMINATION

21. Height (inches)	22. Weight (pounds)	23. Statement of Demonstrated Ability ("Waiver") ☐ YES ☐ NO Defect Noted:	24. WAIVER SERIAL NUMBER

CHECK EACH ITEM IN APPROPRIATE COLUMN	Normal	Abnormal	CHECK EACH ITEM IN APPROPRIATE COLUMN	Normal	Abnormal
25. Head, face, neck, and scalp			37. Vascular system (Pulse, amplitude and character: arms, legs, others)		
26. Nose			38. Abdomen and viscera (Including hernia)		
27. Sinuses			39. Anus and rectum (Digital exam only if clinically indicated or requested)		
28. Mouth and throat			40. Skin		
29. Ears, general (Internal and external canals: Hearing under item 49)			41. G-U system (Pelvic exam only if clinically indicated or requested)		
30. Ear Drums (Perforation)			42. Upper and lower extremities (Strength and range of motion)		
31. Eyes, general (Vision under items 50 to 54)			43. Spine, other musculoskeletal		
32. Ophthalmoscopic			44. Identifying body marks, scars, tattoos (Size & location)		
33. Pupils (Equality and reaction)			45. Lymphatics		
34. Ocular motility (Associated parallel movement, nystagmus)			46. Neurologic (Tendon reflexes, equilibrium, senses, cranial nerves, coordination, etc.)		
35. Lungs and chest (Breasts exam only if clinically indicated or requested)			47. Psychiatric (Appearance, behavior, mood, communication, and memory)		
36. Heart (Precordial activity, rhythm, sounds, and murmurs)			48. General systemic		

NOTES: Describe every abnormality in detail. Enter applicable item number before each comment. Use additional sheets if necessary and attach to this form.

49. Hearing	Right Ear	Left Ear	Audiometer (Threshold in Decibels)	Right Ear 500	1000	2000	3000	4000	Left Ear 500	1000	2000	3000	4000
Voice Test													

50. Distant Vision	51a. Near Vision	51b.	52. Color Vision
Right 20/ Corrected To 20/	Right 20/ Corrected To 20/		☐ Normal
Left 20/ Corrected To 20/	Left 20/ Corrected To 20/		☐ Abnormal
Both 20/ Corrected To 20/	Both 20/ Corrected To 20/		

53. Field of Vision	54. Heterophoria 20′ (In diopters)	Esophoria	Exophoria	Right Hyperphoria	Left Hyperphoria
☐ Normal ☐ Abnormal					

55. Blood Pressure (Sitting, mm of Mercury)	Systolic	Diastolic	56. Pulse (Resting)	57. Urinalysis (If abnormal, give results)	Albumin	Sugar	58. ECG (Date) MM	DD	YY
				☐ Normal ☐ Abnormal					

59. Other Tests Given

60. Comments on History and Findings: AME shall comment on all "Yes" answers in the Medical History section and for abnormal findings of the examination. (Attach all consultation reports, ECGs, X-rays, etc. to this report before mailing.)

FOR FAA USE
Pathology Codes:

☐ **No Significant Medical History** ☐ **No Abnormal Physical Findings**

61. Applicant's Name

62. Has Been Issued — ☐ Medical Certificate ☐ Medical & Student Pilot Certificate
☐ **No Certificate Issued** — Deferred for Further Evaluation
☐ **Has Been Denied** — Letter of Denial Issued (Copy Attached)

63. Disqualifying Defects (List by item number)

64. Medical Examiner's Declaration— I hereby certify that I have personally reviewed the medical history and personally examined the applicant named on this medical examination report. This report with any attachment embodies my findings completely and correctly.

Date of Examination	Aviation Medical Examiner's Name and Address (Including Zip Code)	Aviation Medical Examiner's Signature
__ __ __ M M D D Y Y	VOID	VOID AME Serial Number AME Telephone ( )

FAA FORM 8500-8 (1-91)

17-3 FAA medical application form (back).

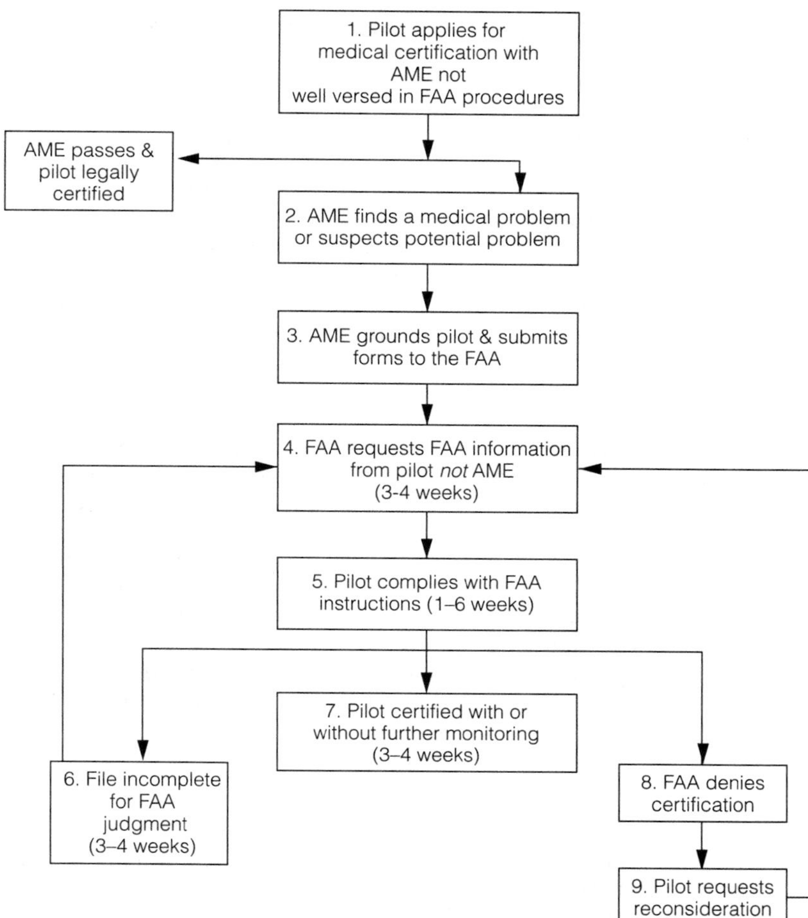

Comments:

❑1. The typical AME is not highly knowledgeable regarding expeditious handling of problem certifications.

❑2. Often AME does not tell pilot significance of findings — allows FAA to be "bad guy" and inform pilot he is grounded.

❑3. Often prematurely submits report to FAA.

❑4. FAA requires additional medical information to reconsider a pilot's fitness.

❑5. Pilot sometimes left on his own to work with specialists and handle additional paperwork required by FAA.

❑6. Because of misunderstanding concerning intent of further evaluation and requirements, FAA requests further data, pilot is back at box 4.

❑8,9. By this point pilot often finds himself trying to coordinate and achieve cooperation from 3 entities — the original AME, any specialists who have entered the picture, and the FAA .

Key to improperly handled sequence is not knowing and properly *anticipating* the FAA needs *in advance* of their requests. Tremendous amounts of time can be lost waiting for the FAA to respond to previous paperwork submission, only to learn more paperwork is necessary.

17-4 Improper certification.

PREFERRED CERTIFICATION SCENARIO

These unnecessary delays do not have to occur. The key to an efficient certification process, whether the pilot is healthy or not, is to work with a knowledgeable AME: a flight surgeon. Such doctors are usually pilots and have a high interest in aviation. They empathize with the pilot's concern over certification and have a reputation amongst the pilot community, as well as the FAA, of working with the pilot on his or her behalf whenever problems arise. Furthermore they are active in aeromedical organizations, might be former military flight surgeons, and are very familiar with up-to-date FAA medical standards. These AMEs anticipate what the FAA requires for further testing, arrange for the evaluations, work up a completed file, discuss your case with one of the FAA doctors, and then submit the entire completed file for FAA review and conclusion (Fig. 17-5).

In less complex situations, an AME-flight surgeon can often be authorized by the FAA, by phone, to certify you based on the results of his or her evaluation. The FAA also respects these AMEs, knows they have a reputation of being thorough and professional, and often relies on their judgment in terms of the pilot being qualified. These AME-flight surgeons can be found by asking around the pilot lounges, looking for the doctor who works with pilots as defined above. Even the FAA in Oke City or the regional flight surgeon knows who the flight surgeons are whom the FAA respects. Call them—you need not identify yourself if you are merely asking for a doctor's name in your area. Despite what their image is made out to be, they are paid to help the pilot.

DENIAL

The most feared word in aviation is *denial*, meaning the same as *grounded*. It's the word used when the FAA feels the pilot has a medical condition that isn't safe. *A denial, contrary to what many think, is never final.* In other words, the FAA will always consider additional information from additional tests and medical opinions. The passage of time without progression of the disorder or the problem is also taken into consideration; therefore, if the pilot truly wants to return to flying, he or she should continue to pursue additional data with the help of a flight surgeon, to prove he or she is safe.

Several medical conditions require mandatory denial at the time they are discovered: a character or behavior disorder severe enough to be unsafe, a psychosis, chronic alcoholism, drug addiction, epilepsy, disturbance of consciousness without satisfactory explanation, heart attack, significant coronary heart disease, and diabetes requiring any medication. It's also important to recognize that this refers to a "history of," which means that if at any time in your life you have had any of these conditions, denial is mandatory.

That's the bad news. The good news is that the federal air surgeon, through the medical regulation FAR 67.19, can "special issue" a certificate, even for the mandatory denials, if it can be proven that the condition, treated or untreated, will not be a safety issue.

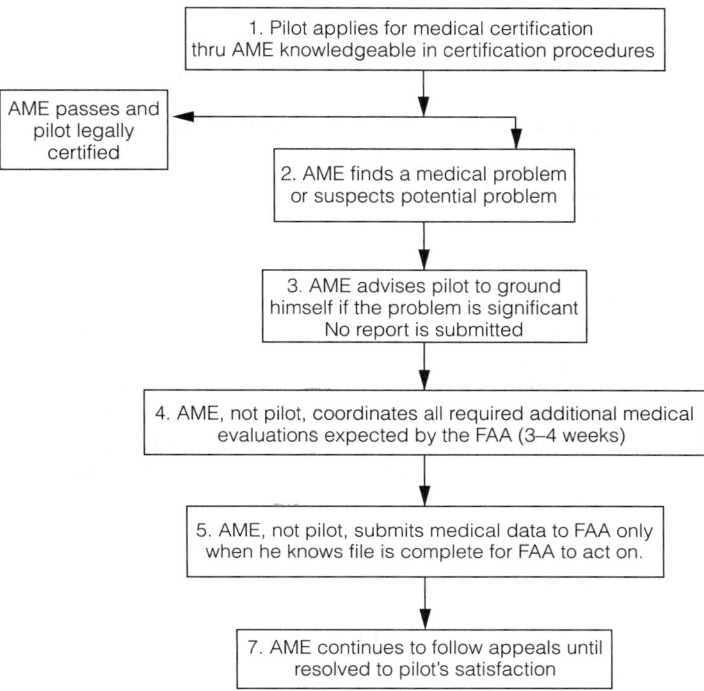

1. Pilot applies for medical certification thru AME knowledgeable in certification procedures
AME passes and pilot legally certified
2. AME finds a medical problem or suspects potential problem
3. AME advises pilot to ground himself if the problem is significant No report is submitted
4. AME, not pilot, coordinates all required additional medical evaluations expected by the FAA (3–4 weeks)
5. AME, not pilot, submits medical data to FAA only when he knows file is complete for FAA to act on.
7. AME continues to follow appeals until resolved to pilot's satisfaction

Comments:

❑1. This is the key to success of control: that the AME is knowledgeable in the certification process and aviation medicine, and is experienced in working with the FAA in protecting the healthy pilot's medical certificate.

❑2. AME will try to anticipate potential problems which can be resolved without compromising your responsibilities as a pilot.

❑3. This is the second key to success: no report is made to the FAA until both the AME and pilot are satisfied with the file. If pilot's condition is not significant, pilot continues to fly.

❑4. AME can make appointments with other specialists and already knows what the FAA will require for further evaluation. There is no need to wait for FAA's direction.

❑5. AME submits entire file to FAA, with cover letter, rather than making the FAA the coordinator of the varied medical reports. AME may also alert FAA about coming file.

❑6. AME can then determine what courses of action to follow once everyone knows what is necessary to satisfy FAA requirements.

❑7. The AME should commit himself to the pilot to assist him thru the entire appeals process and explain why such actions are necessary.

17-5 Proper certification.

RECERTIFICATION AND SPECIAL ISSUANCE

If the pilot is denied for any reason, from failing to submit a completed application to having a serious medical condition, the process is the same. In the hands of a competent and helpful AME, additional tests are accomplished, a file is compiled, and discussions are held with the FAA doctors.

The FAA expects certain minimum tests to be done. Some personal doctors will say that such tests are not necessary; however, in most cases, the FAA will still require these evaluations and will not consider certification until they are done. The FAA will not accept the doctor's unproven opinion that what you have is not significant. Unless there is objective evidence to back up that opinion, the FAA cannot act favorably and there are unnecessary delays. It might take several attempts with new information to convince the FAA that you are safe.

Recall that FAR 61.53 (which deals with your responsibility to be medically qualified every time you fly) is in effect only when you fly. This is a simplistic statement, but it is the key to keeping control during the recertification process. Providing you are working with an AME/flight surgeon and everyone is in the process of evaluating the pilot's condition, then nothing has to be reported to the FAA as long as the pilot doesn't fly—the pilot grounds himself or herself. This prevents premature submission of incomplete files to the FAA that would inevitably result in very lengthy and often unsatisfactory delays.

A responsible pilot's fear is not just that something will be found during a medical exam, it's the feeling that the condition must be reported to the FAA immediately and prematurely, leading to the familiar "war stories" of prolonged and unfair results. If the pilot doesn't fly, and the AME respects you and your commitment, then the pilot and the AME have "breathing room" to work out the problem before anything is reported. When everything has been accomplished and the file is complete, then the AME should contact the FAA and ultimately send in the file.

Finally, if the pilot is willing to accept the responsibility of being familiar with the FAA certification process and works with an AME/flight surgeon while the pilot is healthy and before any problem unexpectedly develops (avoiding crisis management), there is less of a threatening situation whenever the time comes for the FAA or company physical. Waiting for something to happen, going to an AME who gives a "pink and breathing" physical, and being informed only through "crew-lounge" counsel will lead to a bona fide "war story" of a pilot losing his or her medical for unexpected reasons. Look for the competent AME (or flight surgeon) now by checking around. Don't wait until a problem develops.

Recommended reading

Aeromedical Handbook for Aircrews
T.G. Dobie, 1972
National Technical Information Service
Springfield, VA 22151
(Comprehensive physiology)

Aeromedical Training for Flight Personnel (FM 1-301)
U.S. Army, 1983
Government Printing Office
(Physiology, military)

Aviation Weather
FAA and NOAA, 1975
Government Printing Office
(Weather)

Cause Factor: Human
Olaf W. Skjenna, 1981
Minister of National Health and Welfare
Canadian Government Press
(Helicopter flight physiology)

Fit to Fly
British Airline Pilots Association, 1980
Granada
London
(Health and fitness)

Flight Surgeon's Guide (AFP 161-18)
Department of Air Force, 1968
Government Printing Office
(Physiology and aviation medicine)

Flying Is Safe—Are You?
W. Bryce Hansen, 1969
Avikon Ltd.
Edmonton, Alberta, Canada
(Basic physiology)

Fundamentals of Aerospace Medicine
Roy L. Dehart, 1985
Lea & Febilger
Philadelphia
(Aerospace medicine text)

Hearing Conservation
Donald C. Gasaway, 1985
Prentice-Hall, Inc.
Englewood Cliffs, N.J.
(Hearing)

Instructors Manual for Physiological Training
Department of Air Force, 1968
Government Printing Office
(Physiology and military)

Manual of Civil Aviation Medicine
International Civil Aviation Organization, 1974
Montreal, Quebec, Canada
(Aerospace medicine)

Medication and Flying
Stanley R. Mohler, 1992
Boston Publishing Co.
Boston
(Medications)

Merck Manual
Merck & Co., Inc., 1992
Rahway, N.J.
(Medical)

Physiological Training
FAA, 1988
Civil Aeromedical Institute
Oklahoma City

Pilot's Manual Medical Certification
Richard O. Reinhart, M.D., 1996
McGraw-Hill
Professional Book Group
New York
(FAA certification and general health)

The Dangerous Sky
Douglas H. Robinson, 1973
University of Washington Press
Seattle
(History of aviation medicine)

U.S. Naval Flight Surgeon's Manual
U.S. Navy, 1980
Government Printing Office
(Flight physiology)

USAF School of Aerospace Medicine Course Material
Brooks AFB, Texas
(Flight physiology and aerospace medicine)

Human Factors for General Aviation
Stanley Trollip and Richard Jensen, 1991
Jeppesen Sanderson
(Human factors)

Your Body Clock
Hubertus Strughold, 1971
Charles Scribner's Sons
New York
(Jet lag)

Human Factors in Flight
H. Hawkins, 1993
Ashgate Publishing
(Human factors)

Aviation Psychology
Richard S. Jensen, 1989
Gower Publishing
Brookfield, Vermont
(Psychology)

Human Factors in Aviation
Earl L. Wiener, 1988
Academic Press
(Human factors)

Cockpit Resource Management
Earl Wiener, 1993
Academic Press
(CRM)

Index

A

acceleration, 216-221
  civilian flying, 217
  G force, 217, 218, 220-221, **218**, **219**, **220**
  types of, 216-217
accident reports, Beechcraft C99, 42
acclimatization, 185
  hypoxia and, 56-57
acetaminophen, 153
aeroembolism, 72, 73
air quality, cabin, 191-192
airworthiness, medical, 5
alcohol, 4, 161-172
  dependency, 170-172
  effects on performance, 165-166
  effects on the body, 164-165
  hangovers, 166-167
  hypoxia and, 55, 165
  physiological effects, 163-166
  physiology, 162-163
  recognizing a problem drinker, 168-170
  sources, 162
  tolerance, 167-168
  vision and, 110
  withdrawal, 168
alertness management, 199-201
altitude, physiology, 47-79
altitude chambers, 78, **79**
alveoli, 21
anemic hypoxia, 50
angular motion illusion, 131-133, **133**, **134**
anoxia, 47
antihistamines, 151
Armstrong's line, 41
arrhythmias, 157
arteries, 22-24
arthritis, 15
aspirin, 153
atmosphere, 29-41
  composition, 30-32
  division/layers of, 37-40, **39**
  effects on human body, 41
  international standard, 33
  ozone, 31-32, 64
  physical characteristics, 32-33
  physiological divisions, 40-41
atria, 19
Aurora Australis, 40
Aurora Borealis, 40
autokinesis, 137, **137**
automation, 231-238
  definition/description, 232-233
  information processing, 233-234
  physiological impairments, 234-236
  quantifying impairment, 236-238
aviation medical examiner (AME), 8-9, 266
Aviation Safety Reporting System (ASRS), 3

B

barodontalgia, 70, **70**
barometric pressure, 33
barosinusitis, 69
barotitis, 67-69, **67**
bends, 72, 73
biorhythms, 205
black-hole effect, 113
blood, 24-25
  pressure, 256
  tests, 258
bones, 14-15
Boyle's Law, 34-35, 66, **34**, **35**
brain, 11-12
  cerebral cortex, 11
  cerebrum, 11
  concussion, 12
  disorders of, 74
  frontal lobes, 11
  hindbrain, 11
  infections of, 11
  midbrain, 11
  occipital lobes, 11
  parietal lobes, 11
  temporal lobes, 11
bruise, 15

Illustrations are indicated by **boldface**.

C

cabin, pressurized, 62-63, 74
caffeine, 155-158
  content in common beverages, 156
  fatigue and, 209
  side effects, 157-158
  sources, 156-157
  tolerance, 158
calories, 250-251
capillaries, 22
carbohydrates, 249
carbon dioxide, 31
carbon monoxide, 63-64, 192-193
  effects of exposure, 63
  symptoms, 63
  treatment, 63-64
  vision and, 110
cardiopulmonary resuscitation, 241-242
cerebral cortex, 11
cerebrum, 11
certification, medical (*see* medical certification)
Charles' Law, 35
choking, 242
circadian rhythms, 204-206
circulatory system, 19-24
  alcohol and, 164
  heart, 19-21, **20**
  lungs, 21-22, **21**, **22**
  vascular system, 22-24, **23**
cocaine, 173
cockpit/crew resource management (CRM), 2
colds, 145-146
colonitis, 16
concussion, 12
conduction, 176
consciousness, loss of, 12
controlled flight into terrain (CFIT), 44
convection, 176-177
coriolis illusion, 130-131, **130**
crew duty time, 203-204
crew resource management (CRM), 222-230
  definition/description, 223-224
  deterrents to good, 227-228
  skills
    communication, 225
    crew coordination, 226
    decision-making, 225
    judgment, 226
    leadership, 225
  stress management, 229-230
  training, 226-227

D

Dalton's Law, 35-36
decibel scale, vs. sound intensity, 84, **84**
DECIDE model, 46
decompression, 2, 64-65
  acute effects on human body, 65
  explosive, 64-65
  rapid, 65
  subtle, 65
decompression sickness (DCS), 72
  factors affecting severity of, 75-76
  preventing, 74
  scuba diving and, 76-77, **76**
decongestants, 152
dehydration, 4, 18, 188-189, 209
delirium tremen (DT), 168
denitrogenation, 74
Department of Transportation (DOT), 1
depth of field, 98
diabetes, 18
diet (*see* nutrition)
diffusion, 27
disorientation, 4
  definition/description, 117-119
  positional, 120-121
  postural, 119-120
  proprioceptive, 119-120
  spatial, 133, 135-136
  temporal, 121-123
  tolerance to, 140-142
  types of, 119-136
  vestibular, 123-133
dissociation curve, 24, **25**
Divers Alert Network (DAN), 77
drugs, 161, 172-174
  amphetamines, 173
  cocaine, 173
  illicit, 172-174
  marijuana, 172-173
  opiates, 173
  PCP, 173
  testing, 173-174, 262
  types of, 172-173
dysbarism, 66, 72

E

ear, 66-69, 81-83, 124-129, **81**, **124** (*see also* hearing; noise; sound)
  inner, 82-83
  middle, 67-69, 82, **67**
  otolith organs, 125, 129, **124**, **125**, **127**, **128**, **129**
  outer, 81-82

semicircular canal, 125, **126**
effective performance time (EPT), 58, **59**, 57
electrocardiogram (EKG), 260-261
electromagnetic radiation (EMR), 189, **190**
electromagnetic spectrum, 96, **97**
encephalitis, 11
endolymph, 125, **126**
esophagitis, 16
ethanol, 162, 163
evaporation, 178, 180
evolved gas disorders, 71-72
exercise, 252
exercising, 160-161
exosphere, 40
eye, 97-99, **98**, **99** (*see also* vision)
cornea, 97
iris, 98
lens, 97-98
retina, 99
ultraviolet radiation and, 115

F

fainting, 241
fatigue, 4, 208-215, 236
alcohol and, 166
causes of, 209-211
coping with, 213-215
symptoms, 211-213
vision and, 110
fats, 248
fecal matter, 16
Federal Aviation Regulations (FARs)
Part 61.53, 267
Part 67, 267
federal flight surgeon, 267
Federal Register, Vol. 55 No. 228, 1
fiber, 249
fibrillation, 20
first aid, 240-242
flicker, fatigue and, 210
flicker vertigo, 142
flight physiology, definition/description, 4
flight surgeon, 8-9, 266
federal, 267
regional, 267
flu, 145-146
fog, illusions and, 138
Food and Drug Administration (FDA), 148
frontal lobes, 11
frostbite, 187-188

G

G force, 217
negative, 221
symptoms, 218, 220, **218**, **219**
tolerance, 221
gallbladder, 17
gas laws, 34-38, 66, **39**
Boyle's, 34-35, 66, **34**, **35**
Charles', 35
Dalton's, 35-36
Graham's, 36-37, **38**
Henry's, 36, 72, **37**
gases
effects on human body, 72-74
evolved disorders, 71-72
inert atmospheric, 31
trapped, 2, 65-71, **71**
gastritis, 16
gastrointestinal system, 15-17, 70-71
alcohol and, 164
infections, 16
large intestine, 16
small intestine, 16
stomach, 15-16
Graham's Law, 36-37, **38**

H

Hall, Joseph, 10
hay fever, 146
haze, illusions and, 138
headaches, 146-147
hearing (*see also* ear; noise; sound)
audiograms, **91**
loss of, 80
caused by noise, 89-90
conductive, 91-92
presbycucis, 92
sensorineural, 92
measuring, 88
protecting, 92-93
test, 261
heart, 19-21, **20**
arrhythmias, 157
atria, 19
fibrillation, 20
murmur, 21
myocardial infarction, 20
premature ventricular contraction, 157
heat cramps, 182
heat exhaustion, 182-183
heat stroke, 183-184
hemoglobin, 24
Henry's Law, 36, 72, **37**

hindbrain, 11
histotoxic hypoxia, 50-51, 165
hydrochloric acid, 15
hypemic hypoxia, 50
hyperopia, 102, **102**
hyperthermia, 182
hyperventilation, 60-61
  causes, 60
  symptoms, 60-61
  vs. hypoxia, 61
hypoglycemia, 158-159, 164, 251-252
  fatigue and, 210
hypothermia, 185-186
hypoxia, 2, 4, 25, 47, 61-62, 201
  absolute altitude and, 57
  acclimatization and, 56-57
  alcohol and, 55, 165
  ambient temperature and, 57-58
  anemic, 50
  causes, 49
  classification, 48
  duration of exposure and, 57
  factors influencing tolerance to, 54-58
  fatigue and, 210
  histotoxic, 50-51, 165
  hypemic, 50
  hypoxic, 25, 49-50, **26**
  physical activity and, 57
  prevention, 58-59
  rate of ascent and, 57
  smoking and, 55-56, **55**
  stages, 51-53, **52**
    compensatory, 52-53
    critical, 53
    disturbance, 53
    indifferent, 51-52
  stagnant, 50
  symptoms, 53-54
  treatment, 59-60
  types of, 49
  vision and, 109
  vs. hyperventilation, 61
hypoxic hypoxia, 25, 49-50, **26**

I

ibuprofen, 153-154
ileitis, 16
ileum, 16
illusions, 136-139
  angular motion, 131-133, **133**, **134**
  autokinetic, 137, **137**
  coriolis, 130-131, **130**
  fog and haze, 138
  landing visual, 138-139, **139**, **140**, **141**
  leans, 131, **132**
  oculogravic, 131, 138
  oculogyral, 137
  rotational, 131-133, **133**, **134**
  tolerance to, 140-142
  vection, 133
  visual, 4, 111-112
  visual-cue, 138
  water refraction, 138
impairment, 6-8
impulse noise, 85-86
incapacitation, 6-8
  distraction, 7
  partial, 7
  recognized, 8
  subtle, 7
  sudden, 7
  total, 7
  unrecognized, 8
inert gases, 31
injuries, 147
inner ear, 82-83
insomnia, 199
  symptoms, 201-202
insulin, 18
International Civil Aviation Organization (ICAO), 5
international standard atmosphere (ISA), 33
intestines, 16
ionosphere, 40

J

jejunum, 16
jet lag, 4, 206-208, **208**

K

kidneys, 17-18
kidney stones, 18

L

landing, visual illusions and, 138-139, **139**, **140**, **141**
Lauber, John, 237
leans illusion, 131, **132**
ligaments, 14
light, 99-101
light spectrum, 96, **97**
liver, 17
loss of consciousness (LOC), 12

M

management (*see* crew resource management; stress management)
marijuana, 172-173

medical airworthiness, 5
medical certification, 264-276
applications, 266, **271**, **272**
aviation medical examiner, 266
denial, 274
FARs, 267
federal flight surgeon, 267
flight surgeon, 266
Form 8500, 268, **269**
limitations, 268
process, 270-276
recertification and special issuance, 275-276
regional flight surgeon, 267
waiver, 268
medical emergencies, 239-245
cardiopulmonary resuscitation, 241-242
choking, 242
fainting, 241
first aid, 240-242
onboard medical kits, 242
survival techniques, 243-245
transporting ill passengers, 243
medical examination, 252-263
blood tests, 258
drug testing, 262
electrocardiogram, 260-261
FAA and company flight physicals, 262-263
hearing test, 261
medical history, 253-255, **254**
physical exam, 255-258, **257**
results, 258-260
spirometry test, 261
urine tests, 258
vision test, 261
X-rays, 261
medication, 4, 69, 148-155
acetaminophen, 153
analgesics, 153-154
antihistamines, 151
aspirin, 153
cough suppressants, 154
decongestants, 152
fatigue and, 210
ibuprofen, 153-154
ingredients, 150, 154
marketing claims, 149-150
nose sprays, 152-153
pain relievers, 153-154
phenylpropanolamine, 154-155
prescription, 160
sleep aids, 151-152, 202
Melancthon, Philip, 10
meningitis, 11
menses (menstrual cycle), 18-19
mesosphere, 40
metabolic system, 17-19
diseases, 18
gallbladder, 17
kidneys, 17-18
liver, 17
pancreas, 18
thyroid, 18
midbrain, 11
middle ear, 82
middle ear block, 67-69, **67**
prevention, 69
symptoms, 68
treatment, 68-69
motion sickness, 143
symptoms, 143
treatment, 143
muscles, 14
musculoskeletal system, 14-15
effects of gases on, 73
myocardial infarction, 20
myopia, 102, **102**
empty-field, 112
space, 112
myositis, 15

N

National Plan for Aviation Human Factors, 1
nervous system, 11-14
brain, 11-12, 74
peripheral, 12-14
spinal cord, 12
neuritis, 13
nitrogen, 30-31
noise, 4, 85-87 (*see also* ear; hearing; sound)
fatigue and, 86-87, 209-210
hearing loss caused by, 89-90
impulse, 85
OSHA standards for exposure to, 89
sources, 88-89
steady-state, 85
threshold shift, 90
types of, 85-86
nonionizing radiation, 190-191
nose, sinus block, 69
nutrition, 247-251
calories, 250-251
carbohydrates, 249
fats, 248
fiber, 249
proteins, 248
vitamins and minerals, 250

O

occipital lobes, 11

oculogravic illusion, 131, 138
oculogyral illusion, 137
orientation, 116-143
outer ear, 81-82
oxygen
atmospheric, 30
human body and, 25, 47-48
masks for, 62
regulators, 62
supplemental, 62
support systems, 62-63
oxygen dissociation curve, 25
ozone, 31-32, 64

P

pancreas, 18
paresthesias, 73
pepsin, 15
peripheral vision, 134
peristalsis, 15, 16
photopic vision, 99
Physician's Desk Reference (PDR), 150
pilots, attitude toward illness, 147-148
positional disorientation, 120-121
postural disorientation, 119-120
pregnancy, 19
premature ventricular contraction (PVC), 157
pressurization, 62
cabin, 74
proctitis, 16
proprioceptive disorientation, 119-120
proteins, 248
pulmonary function test, 261

R

radiation, 177-178, 189-191
electromagnetic, 189, **190**
ionizing, 191
nonionizing, 190-191
ultraviolet, 115, 190-191
refraction, 99
regional flight surgeon, 267
respiratory system, 24-28, 48-63
alcohol and, 164
physiology of, 25-27, **26**, **27**, **28**
Robinson, Douglas, 29
rotational illusion, 131-133, **133**, **134**

S

saliva, 15
scuba diving, decompression sickness and, 76-77, **76**
seasonal affective disorder (SAD), 204
serum, 24
SHEL model, 5, **6**
sinus block, 69
situational awareness, 7, 42-46, 116, 118-119
causes/clues of lost, 45
definition/description, 43-44
performance components, 44-45
prevention of lost, 45-46
skeletal system, 14-15
sleep, 195-204
alertness management, 199-201
disorders, 199
factors affecting quality of, 201
fatigue and, 209
how to, 202
insomnia, 199, 201-202
medication for, 151-152, 202
REM, 197-198
stages of, 195-197, **196**
variations to normal, 198-199
smoking, 4, 159
hypoxia and, 55-56, **55**
snow, blowing, 113
sound, 83-85 (*see also* ear; hearing; noise)
conversational range, 85
decibel scale vs. intensity, 84, **84**
duration, 85
frequency, 83
intensity, 84, **84**
perception, 87
spatial disorientation, 133, 135-136
spinal cord, 12
spirometry test, 261
stagnant hypoxia, 50
steady-state noise, 85-86
sterile-cockpit rule, 8
stomach, 15-16
stratosphere, 40
stress management, 229-230
sunglasses, 113-114
survival techniques, 243-245

T

teeth, 70, **70**
temperature, 175-189
atmospheric, 33
cold, 185-188
frostbite, 187-188
hypothermia, 185-186
loss of body heat, 186-187
windchill, **187**
fatigue and, 210
heat, 176-178
comfort factors, 178
conduction, 176
convection, 176-177
coping with extreme, 181-184
evaporation, 178, 180

generating body, 179
hyperthermia, 182
radiation, 177-178, **177**
human body and, 178-181
temporal disorientation, 121-123
temporal lobes, 11
tendinitis, 15
tendons, 14
threshold shift, 90
thyroid, 18
time of useful consciousness (TUC), 57, 58
toxic fumes, 192-193
trapped gases, 2, 65-71, **71**
tropopause, 39
troposphere, 37

U

ultraviolet radiation (UV), 115, 190-191
urine, 18
tests, 258

V

Valsalva maneuver, 69
vascular system, 22-24, **23**
vasodilatation, 163
vection illusion, 133
veins, 22, 24
venules, 22
vertigo, 123
flicker, 142
vestibular disorientation, 123-133
illusions, 130-133, **130**, **132**, **133**
vestibular system, alcohol and, 165
vibration, 4, 93-94
vision, 95-96 (*see also* eye)
acuity variations, 101-102
alcohol and, 110, 165
areas of, **100**
carbon monoxide and, 110
cockpit displays and, 235
corrective lenses, 103
depth of field, 98
depth perception, 112
factors affecting acuity, 108-110
fatigue and, 110, 210
geometric perspective, 112
hyperopia, 102, **102**
hypoxia and, 109
illusions (*see* illusions)
light spectrum and, 96, **97**
mesopic, 104
motion parallax, 112
myopia, 102, **102**
night, 103-104, **106**
adapting to, 105-106
preserving, 106-108
techniques for improving, 108, **109**
peripheral, 134
photopic, 99, 104
refractive error, 102-103
scanning techniques, 111
scotopic, 104
space myopia, 112
sunglasses, 113-114
test, 261
types of, 104, **105**
visual-cue illusion, 138
vitamins, 250

W

water refraction, 138
water vapor, 31
whiteouts, 113
windchill, **187**

X

X-rays, 261

About the author

Richard O. Reinhart, M.D., is the president of Human Factors Resources, Inc., a Minneapolis-based company that provides educational resources and training materials in human factors and flight physiology as well as advice on FAA medical certification and medical management for the aviation community.

He is an instrument-rated commercial pilot, senior FAA aviation medical examiner (AME), and retired U.S. Air Force/Air National Guard flight surgeon. Reinhart was the aeromedical adviser to Republic Airlines for 10 years and currently advises several corporate flight departments and individual pilots.

He is medical editor for *Business & Commercial Aviation* magazine, author of a monthly column for NBAA's *Digest*, and he publishes *The Airworthy AirCrew*, a monthly human-factors newsletter.

Reinhart is also active in the Aerospace Medical Association, National Business Aircraft Association, Helicopter Association International, Flight Safety Foundation, University Aviation Association, and Airline Medical Directors Association. He has participated in human-factors and crew resource management workshops and presentations for Embry-Riddle Aeronautical University and NBAA.

Other Bestsellers of Related Interest

Advanced Aircraft Systems: Understanding Your Airplane
David Lombardo
This book explains how to use and maintain today's state-of-the-art general aviation aircraft systems. Engine components, the electrical system, cockpit instrumentation, hydraulics, fire protection, de-icing equipment—Lombardo thoroughly examines more than 30 essential topics, fully explaining the theory, terms, and practical applications of each.
ISBN 0-8306-0823-0, #155265-0 $21.95 Paperback

Cockpit Resource Management: The Private Pilot's Guide
Thomas P. Turner
A comprehensive guide to cockpit resource management. The key to becoming a safer, more proficient flier in the 1990s. For both the private and professional pilot.
ISBN 0-07-065604-5 $19.95 Paperback,
ISBN 0-07-065603-7 $32.95 Hardcover,

How to Order

Call 1-800-822-8158
24 hours a day,
7 days a week
in U.S. and Canada

Mail this coupon to:
McGraw-Hill, Inc.
P.O. Box 182067
Columbus, OH 43218-2607

Fax your order to:
614-759-3644

EMAIL
70007.1531@COMPUSERVE.COM
COMPUSERVE: GO MH

Shipping and Handling Charges

Order Amount	Within U.S.	Outside U.S.
Less than $15	$3.50	$5.50
$15.00 - $24.99	$4.00	$6.00
$25.00 - $49.99	$5.00	$7.00
$50.00 - $74.49	$6.00	$8.00
$75.00 - and up	$7.00	$9.00

EASY ORDER FORM—SATISFACTION GUARANTEED

Ship to:
Name ______
Address ______
City/State/Zip ______
Daytime Telephone No. ______

Thank you for your order!

ITEM NO.	QUANTITY	AMT.
	Shipping & Handling charge from chart below	
	Subtotal	
	Please add applicable state & local sales tax	
	TOTAL	

Method of Payment:
☐ Check or money order enclosed (payable to McGraw-Hill)
☐ DISCOVER ☐ AMERICAN EXPRESS Cards
☐ VISA ☐ MasterCard

Account No. ______

Signature ______ Exp. Date ______
Order invalid without signature

In a hurry? Call 1-800-822-8158 anytime, day or night, or visit your local bookstore.

Key = BC95ZZA